SOLID-STATE NMR

SOLID-STATE NMR:

BASIC PRINCIPLES & PRACTICE

DAVID C. APPERLEY, ROBIN K. HARRIS
& PAUL HODGKINSON

MP MOMENTUM PRESS

MOMENTUM PRESS, LLC, NEW YORK

First published by Momentum Press®, LLC
222 East 46th Street, New York, NY 10017
www.momentumpress.net

ISBN-13: 978-1-60650-350-8 (hard back, case bound)
ISBN-10: 1-60650-350-2 (hard back, case bound)
ISBN-13: 978-1-60650-352-2 (e-book)
ISBN-10: 1-60650-352-9 (e-book)

DOI: 10.5643/9781606503522

Cover design by Jonathan Pennell
Interior design by Exeter Premedia Services Private Ltd.,
Chennai, India

10 9 8 7 6 5 4 3 2 1

Printed in the United States of America

CONTENTS

PREFACE

Nuclear magnetic resonance (NMR) has proved to be a uniquely powerful and versatile spectroscopy, and no modern university chemistry department or industrial chemistry laboratory is complete without a suite of NMR spectrometers. The phenomenon of nuclear spin may seem an odd basis for an analytical tool, but it is the relative isolation of the nuclear spin from its surroundings that makes it an ideal noninterfering probe of the electronic environment. Different sites are clearly identified by their chemical shifts, while J couplings in ^1H spectra provide connectivity information. The combination of these two complementary interactions, plus the formidable array of different NMR experiments developed since the arrival of Fourier transform NMR in 1966, has revolutionized the practice of chemistry.

While the original discovery of NMR involved both solids and liquids, the application of NMR to materials in solid form developed at a markedly slower rate than its solution-state counterpart until relatively recently. The most obvious explanation for this difference is the fact that molecular mobility in isotropic solutions averages anisotropic interactions (such as shielding) to their isotropic values. In particular, for solids it is necessary to consider a number of NMR interactions (notably the dipolar and quadrupolar interactions) that can generally be ignored for the solution state. These can significantly reduce spectral resolution and complicate interpretation of spectra. On the other hand, the direct effect of these interactions means that solid-state NMR is potentially a much richer information source than NMR of solutions.

A second, perhaps less appreciated, reason for the relatively tardy progress of solid-state NMR is the nature of the solid state itself. Solids, and especially the samples the solid-state NMR spectroscopist is asked to deal with, are rarely the simple monocrystals beloved by diffraction crystallographers. Solid-state NMR may well give broad, featureless lines for amorphous or heterogeneous samples, but this reflects the underlying nature of the system—the chemical shift of a given nucleus may be a distribution and not a single value as it would be if the sample were dissolved. As a result, solid-state NMR may be providing *too much* information, so that the art of solid-state NMR spectroscopy lies in finding the right approach to refining the information content.

It is this complex interaction between the nature of the sample and the multiple anisotropic NMR interactions that makes solid-state NMR a challenging technique to master, and the treatments in undergraduate texts do little to dispel the image of a dark and troublesome technique. The purpose of the current text is to provide a bridge between the familiar world of isolated molecules relevant to solution-state NMR and the subtle world of solid materials. We hope that this introduction and survey will be of value to both established researchers wishing to learn what solid-state NMR can (and cannot) do for their systems and graduate students starting work in this area. We have deliberately avoided detailed mathematical treatments in the early chapters, concentrating on providing that essential qualitative "feel" for the different aspects of solid-state NMR. The theoretical treatment, which will be of most relevance to those starting a career in the field, is delayed to chapter 4. This survey of the theoretical tools used in solid-state NMR should allow the reader to tackle the more specialized literature with increased confidence. The advanced experiments described in chapters 5–8 can only be completely understood in terms of the underlying theory, but again a basic feel for the experiments and what information they provide should still be clear without such knowledge.

Solid-state NMR is now too vast and varied to try to present a comprehensive review. Our goal is rather to provide a survey of the technique and the essential tools for exploring further, together with a practical guide on the application and use of solid-state NMR. A selection of review articles is listed at the end of each chapter for those wishing to explore individual topics in more detail. Similarly, to keep the text to a manageable size, we are assuming a basic familiarity with ^{1}H and ^{13}C solution-state NMR and the Fourier transform NMR experiment (see the "Further reading" for chapter 1 for some excellent introductory texts).

We have tried to emphasize the interactions between solid-state NMR and other techniques. A very significant one of these is the cross-fertilization between the solid-state and solution-state branches of NMR. The recent past has seen a proliferation of multidimensional techniques, often initially developed in solution, being applied to solid-state NMR. Solids are intrinsically complex materials and we cannot expect a single technique to provide all the answers, so we also point out where the interpretation of NMR results is aided by other experimental techniques, such as X-ray diffraction, or tools from theoretical chemistry, such as *ab initio* calculation of chemical shifts. Moreover, we emphasize the additional information (much of it complementary to that obtained by diffraction measurements) that NMR brings to our knowledge of solid-state structure.

In summary, making the best use of solid-state NMR involves a good understanding of the materials under study, familiarity with a broad range of techniques, and a sense for the complementary information that can be provided by other techniques. This may seem a daunting challenge for a newcomer, but the subtle interactions between sample, equipment, and theory are key as to why solid-state NMR is such an intellectually rewarding field in which to work.

Many people have contributed to the appearance of this book. Members of our research group have read and commented on the various chapters as they were written, for which we thank them. We are also very grateful for detailed comments on chapter 6 by Dr. Sharon Ashbrook. A number of research colleagues in other locations have supplied the basis for some of the diagrams, as is acknowledged at appropriate places, and we thank these friends for such help. Finally, we thank our publishers, Momentum Press, for the attractive style and efficient production of the book.

David C. Apperley, Robin K. Harris, Paul Hodgkinson
October 2011, Durham

ABOUT THE AUTHORS

David C. Apperley studied chemistry at the University of East Anglia, Norwich, and gained a Ph.D. for research on dipolar coupling in solids from the Open University, Milton Keynes, in 1986. He further developed his interest in solid-state NMR while working as a senior experimental officer in the Durham Solid-State NMR Research Service, at first in the Industrial Research Laboratories and now in the Department of Chemistry, Durham University. For many years he has been the manager of the facility, which serves both industry and other universities (the latter operation funded by the Engineering and Physical Sciences Research Council). His role with the NMR Service is to provide access to, training in, and interpretation of results from solid-state NMR measurements for scientists in industry and in the UK universities. As well as providing support to organic, organometallic, inorganic, and physical chemists, he specializes in the experimental application of solid-state NMR techniques to characterization or problem solving in a wide range of solid (and sometimes not so solid) materials, including pharmaceuticals, catalysts, ceramics and glasses, polymers (synthetic and natural), soils (and related materials), and household products. He has coauthored 150 publications in the scientific literature.

Robin K. Harris is an emeritus professor at the University of Durham, UK, where he previously served as professor of chemistry, head of the Physical and Theoretical Chemistry section, and chairman of the Chemistry Department. He obtained his first degree from Cambridge University and undertook research in NMR there, supervised by Norman Sheppard, for his Ph.D. After 2 years as a postdoctoral fellow at the Mellon Institute, Pittsburgh, he joined the (then new) University of East Anglia (Norwich, UK) as a lecturer, winning promotion to a readership and then obtaining a personal chair. He moved to the University of Durham in 1984. His early research was on solution-state NMR, but, starting in 1976, he has carried out pioneering research in solid-state NMR, using cross-polarization and magic-angle spinning, for a wide variety of elements and in an extensive range of chemical systems, including organometallics, polymers, ceramics, and inorganics. In the last 20 years, much research has been directed toward problems involving pharmaceutical compounds and systems, especially relating to polymorphism and quantitative studies. Current interests center around the concept of "NMR crystallography" and include both sophisticated NMR experimental methods and computations of chemical shifts using crystallographic repetition. Professor Harris was awarded an Sc.D. degree by Cambridge University for his research in 1978. He is a Fellow of the Royal Society of Chemistry and has won its awards in Chemical Instrumentation (1985) and in Analytical Spectroscopy (1998). He was the director of the UK National Research Service in Solid-State NMR (1986–2004). He has been an author of over 500 research articles, mostly on NMR, and has acted as author or editor of several books on the subject. He is the senior editor-in-chief of the *Encyclopedia of Magnetic Resonance.*

Paul Hodgkinson studied chemistry at Queen's College, Oxford, completing his Ph.D. in 1995 on the sampling of NMR data. His interests in solid-state NMR developed during postdoctoral research at UC Berkeley (Royal Society/NATO fellowship with Professor Alex Pines) and at the Ecole Normale Supérieure de Lyon (Marie-Curie fellowship with Professor Lyndon Emsley). He was appointed to a research fellowship in the Chemistry Department, University of Durham, UK, in 1998 and is currently a reader in magnetic resonance at Durham, where he directs the Durham Solid-State NMR Research Service. His research combines interests in technique development and methodology in solid-state NMR as well as applications to chemical problems. His group develops NMR theory and numerical simulation software to explore the dynamics of large coupled spin systems and applies solid-state NMR in the area of structural chemistry, particularly of systems with mobility such as soft solids and solvates, and to pharmaceutical solids. A particular interest is in combining information from diffraction-based experiments, NMR and computation of NMR parameters (using DFT codes), and dynamics (molecular dynamics simulations). He has authored/coauthored over 60 research articles in the area of NMR.

CHAPTER 1

INTRODUCTION

1.1 THE UTILITY OF NMR

This chapter gives a brief introduction to solid-state nuclear magnetic resonance (NMR). However, in order to understand its applications, practitioners need to be aware of a number of facets of solid-state structure, so these are described. The final section introduces the three key techniques of high-resolution NMR of solids and the types of nuclei for which different approaches are needed.

NMR spectroscopy is undoubtedly one of the most powerful techniques for determining molecular-level structure and dynamics. Three main factors contribute to this situation. Firstly, it can be applied to the vast majority of samples. Secondly, nearly all elements have spin-active nuclides that can be accessed—and the resonance regions do not generally overlap (i.e., the spectra are isotope specific). Thirdly, under suitable experimental conditions the resolution is extremely high so that small differences in the electronic environment of atoms result in observably different resonance frequencies.

The first successful NMR experiments on condensed phases (conducted by physicists in late 1945) encompassed both solids and solutions. However, it soon became clear that (i) resolution for solutions was orders of magnitude better than for solids and (ii) the solution-state spectra gave molecular structure information of great value to chemists. This stimulated the production of commercial spectrometers for solution-state work. For several decades, the overwhelming majority of NMR applications were carried out for solutions by chemists. The use of solid-state NMR remained the preserve of a handful of physicists and physical chemists.

In order to understand why NMR of solids is so different from solution-state NMR, it is necessary to discuss the nature of the solid state and the various factors affecting NMR spectra, as in the sections below. Many books describe solution-state NMR (see Further reading), and graduate chemists will certainly have received lectures on the topic, so the present book assumes a certain amount of background knowledge. The current chapter will discuss the nature of the solid state and will introduce the ways in which solid-state NMR is distinguished from its solution-state analog, in spite of the fact that the same factors are at work.

NMR spectra are generally obtained from samples in strong magnetic fields by recording the response to radiofrequency (RF) radiation, which is normally applied in the form of short pulses. The detailed behavior of the nuclear spins depends on the RF pulsing regime, on the spin properties, and on the interactions of the spins. Thus, one must evaluate the terms governing the energy of the spins in the magnetic field and subject to irradiation. The Zeeman effect of the spin magnets in the applied magnetic induction field (B_0—conventionally in the z direction) almost always forms by far the largest contribution to the energy. The full Zeeman energy, E_Z, involves a sum over all the nuclear species in the sample. For a given spin j (ignoring shielding), it is expressed as:

$$E_Z = \left(\frac{\gamma_j}{2\pi}\right) h m_j B_0 \qquad 1.1$$

where h is Planck's constant, γ_j is the magnetogyric ratio of the nucleus concerned (i.e., a measure of its magnetic strength), and m_j is the spin component quantum number of nucleus j.

Magnetic fields arising from RF radiation form small perturbations in energy and cause transitions. The remaining interactions, which are all internal to the sample and therefore give chemical information, are of four main types,[1] one of which (shielding) modifies the Zeeman effect, while the other three involve spin couplings (indirect, dipolar, and quadrupolar) and mostly lead to splittings in the spectra. These four interactions give rise to energies designated here as E_S, E_J, E_D, and E_Q, respectively. Thus the total energy of an ensemble of spins, E_{NMR}, is the sum of the several effects:

$$E_{NMR} = E_Z + E_{RF} + E_S + E_J + E_D + E_Q \qquad 1.2$$

where E_{RF} refers to the interaction of the spins with applied radiofrequencies. The "internal" interactions, namely E_S, E_J, E_D, and E_Q, are discussed in separate sections in chapter 2. *The important feature to note here is that these internal interactions all depend on the orientation of the relevant molecular-level fragment in the applied magnetic field of the NMR experiment (i.e., they are anisotropic).* Thus for isotropic solutions, where the molecules are rapidly and chaotically tumbling, the relevant energies are averaged over all orientations. However, for solids, with relatively static molecules (and certainly not the complete motional disorder of solutions), the internal interactions depend strongly on the orientation of the solid in the magnetic field, giving a significantly more complex situation than for solutions.

An extra term is required in equation 1.2 when paramagnetic systems are involved because the unpaired electrons can couple with nuclei. While this book concentrates on diamagnetic systems, which attract the vast majority of NMR studies of solids, section 7.5 discusses some of the special effects seen in paramagnetic systems.

---

[1] Spin-rotation interactions, which do not generally affect spectra of solids, will be ignored herein.

1.2 A PREVIEW OF SOLID-STATE NMR SPECTRA

In order to give a taste of what is to come in this book, a few examples of spectra are given at this point. Figure 1.1 compares ^{13}C spectra for a solution of alanine and for a microcrystalline sample, with low-power proton decoupling for the former, as normally used for solution-state spectra. The overarching reason for the difference is, as described above, the extensive averaging of various interactions in the solution state, caused by rapid random isotropic tumbling of the molecules. Thus the effects of anisotropies in the interactions (i.e., of variations with the molecular-level or sample orientation in the applied magnetic field B_0) are removed from the spectra, causing considerable simplification (and therefore high resolution)—but simultaneously losing information. Some solid-state situations give better resolution, for instance, molecular solids when the molecules are highly mobile, as is the case for plastic crystals such as adamantane (see inset 7.8). Inorganic compounds lacking hydrogens can give better-defined spectra, as for the ^{207}Pb spectrum of lead nitrate, shown in figure 1.2. Figures 1.1 (upper trace) and 1.2 (upper trace) are for microcrystalline samples, yielding what are referred to as *powder-pattern spectra*. It is possible, though unusual, to record NMR spectra of single crystals (a rather large crystal is required!). A single crystal of lead nitrate would give a relatively sharp singlet ^{207}Pb spectrum rather than a powder

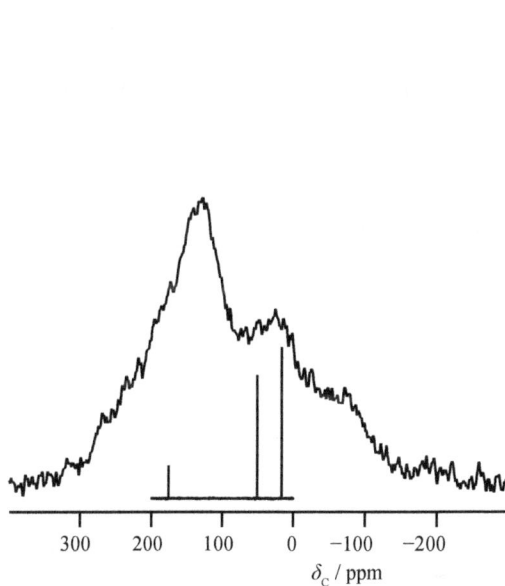

Figure 1.1. Carbon-13 spectra of alanine in the solid state (upper trace) and in solution (lower trace). The latter was obtained using proton decoupling, as is usual for solutions. The former involved no special techniques.

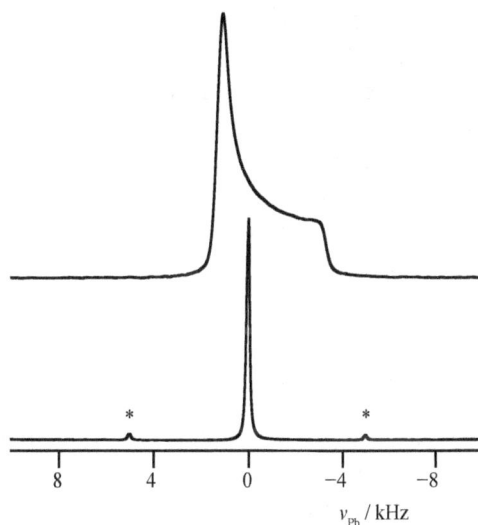

Figure 1.2. Lead-207 NMR spectra of microcrystalline lead nitrate under static (upper trace) and magic-angle spinning (lower trace) conditions. The former is an example of a powder pattern. The latter is discussed in section 1.5. Magic-angle spinning (including spinning sidebands, indicated in this figure by asterisks) is discussed in section 2.7.

pattern, but its chemical shift will depend on the orientation of the crystal in the magnetic field, as discussed in sections 2.3 and 8.2.4. For a microcrystalline powder, there will be a random (statistical) distribution of orientations, so the spectrum consists of a summation of many such sharp lines at different shifts, which overlap to form the powder pattern.

1.3   THE SOLID STATE

1.3.1   INTRODUCTION

The principal way in which solids differ from fluids lies in the relative lack of molecule-level motion. In isotropic solution, molecules and ions tumble rapidly, randomly, and chaotically. Therefore, any properties that depend on molecular orientation become averaged to their isotropic values. The internal NMR interactions of importance come into this category. Such interactions each involve products of two vectors, so they are classed as *tensors* (see chapter 2). In fact, two of the most important features, namely the electric quadrupolar and magnetic dipolar interactions, are averaged to zero by isotropic tumbling and therefore do not affect the NMR transition frequencies of solutions. Simple interpretations of spectra can ignore their existence, though they do affect relaxation times. Moreover, the other two important interactions (shielding, which gives rise to chemical shifts, and indirect coupling) are simplified so that for solutions a single *scalar* chemical shift results for each chemical site and a scalar coupling constant (symbol J) exists for every pair of nuclear spins.

There are many types of solids and solid systems. The simplest (and most commonly met with in chemistry research) is the crystalline state. In "ideal" crystalline samples, the atoms (whether in molecular or in framework solids) are effectively locked into positions. The structures as a whole are described by unit cells wherein the atomic positions can be determined by diffraction methods. The orientation of a crystal in B_0 will determine the resonance frequency for each NMR-active nucleus.

1.3.2   SYMMETRY IN THE CRYSTALLINE STATE

The concept of a unit cell underlies the translational symmetry inherent in crystalline solids. NMR spectra reflect the contents of the unit cell so that resonances can be interpreted in terms of only the unit cell. Of course, atoms in neighboring unit cells will influence the chemical shift of a nucleus in a given cell. However, such influences die away very rapidly with distance. NMR thus responds to short-range effects, in contrast to diffraction techniques, which largely rely on long-range order.

For NMR purposes, it is important to distinguish between the effects of different symmetry elements on the equivalence or otherwise of molecules or atomic groupings. For rigid fully-ordered structures, translational symmetry means that each unit cell is identical and therefore gives the same NMR spectrum, that is, spectra directly reflect the contents of the unit cell. However, there may be several chemically identical sites for a given nucleus, related by symmetry other than translation. Consider molecular systems (similar matters arise for framework structures). Molecules related by a center of symmetry (figure 1.3(a)) are fully equivalent and give identical spectra. However, molecules related by other symmetry elements, such as planes or axes of symmetry (figure 1.3(b)), may give different spectra in some experiments (see below) and so are only conditionally equivalent. In other situations, there may be two or more whole molecules in a unit cell which are not related by symmetry (figure 1.3(c)) and are therefore classed as *independent*. In fact, the *minimum* group of atoms which, when repeated by *all the symmetry* of the crystal, gives the contents of the whole unit cell, forms what is known as the *crystallographic asymmetric unit*. Figure 1.4 presents a practical case, that of the α

(a) $Z = 2$, $Z' = 1$ (b) $Z = 2$, $Z' = 1$ (c) $Z = 2$, $Z' = 2$

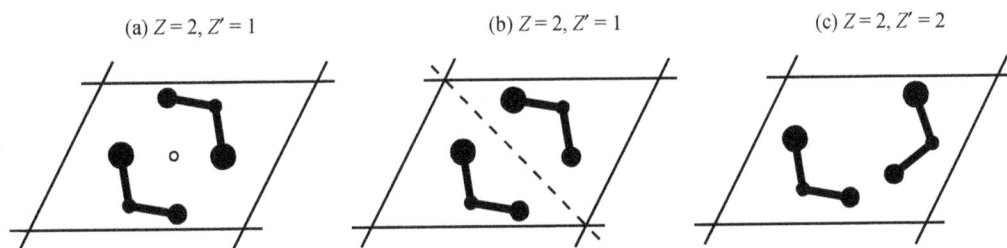

Figure 1.3. Schematic illustration in two dimensions of (a) fully equivalent (under the influence of a center of symmetry), (b) conditionally equivalent (under the influence of a mirror plane), and (c) nonequivalent molecules of the same chemical species in a unit cell. The symbol Z is conventionally used for the number of molecules in the unit cell, whereas Z' refers to the number of molecules in the crystallographic asymmetric unit (see the text).

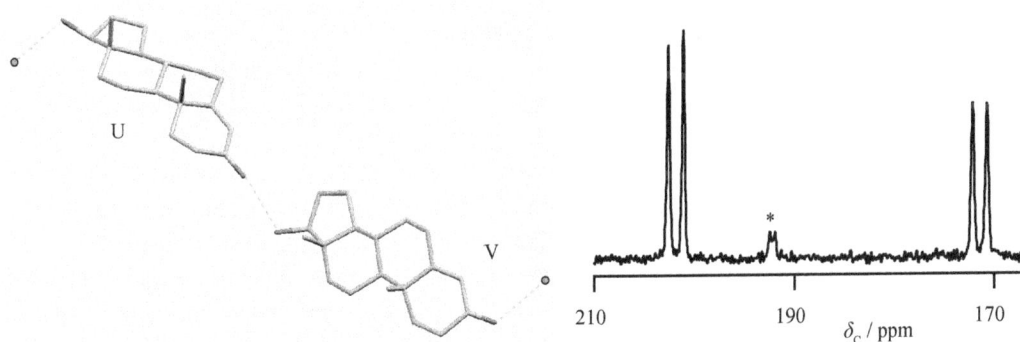

Figure 1.4. Left: The two independent molecules (labeled U and V) in the asymmetric unit of the α form of the steroid testosterone. These have different environments and slightly different geometries. Right: Carbon-13 spectrum of α-testosterone (high-frequency region only), obtained under conditions of proton decoupling and magic-angle spinning (see section 1.5) to show crystallographic splittings. The asterisk indicates a spinning sideband (see section 2.7).

form of testosterone, for which the asymmetric unit consists of two molecules, labeled U and V. There are analogous situations for network structures of solids; again, one can define an asymmetric unit.

1.3.3   EFFECTS OF CRYSTAL STRUCTURE ON NMR

As stated above, solid-state NMR spectra reflect the situation of all the nuclei in the unit cell (in contrast to solution-state spectra, which reflect the situation of the isolated molecule). There are at least four distinguishable ways in which the internal structure of a static crystal will give rise to a larger number of resonances than for a solution of the same compound:

- For a static single crystal, atomic sites which are in identical chemical positions will, in general, be in different orientations with respect to B_0 and will therefore give rise to different resonance frequencies. Then more peaks are obtained than for the solution state, forming *crystallographic splittings*. However, such splittings may be eliminated by the technique of *magic-angle spinning* (MAS), discussed later (section 1.5).

- Independent molecules in an asymmetric unit will differ in both their intermolecular (packing) and intramolecular (geometry) arrangements. Therefore the environment of chemically analogous spins in (say) two independent molecules will be different, and so they will give rise to different resonance frequencies—another (more commonly observed) cause of crystallographic splittings. This type of splitting cannot be eliminated by MAS. A partial ^{13}C spectrum of α-testosterone is shown in figure 1.4 to illustrate this phenomenon. It is difficult to disentangle experimentally the influences of packing and geometry on NMR spectra.

- Moreover, a molecule that has symmetry in its isolated state (as in solution) may well lose its symmetry in the crystalline state. Thus, for instance, if the two sides of a paraphenylene group (related by symmetry in the isolated molecule) have different local environments in the crystal, the relevant spins will have different chemical shifts. In such a case, the whole paraphenylene group is (part of) the asymmetric unit, though if the symmetry of the isolated molecule is retained, the asymmetric unit may be half the molecule (plus any atom on the effective mirror plane or symmetry axis relating the two sides of the ring). The two situations are illustrated schematically in figure 1.5.

    The pharmaceutical drug barbital provides an example (figure 1.6). In this case the intramolecular lack of symmetry induced by the molecular environment in the crystal also extends to the two ethyl groups in some forms. The structure of polymorphic form III (figure 1.6(a)) consists of molecular chains in which the two halves of the ring are related by symmetry, as are the two ethyl groups, so the asymmetric unit is half a molecule. This means that only single resonances are seen for the methyl and methylene carbons (as would be the case for a solution-state spectrum). However, form I has a double-chain structure (figure 1.6(b)) in which there is no such relationship, so an entire molecule is the asymmetric unit. Thus the methyl and methylene signals are split into two. Note, however, that the ring carbon signal is unsplit.

Solution-state

Solid-state

general case

symmetrical case

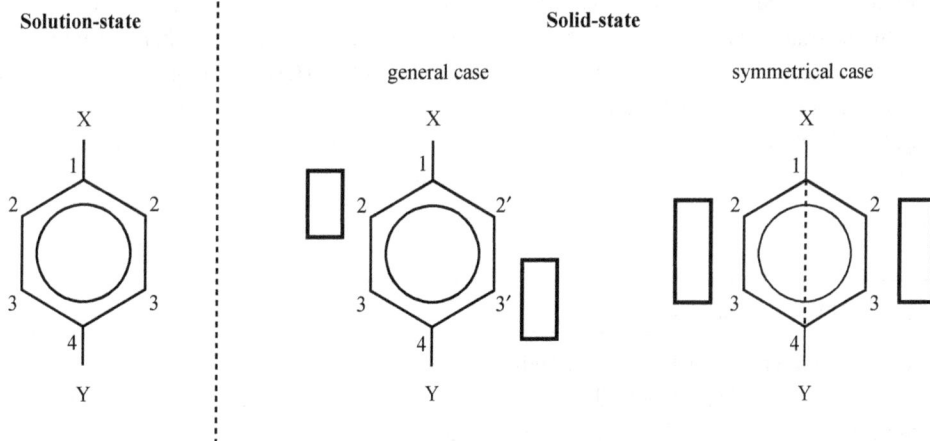

Figure 1.5. Schematic diagram to show the equivalences and inequivalences for carbon nuclei in a phenylene ring in different situations. The rectangles beside the aromatic rings are simply indicative of the local environment. In the general case (only) two resonances will be seen for the two carbons ortho to substituent X (and similarly for those ortho to Y). The dotted line in the right-hand case refers to a twofold rotation axis or a plane of symmetry.

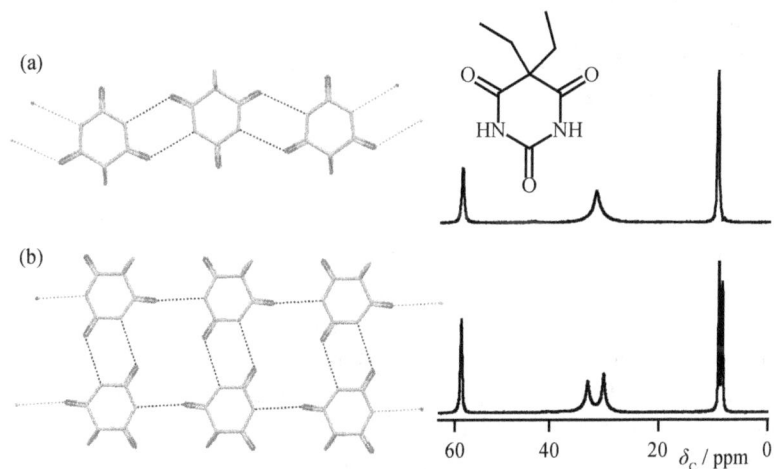

Figure 1.6. Left: Structures of barbital polymorphs, showing the molecular chains that affect the relationships between the two halves of the ring. Right: The corresponding ^{13}C spectra (omitting the C=O signals). (a) Barbital III. (b) Barbital I. The three signals (or groups of signals) are assigned to (right to left) the methyl, methylene, and ring carbons.

- Finally, molecular-level motion, which is fast on the *NMR timescale* (see inset 1.1) in the solution state, may well be slowed in the crystalline state, causing further nonequivalences to be revealed. Thus the methyl carbon nuclei of a $C-C(CH_3)_3$ group may be effectively equivalent in solution because of rapid internal motion about the C–C bond, therefore giving a single ^{13}C signal, but when the t-butyl group is in a fixed position in a crystal, the environments of the methyl groups may differ, resulting in three ^{13}C signals.

Inset 1.1. NMR timescales

There are a number of timescales involved in the NMR experiment:

(a) The *acquisition time* to record a single *free-induction decay* (from tens of milliseconds to several seconds; see inset 3.2). It is this range that is generally referred to as the *NMR timescale*. If molecules are essentially static (apart from vibrational motion) over such times, then the spectrum will be a superposition of those for all molecular positions (e.g., conformations) present. However, if there is rapid motion, that is over times shorter than this, an averaged spectrum will be obtained (see section 7.3). Typically, the critical motional rates are those comparable to chemical shift differences between corresponding nuclei for the different molecular positions.

(b) The total *accumulation time* taken to record a spectrum, which may be from minutes to hours or even days, depending on the number of free induction decays that are recorded. If the compound is stable over this time, its spectrum will be uniquely obtained. If reaction occurs, the spectrum will be a superposition of all the compounds present.

(c) Times relating to relaxation, ranging from microseconds to seconds (or even up to hours in some cases; see section 2.8).

It is actually a misconception to think that atoms and molecules in solids are static. There is, for example, always vibrational motion. Moreover, many crystals have much more motion. Whole molecules or parts of them may be rotating rapidly. Thus C-methyl groups commonly rotate rapidly on the NMR timescale about the $C-CH_3$ axis at ambient temperatures, rendering the protons equivalent. Phenyl and phenylene groups often undergo rapid 180° ring-flips about their axes. Ring inversion may occur for cyclohexyl groups. Such processes lead to averaging of resonances for solids just as they do for solutions, though usually the motional rates are much higher in the latter state. Significant motion of molecules as a whole leads to the *plastic* state. An example is adamantane, which has a roughly spherical molecular shape (see inset 7.8) and so can readily rotate isotropically in its crystalline form (though the molecules do not translate). In other cases, atoms, molecules, and ions can translate through the crystal lattice, as happens in ion conductors such as lithium-based battery materials. Systems with channel or tunnel structures, such as zeolites and urea complexes, may contain mobile guest molecules. In all these cases, there may be substantial averaging of the NMR spectrum.

1.3.4 TYPES OF SOLIDS

There are a number of types of solid state that are not crystalline or not perfectly crystalline. Some solids exhibit *disorder*. This may take a number of forms; in particular, it may be static or dynamic (*spatial* or *temporal* disorder). In either situation, diffraction techniques report *occupancy factors* at the possible sites, implying that there are fractional atoms there. In the former case, adjacent "unit cells" may have some atoms in different positions, which do not change with time. In the latter case, there may be exchange (over a particular timescale) of atoms between two or more possible positions for any given unit cell; the rate of exchange will determine the appearance of the NMR spectrum (see section 7.3). An example is the 3:2 complex of phenol and triphenylphosphine oxide (TPPO). A partial structure is shown in figure 1.7. The phenol molecule is rotationally disordered (and was thus solved in a $P\bar{1}$ space group), with arrows indicating two notional oxygen half-atoms. With this molecule fixed in one of its two arrangements, the phosphorus atoms of the two TPPO molecules are nonequivalent. At the lowest temperature this is the case: the molecules are static on the NMR timescale, so two ^{31}P signals appear. However, as the temperature increases, rotation of

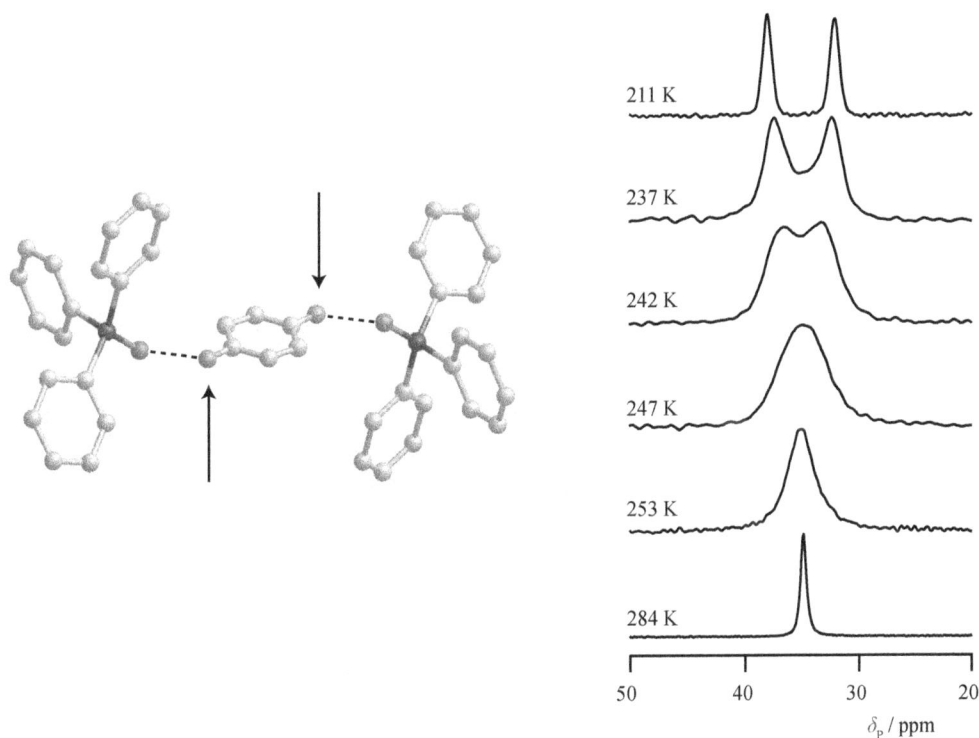

Figure 1.7. Left: Partial structure of the unit cell contents of a 3:2 adduct of phenol and triphenylphosphine oxide. The other two phenol molecules in the unit cell, which are not disordered, are omitted for clarity. Right: Phosphorus-31 spectra as a function of temperature.

the phenol molecule becomes rapid, causing the signals to merge because an average situation is observed by NMR. The NMR proves that the disorder is dynamic (temporal, not spatial).

The different sites in disordered systems do not need to occur with rational occupancy ratios. Many inorganic crystalline systems exist in which there is a fractional excess of some types of atom (over and above the nominal stoichiometric amount). In other cases, there is a deficiency of some types of atom. Such systems are known as defect structures, and their nature has consequences on the NMR spectra (e.g., on relative intensities of resonances).

Yet other structures are classified as *incommensurate*. They contain different chemical entities with translational repetition distances that do not bear any rational relationship. An example is discussed in section 1.4.

Many pure (homogeneous) systems are amorphous. These may be glassy (i.e., relatively rigid at the molecular level) or rubbery (much more mobile). Some organic systems readily produce a supercooled (rubbery) state when cooled rapidly to below their freezing points. As temperature is lowered further, they may pass into the glassy condition at the *glass transition temperature*, T_g. All amorphous systems contain the same atoms and chemical groups in a range of local environments, and so they give rise to a corresponding range of NMR transition frequencies, thus causing line broadening, as shown in figure 1.8. Actually, apparently amorphous

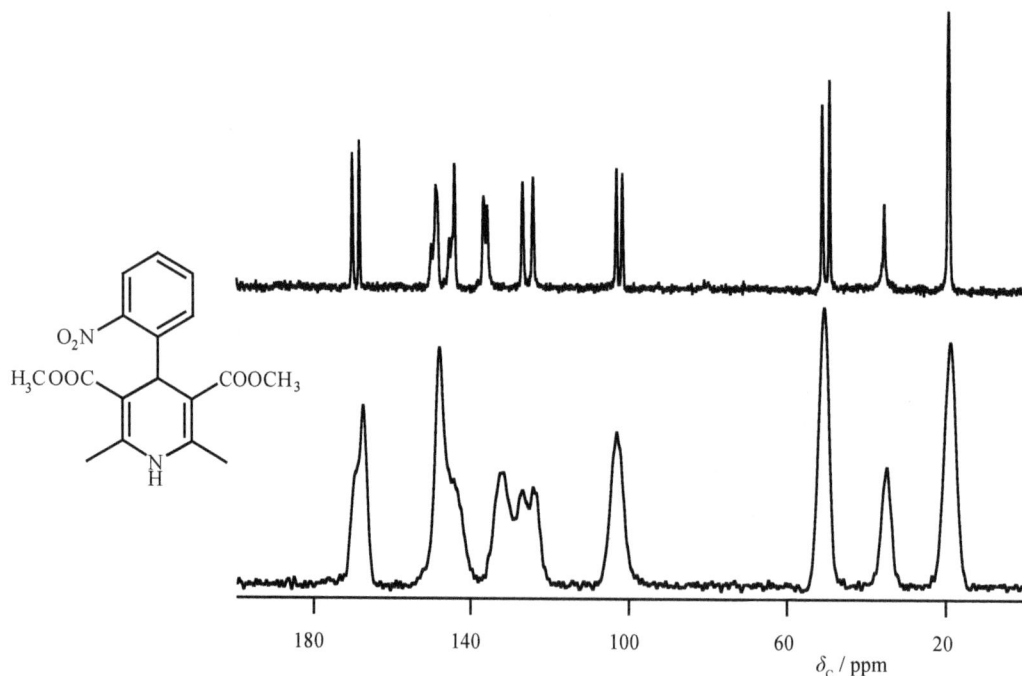

Figure 1.8. Carbon-13 NMR spectra of the organic compound nifedipine, obtained under conditions of high-power proton decoupling and magic-angle spinning, showing the line broadening for the amorphous state (lower spectrum) compared with the crystalline state (upper spectrum).

polymers and fibers (including naturally occurring systems such as silk) can have a degree of order in terms of a *preferred orientation*. Thus polymers drawn through a hole may have chains preferentially oriented along the draw direction. Such preferences can be studied by NMR using static samples.

Moreover, even chemically homogeneous solids may be physically heterogeneous, some with relatively simple biphasic structures (for instance, many semicrystalline homopolymers have amorphous and crystalline domains). Chemical heterogeneity is exceedingly common (e.g., wood and other natural products), resulting in highly complex structures and correspondingly crowded NMR spectra. Crystalline compounds may have many defects—and naturally the nuclear spins near such defects or at the surfaces of microcrystalline particles have different environments to those in the core of a perfect single crystal and so give rise to different chemical shifts in an NMR experiment.

A final type of physical system that needs special consideration is the liquid crystalline state, which can be considered as a half-way house between the isotropic solution and crystalline solid states. Such systems have considerable mobility analogous to that of a liquid but possess order analogous to that of a solid in some dimensions. The behavior of liquid crystals in NMR experiments is a specialized topic which will not be discussed in this book. Suffice it to say here that dipolar and quadrupolar interactions are not averaged to zero for liquid crystals and therefore influence the resonance frequencies.

Thus questions of both spatial order/disorder and temporal disorder (mobility) strongly affect the experimental methods used in solid-state NMR and/or the quality of the spectra obtained. Figure 1.9 illustrates the relationships between the two types of disorder for a variety of condensed-phase situations.

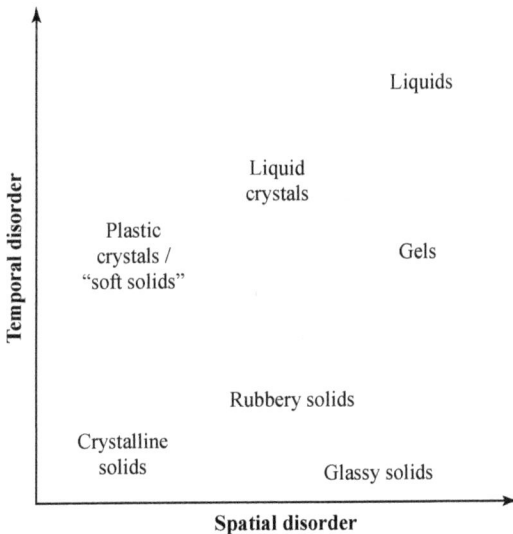

Figure 1.9. Classification of different types of condensed-phase matter according to their spatial and temporal disorder.

1.4 POLYMORPHISM, SOLVATES, CO-CRYSTALS & HOST:GUEST SYSTEMS

Chemical compounds frequently exist in more than one crystalline "form." This may affect only the external appearance or *habit* of the crystals, with no change in the underlying structure. However, in many cases, the internal structure differs in some way, resulting in the phenomenon of *polymorphism* (the corresponding situation for structures of elements is known as allotropy). Carbon is a classical example of allotropy since it can exist as diamond or graphite or the series of fullerenes such as C_{60} (with the soccer football shape). Polymorphism is, in fact, ubiquitous in chemistry, occurring for organic and inorganic compounds, as well as polymers and natural products (e.g., cellulose). The example of barbital has already been mentioned (see figure 1.6). The classification of polymorphs represents some difficulties, but here network systems (such as the various forms of silica) and molecular polymorphs, which are of considerable importance in the pharmaceutical industry, can be distinguished.

All properties of polymorphs for a given compound differ, at least in principle. For instance, dissolution rates of the polymorphs and their equilibrium concentrations will differ. However, their solutions will behave identically. Polymorphs are obtained in a variety of ways (for instance, by crystallization from different solvents). A pair of polymorphs can be either monotropic or enantiotropic. In the former case only one of the forms can be stable (under a range of temperatures and pressures), whereas the other is always metastable (i.e., not thermodynamically stable but often with a long lifetime). In enantiotropic cases, however, the two (or more) forms are each stable, under different conditions, and there will be a transition temperature at which they will interconvert (depending on the pressure). The situation may be expressed in a phase diagram (see figure 1.10), which shows the regions of pressure–temperature space for the various states of a pure compound. However, transition rates between polymorphs can be extremely slow so that it is common to be able to obtain several polymorphs (even in monotropic cases) at, say, ambient temperature, which

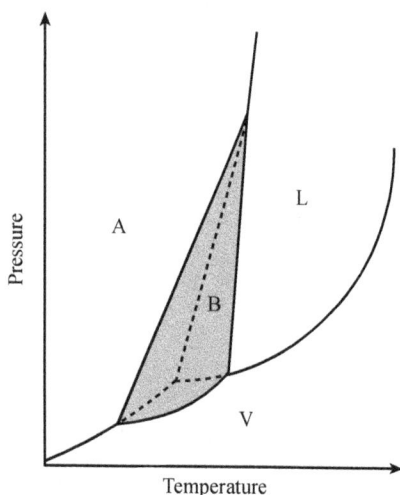

Figure 1.10. Schematic phase diagram for an enantiotropic system. The regions of stability for the two solid forms, A and B (the latter in gray), are indicated. The dashed lines show the boundaries of form A existence in the absence of form B (e.g., if transitions from A to B are slow). L = liquid phase; V = vapor phase.

clearly makes their study easier. For more information on phase diagrams and the thermodynamics of polymorphic systems, see Further reading.

In addition, many organic compounds form stoichiometric solvates or co-crystals, for example as clathrates, in which a host system accepts a guest, either with a change of crystal structure or with retention of the host structure with minimal adjustment to the presence of the guest. Sulfathiazole, for instance, has been shown to participate in over a hundred solvates. When the guest molecule is solid in its pure state, the term *co-crystal* is frequently used instead of solvate. The most important class of solvates involves water as the guest, in which case the term *hydrate* is appropriate. The water molecules often act by hydrogen bonding to hold the structure together. In some cases, hydrates with different stoichiometries (and different crystal structures) exist.

NMR proves to be one of the best ways of characterizing polymorphs and solvates, though it is always wise to use it in conjunction with other techniques such as vibrational spectroscopy and X-ray diffraction, together with thermal methods.

In many cases, the basic structure of the host compound contains cages or channels that can be occupied by a range of guest molecules. Zeolites, for instance, often contain both cages and channels, which result from the use of organic molecules as templates during synthesis. Another example is provided by the complexes of small organic molecules with urea. The urea provides a host structure (unstable in the absence of guest molecules!) containing linear tunnels into which elongated molecules

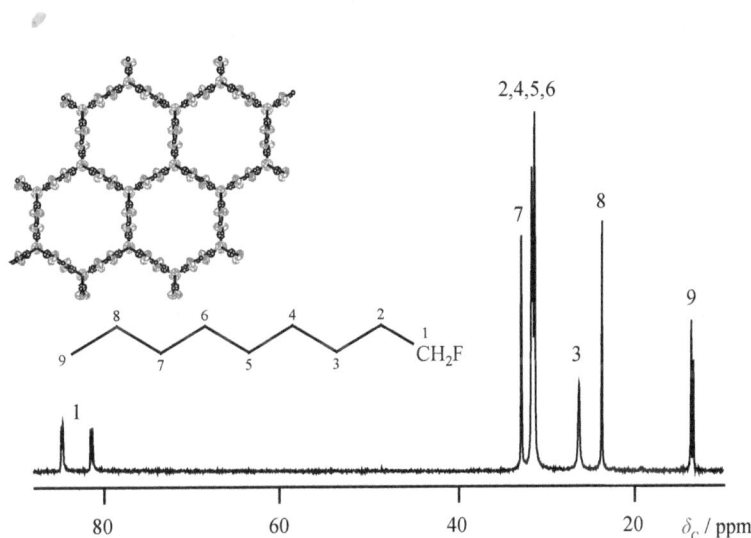

Figure 1.11. Top left: Structure of the inclusion compound between urea and 1-fluorononane as determined by X-ray studies at ambient temperature, viewed down the tunnels formed by the urea host. Note that the disordered guest molecules do not give rise to Bragg scattering and so the tunnels appear to be empty in the reported structure. Bottom: The low-frequency part of the ^{13}C spectrum (showing the signals arising from the guest molecule only), obtained under conditions of high-power proton decoupling and magic-angle spinning (see section 1.5).

such as alkanes and their simple derivatives can fit. However, the repeat distance of the urea host in the tunnel direction is not in general the same as the repeat distance of the guest molecules, which is governed by their length. Such structures fall into the class of incommensurate systems mentioned in section 1.3.4. Frequently, the guest molecules in such complexes are highly mobile in the host tunnels and high-quality spectra can therefore be readily obtained (figure 1.11). Diffraction techniques struggle to precisely locate the guest molecules in the tunnels because of their high degree of dynamic disorder.

Sometimes guest molecules can be incorporated into host crystal structures for a wide range of guest:host ratios, that is, in non-stoichiometric fashion. Such situations are highly likely when suitably sized channels or tunnels are present in the host chemical structure, allowing ready ingress or egress of guest molecules. For instance, the compound sildenafil citrate (the active principle of Viagra®) readily accepts water molecules—to a variable extent depending on the humidity of the environment.

Chiral molecules may form crystals that each contain only one enantiomer, but in such cases usually as a mixture of crystals of the two forms (a *conglomerate*). The NMR spectra of the two pure enantiomers (and of the corresponding conglomerate) will be identical. Alternatively, crystals of an internal *racemate* may form, with the enantiomers in a unit cell related by a center or a plane of symmetry. The spectrum of a racemate will in general differ from that of the related enantiomers, thus providing a powerful way of distinguishing between the two situations.

1.5 NMR OF SOLIDS & THE PERIODIC TABLE

As discussed above, the nature of the crystalline state implies that NMR spectra will be more complex than those of isotropic solutions. Additional considerations arise when microcrystalline powders are examined by NMR (as is the normal case), since each crystallite will be separately oriented in the applied magnetic field. This results in powder-pattern spectra for static samples, as illustrated earlier. Three particular techniques, which are discussed in more detail later, are of considerable importance in obtaining high-quality, well-resolved spectra, especially from powdered samples. Two of these relate to improvements in resolution, while the third is concerned with increasing signal intensities:

High-power proton decoupling (HPPD). A typical problem for observing spectra (of, say, ^{13}C or ^{15}N) of organic or organometallic compounds arises because they contain many protons. Heteronuclear dipolar interactions such as ($^{13}C,^1H$) or ($^{15}N,^1H$) can be very strong (up to ~50 kHz), which causes extensive line broadening. The difficulty may be overcome by *high-power* heteronuclear proton decoupling (see sections 3.3.7 and 5.1.2) analogous to proton decoupling for solution-state NMR, though the latter requires only low-power irradiation since it is required merely to eliminate "scalar" (J) coupling (which is much weaker than dipolar coupling for solids). High-power decoupling for solids eliminates scalar as well as dipolar coupling.

Magic-angle spinning (MAS). Another problem of spectral resolution arises because of the existence of shielding anisotropy, which, as mentioned earlier, implies that identical nuclei in different crystalline particles (i.e., at different orientations to B_0) will have different chemical shifts, resulting in line broadening (up to hundreds of parts per million for ^{13}C and even more for heavy-metal nuclei). Very rapid sample spinning about an angle of 54°44′ (the "magic" angle) suffices to average orientation effects and to produce relatively sharp lines. It may be noted that in principle this also averages dipolar interactions, but higher rotation rates are usually required, which are not readily achieved.

Cross polarization (CP). The third problem, which affects signal intensities rather than resolution, is that relaxation times for dilute spins in solids tend to be long (generally tens to hundreds of seconds, but in some cases as much as hours), rendering the usual pulse-and-acquire technique for recording free induction decays very inefficient (i.e., requiring long inter-pulse delays; see section 3.3.1). This difficulty is ameliorated by the use of CP from protons to the dilute spins (achieved by a pulse sequence as described in section 3.4).

The difficulties involved in obtaining good-quality spectra of powdered solids also depend on the type of nuclear spin involved. Since the properties of nuclides (spin quantum number I, natural abundance x, magnetogyric ratio γ, and quadrupole moment Q) are all-important for obtaining NMR spectra, they are listed in appendices A and B. Roughly speaking, four categories may be distinguished:

"Dilute" or "rare" spin-$\frac{1}{2}$ nuclides. In these cases, all three problems mentioned above are important. Some atoms are "dilute" because their natural isotopic abundances are low. Typical cases are ^{13}C, (1.07%) and ^{15}N (0.368%). These can be made more abundant by isotopic enrichment, which is commonly practiced for biochemical NMR studies. However, it is not only *isotopically* dilute nuclides that fall into the "dilute" category. A nucleus such as ^{31}P, which is present in 100% natural abundance, may be *chemically* dilute, as, for example, for a large biomolecule containing a single phosphorus atom. In fact, one can also envisage a *physically* dilute situation as for molecules in an inert matrix, though this would probably not be a crystalline situation. In practice in most situations nearly all spin-$\frac{1}{2}$ nuclei fall into the "dilute" category as far as the application of NMR techniques is concerned. Dilution limits sensitivity, so CP is usually vital, but for reasons discussed above so are HPPD and MAS. Such experiments form what is usually termed CPMAS NMR (the high-power proton decoupling being assumed). The combination of CP with MAS and high-power proton decoupling was first successfully implemented (for ^{13}C in organic systems) in 1976, which may be regarded as the birth of modern solid-state NMR, though the three techniques had already been known individually for several years.

"Abundant" spin-$\frac{1}{2}$ nuclides. Typically, such cases involve protons (because these are not only isotopically abundant, but also form a high proportion in number of atoms in most organic and organometallic compounds). For highly fluorinated compounds, ^{19}F nuclei may also act as "abundant" spins. In relatively rare situations, other nuclei may fall into the abundant category, for example, by isotopic enrichment. These situations involve strong *homo*nuclear dipolar interactions, which cause very substantial line broadening. In principle, MAS can overcome this problem also, but until the beginning of the 21st century, the required spin rates were not available, and the "classical" technique for overcoming this problem is so-called *multiple-pulse decoupling* (see section 5.5), which is a technically demanding method of homonuclear decoupling. Of course, if protons are chemically dilute for some reason, this difficulty does not arise and MAS at modest rates may suffice. Heavy deuteration can be used to artificially dilute the proton content of compounds.

Quadrupolar nuclides with half-integral spin quantum numbers. When nuclides have spin quantum numbers $> \frac{1}{2}$, they have not only magnetic dipole moments but also electric quadrupole moments, which interact with electric field gradients at the nuclei to give quadrupolar coupling. In fact the majority of the elements in the periodic table have at least one quadrupolar nuclide. The quadrupolar coupling constants can be very large (up to several hundred megahertz), so it is usually not possible to eliminate these by MAS. Fortunately, when the spin quantum number is an odd multiple of $\frac{1}{2}$, the central $-\frac{1}{2} \leftrightarrow \frac{1}{2}$ transition is unaffected by first-order quadrupolar coupling, as will be explained further in section 2.6. The broadening influences of proton dipolar coupling and chemical shift anisotropy can be removed and spectra involving this transition sharpened by using high-power proton decoupling and MAS respectively.

Quadrupolar nuclei with integral spin quantum numbers. Such nuclei are in principle the most difficult to deal with, since they have no central transitions. Fortunately, there are very few such nuclides, the principal ones being ^{2}H, ^{6}Li, ^{10}B, and ^{14}N. In each case, there is an alternative nuclide. However, it is sometimes valuable to have spectra of one or other of these four (though not normally of ^{10}B). In fact, the quadrupole moments of ^{2}H and ^{6}Li are small, so their spectra present no problem and only ^{14}N can be said to still represent a difficulty (see section 2.6).

FURTHER READING

GENERAL TEXTS ON NMR

"*Nuclear magnetic resonance: Concepts and methods*", D. Canet, John Wiley & Sons Ltd. (1996), ISBN 0 471 94234 0.

"*Understanding NMR spectroscopy*", J. Keeler, John Wiley & Sons Ltd. (2005), ISBN 0 470 01786 4.

"*Magnetic resonance in chemistry and medicine*", R. Freeman, Oxford University Press (2003), ISBN 0 19 926225 X.

GENERAL TEXTS ON THE SOLID STATE

"*Crystalline solids*", D. McKie & C. McKie, Thomas Nelson & Sons Ltd. (1974), ISBN 0 17 761001 8.

"*Fundamentals of crystallography*", Ed. C. Giacovazzo, Oxford University Press (1992), ISBN 0 19 8555 79 2.

"*Molecular crystals*", 2nd edition, J.D. Wright, Cambridge University Press (1995), ISBN 0 521 46510 9.

"*Solid state chemistry*", L. Smart & E. Moore, Chapman & Hall (1992), ISBN 0 412 40040 5.

"*Basic solid-state chemistry*", A.R. West, John Wiley & Sons Ltd. (1988), ISBN 0 471 91797 4.

"*Polymorphism in the pharmaceutical industry*", J. Bernstein, Oxford University Press (2002), ISBN 0 19 850605 8.

SPECIAL TEXT ON LINKING NMR AND THE SOLID STATE

"*NMR crystallography*", Eds. R.K. Harris, R.E. Wasylishen & M.J. Duer, John Wiley & Sons Ltd. (2009), ISBN 978 0 470 69961 4.

CHAPTER 2

BASIC NMR CONCEPTS FOR SOLIDS

2.1 NUCLEAR SPIN MAGNETIZATION

This chapter is designed to give readers an understanding of the nature of tensor properties, in particular those that affect NMR. The present section will cover some basic aspects of the behavior of nuclear spins. The material can be found in many introductory NMR textbooks and it will be familiar to anyone with an undergraduate knowledge of NMR. For this reason, the section is short and the reader is encouraged to explore other texts for additional details.

A moving (spinning) charge, such as that associated with a nucleus of non-zero magnetic quantum number, I, generates a magnetic moment. In the presence of a strong magnetic field, as used in an NMR experiment, this nuclear magnetic moment is quantized into $2I + 1$ directions; that is, there are different spin states defined by the magnetic component quantum number, m_I. The discussion at this point is restricted to spin $I = \frac{1}{2}$ nuclei so that there are just two such states. The low-energy state (when γ is positive) has a component of magnetic moment aligned with the magnetic field and is labeled α ($m_I = +\frac{1}{2}$) while the high-energy state (with a component opposed to the field) is designated β ($m_I = -\frac{1}{2}$). The energies of these states are governed by the Zeeman effect, as mentioned in chapter 1 and expressed in frequency terms in equation 2.1:

$$h^{-1}E_Z = -\frac{\gamma m_I B_0}{2\pi} \qquad\qquad 2.1$$

The transition between these states is the basis of NMR spectroscopy. The energy difference between them is

$$h^{-1}\Delta E_Z = \frac{\gamma B_0}{2\pi} \qquad\qquad 2.2$$

and this equates to a resonance frequency (v_{NMR}) through

$$\Delta E_Z = h v_{NMR} \qquad 2.3$$

so that

$$v_{NMR} = \frac{\gamma B_0}{2\pi} \qquad 2.4$$

This frequency is often referred to as the *Larmor* frequency. In order to understand or even describe (in pictures) the effect of an NMR experiment, it is helpful to introduce some tools. Key to these is the behavior of a bulk sample (as opposed to the quantum mechanical treatment of individual nuclei). This description of NMR comes from the classical physics of electromagnetism. For an ensemble of nuclear spins, such as is contained in a typical NMR sample, the population (number of nuclei) in each spin state is p_α and p_β, respectively. At thermal equilibrium, there is a Boltzmann distribution between the two states:

$$\frac{p_\beta}{p_\alpha} = e^{-h v_{NMR}/kT} \qquad 2.5$$

where k is the Boltzmann constant and T is the absolute temperature. This ratio depends both on the applied magnetic field and on the nuclide in question (see inset 2.1). Equation 2.5 predicts that there will be slightly more nuclei (magnetic moments) aligned with the applied magnetic field than opposing it. Thus, there will be a net (bulk) magnetization aligned with the field. The magnitude, M_0, of this bulk magnetization for a sample containing N nuclei is given by

$$M_0 = \frac{N(\gamma h)^2 B_0}{16\pi^2 kT} \qquad 2.6$$

It is the behavior of this bulk magnetization that provides a convenient way of describing an NMR experiment (see section 3.2).

The resulting spectra are influenced by the interactions mentioned in section 1.1. Consequently, a more detailed description of the "internal" interactions, which ultimately give useful chemical information about the sample, is now given.

Inset 2.1. NMR and sensitivity

For ^{13}C, from equations 2.4 and 2.5 with $\gamma = 6.728 \times 10^7$ rad s^{-1} T^{-1} at a magnetic field of 9.4 T and at a temperature of 294 K, $p_\beta/p_\alpha = 0.999983$. Put another way, for every 1 million nuclei in the β state, there are only 17 more in the α state. This is the reason why NMR is, as analytical techniques go, relatively insensitive.

2.2 TENSORS

All the internal interactions involved in NMR are orientation dependent, so they are properly expressed as *tensor* properties. While this section will be written in terms of a general tensor, R, the shielding tensor is chosen when a specific example is needed. A tensor may be thought of as the link between two vectors. Thus the shielding tensor, σ, links the magnetic field arising from shielding of the nucleus by the electrons, B_S, to the applied magnetic field, B_0:

$$B_S = -\sigma B_0 \qquad\qquad 2.7$$

Since vectors have three components each, any tensor R (in its Cartesian representation) involves a 3×3 matrix, that is, nine components:

$$\begin{pmatrix} R_{xx} & R_{xy} & R_{xz} \\ R_{yx} & R_{yy} & R_{yz} \\ R_{zx} & R_{zy} & R_{zz} \end{pmatrix} \qquad\qquad 2.8$$

Thus, if we suppose that B_0 is directed along z, then the shielding field has the following components:

$$\begin{aligned} B_{Sx} &= -\sigma_{xz} B_0 \\ B_{Sy} &= -\sigma_{yz} B_0 \\ B_{Sz} &= -\sigma_{zz} B_0 \end{aligned} \qquad\qquad 2.9$$

In other words, the shielding field is *not* in general parallel to the applied field (figure 2.1).

However, tensors are often symmetric, that is, $R_{ji} = R_{ij}$ and, even when this is not the case, the antisymmetric contributions have little influence on NMR spectra and are usually ignored. In such circumstances, there are only six distinct components to a tensor, and it is possible to choose axes (forming the *principal axis system,* PAS) in which R is diagonal:

$$\begin{pmatrix} R_{XX} & 0 & 0 \\ 0 & R_{YY} & 0 \\ 0 & 0 & R_{ZZ} \end{pmatrix} \qquad\qquad 2.10$$

The terms R_{XX}, R_{YY}, and R_{ZZ} are called the *principal components* of the tensor (and the use of capital X, Y and Z indicates this). They provide information of interest, as does the orientation of the PAS in an axis system fixed in the molecule or crystal (involving the other three necessary parameters). The

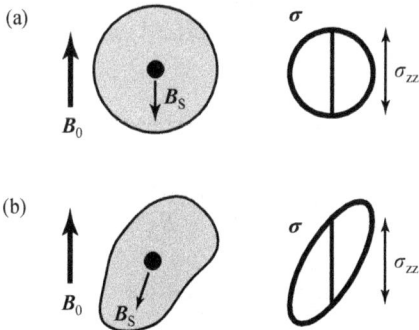

Figure 2.1. Schematic relationship of a shielding field to the applied field: (a) isotropic site; (b) anisotropic site. Shielding ellipsoids are shown on the right—ellipsoids indicate the magnitude of the tensor in different orientations.

transformation of a tensor from a general frame of reference to the PAS (or vice versa) is carried out by a process of rotation (see section 4.6 and Further reading).

Actually, instead of R_{XX}, R_{YY}, and R_{ZZ} being used, it is more common to define three alternative quantities. One of these is the *isotropic average*:

$$R_{iso} = \tfrac{1}{3}(R_{XX} + R_{YY} + R_{ZZ}) \qquad 2.11$$

which is the quantity that would be observed in a solution-state experiment, since it is invariant to rotational transformation. In many high-resolution spectra of solids, this is also the only parameter that is measured. The other two parameters are the *anisotropy* and the *asymmetry*, which are defined (together with the choice of axis labeling) in inset 2.2.

Inset 2.2. Definitions of tensor anisotropy and asymmetry

In the case of shielding, there are two alternative definitions (with different symbols) for anisotropy:

$$\Delta R = R_{ZZ} - \tfrac{1}{2}(R_{XX} + R_{YY}) \qquad 2.12$$

$$\zeta = R_{ZZ} - R_{iso} \qquad 2.13$$

The relation between them is $\zeta = 2\Delta R/3$. Either definition may be used for shielding. However, ΔJ is the normal form for indirect coupling.

The asymmetry, η, is a dimensionless quantity lying between 0 and 1. It is defined by:

$$\eta = (R_{YY} - R_{XX})/\zeta \equiv 3(R_{YY} - R_{XX})/2\Delta R \qquad 2.14$$

Actually, as can be seen, the asymmetry is also a type of anisotropy. This definition is common to shielding, indirect coupling, and quadrupolar coupling. When a nucleus is at a site of axial symmetry, R_{XX} and R_{YY} are necessarily equal, so η is zero, and this is normally the case for dipolar coupling. In such a situation, the notations $R^{\parallel}$ and $R^{\perp}$ may be used for the unique (Z) component and the ones perpendicular to it respectively. The term $\tfrac{1}{3}(R^{\parallel} + 2R^{\perp})$ is the isotropic average (as defined in equation 2.11), while $(R^{\parallel} - R^{\perp})$ is the anisotropy ΔR.

Ordering of X, Y, and Z is defined herein (for the NMR tensors listed in table 2.1, except the quadrupolar coupling (see page 29)) by:

$$|R_{ZZ} - R_{iso}| \geq |R_{XX} - R_{iso}| \geq |R_{YY} - R_{iso}| \qquad 2.15$$

This choice looks a little odd until it is realized that the X, Y, and Z components are in alphabetical order in the spectrum (see figure 2.2), either $R_{zz} \geq R_{yy} \geq R_{xx}$ or $R_{zz} \leq R_{yy} \leq R_{xx}$ (depending on the system studied).

The asymmetry becomes 1 when the Y component is midway between the X and Z components.

The following sections give details of the four tensor properties most important for solid-state NMR, namely shielding, indirect coupling, dipolar coupling, and quadrupolar coupling. Table 2.1 gives some of the general properties of these tensors. The equations in this book are written such that the coupling parameters (and their anisotropies) appear as frequencies (other texts use angular frequencies).

Table 2.1. Properties of tensors involved in NMR

Tensor	General label	Tensor symbol	Isotropic value	Symmetric?	Axially symmetric?
Shielding	σ	$\boldsymbol{\sigma}$	σ_{iso}	No	Not in general
Indirect coupling	J	$\boldsymbol{J}$	J_{iso}	No	Not in general
Dipolar coupling	D	$\boldsymbol{D}$	0	Yes	Yes[a]
Quadrupolar coupling	Q	$\boldsymbol{q}$[b]	0	Yes	Not in general

[a] In certain circumstances involving motional averaging (see section 8.2), these tensors will not be axially symmetric.

[b] Electric field gradient.

2.3 SHIELDING

When a molecule is placed in a magnetic field, the nuclei are magnetically shielded by the presence of electrons, resulting in a shielding field as expressed in equation 2.7. Shielding leads to the well-known phenomenon of *chemical shift*. Isotropic chemical shifts have undoubtedly been the most important items of information in solution-state NMR, allowing chemical structures to be determined. Isotropic chemical shifts can also be obtained for solids, as described in this book, and are similarly used, though there are also special features that apply only to the solid state.

In general, however, electrons are not spherically distributed around any given nucleus, so therefore such a field will be anisotropic. Thus, shielding will be a tensor property, $\boldsymbol{\sigma}$. It is a single-spin quantity since the primary field, $\boldsymbol{B}_0$, is external to the system. The lead nucleus in lead nitrate, for instance, is differently shielded when the local symmetry axis at the lead nucleus is perpendicular to $\boldsymbol{B}_0$ than when it is parallel to $\boldsymbol{B}_0$ (see figure 1.2). These extreme tensor components are labeled $\sigma^{\perp}$ and $\sigma^{\parallel}$, respectively. When the molecular axis is at an angle θ to $\boldsymbol{B}_0$, the shielding in the z ($\boldsymbol{B}_0$) direction can be shown to be:[1]

$$\sigma_{zz}(\theta) = \tfrac{1}{3}(\sigma^{\parallel} + 2\sigma^{\perp}) + \tfrac{1}{3}(3\cos^2\theta - 1)(\sigma^{\parallel} - \sigma^{\perp}) \qquad 2.16$$

---

[1] Note the use of lower-case zz to indicate that σ_{zz} in this equation does not refer to the ZZ principal component.

Since the shielding field, B_S, is orders of magnitude less than B_0, any components of B_S perpendicular to B_0 can be ignored, so the total field considered is:

$$B_0 (1-\sigma_{zz})$$

2.17

Equation 2.1 must therefore be modified by the shielding energy, E_S, and the energy of an ensemble of spins of type j in B_0 therefore becomes:

$$h^{-1}(E_Z + E_S) = \frac{\gamma_j}{2\pi} B_0 \{1-\sigma_{zz}^j(\theta)\}$$

2.18

for a given orientation θ, assuming shielding is the only important interaction other than the Zeeman effect. The transition (Larmor) frequency is correspondingly modified from equation 2.4, thus giving rise to a chemical shift that depends on the electronic environment of the nucleus in question. The total spin energy of a sample will involve summation over all types and numbers of spins, j.

For a polycrystalline sample, there will be a complete spread of angles θ, so a range of resonances, constituting a *shielding powder pattern*, will be observed. Such a spectrum for an axially symmetrical case was shown in figure 1.2. The resonance frequency depends on $(3\cos^2\theta - 1)$ and therefore becomes a measure of θ so that any preferred orientation (such as occurs in a drawn polymer sample) can be observed as a departure from the expected intensity distribution, recognized qualitatively and determined quantitatively. The intensity of the signal at a particular frequency will depend on the probability of the relevant angle θ. Clearly, $\theta = 90°$ is much more common than $\theta = 0°$ (think of the many directions from the center of the earth to the equator versus the unique directions to the poles), so the intensity at $\sigma^\perp$ will be much higher than that at $\sigma^\|$, as seen schematically in figure 2.2(a) (and experimentally in the upper spectrum of figure 1.2). The more intense edge of the powder pattern will depend on whether $\sigma^\perp > \sigma^\|$ or $\sigma^\| > \sigma^\perp$, which will vary with the chemical system in question.

For a nucleus at a site of cubic symmetry (as for ^{19}F in crystalline CaF_2, for example), the three tensor components will be equal, so the shielding will be independent of the sample orientation and a single relatively sharp resonance will be seen.

In the more general case, there will be three different principal components of the shielding tensor, as mentioned in section 2.2 above, and the powder pattern will have, correspondingly, three turning points as in the three lower spectra in figure 2.2. The relevant equation for the effective shielding (to be used in equation 2.18 for deriving energy-level diagrams and hence transition frequencies) will be:

$$\sigma_{zz}(\theta) = \sigma_{iso} + \frac{1}{3} \sum_{j=X,Y,Z} (3\cos^2\theta_j - 1)\sigma_{jj}$$

2.19

where θ_j is the angle between the shielding principal axis j and the magnetic field. This equation may be rewritten in terms of the anisotropy, ζ, and asymmetry, η, as:

$$\sigma_{zz}(\theta,\varphi) = \sigma_{iso} + \frac{1}{2}\zeta(3\cos^2\theta - 1 - \eta\sin^2\theta\cos 2\varphi)$$

2.20

where θ and φ are the spherical angles defining the orientation of the Z principal axis of shielding in the magnetic field B_0.

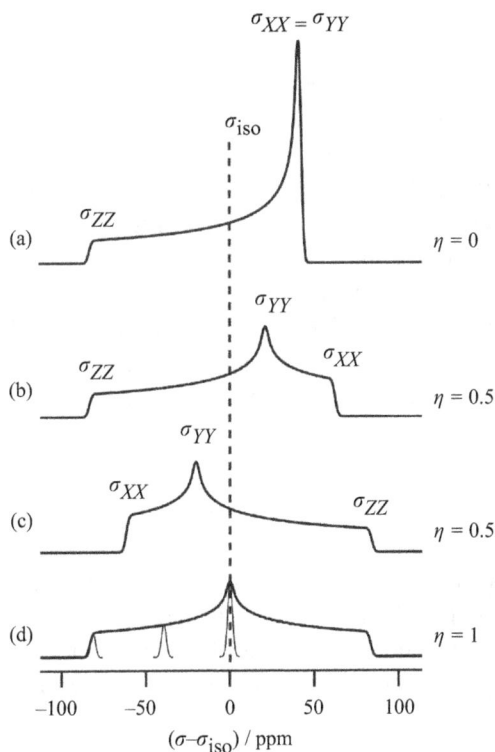

Figure 2.2. Schematic powder patterns of shielding for (a) an axially symmetric case with negative shielding anisotropy, (b) a general case with negative shielding anisotropy, (c) an analogous case with positive shielding anisotropy, and (d) a case with an asymmetry of one. The bottom spectrum also shows the narrow lines obtained from three particular crystallite orientations, with the X and Y tensor components aligned along $\boldsymbol{B}_0$ and for an intermediate situation. Note that as the scale uses the normal convention of frequency increasing right to left, the orientation of the lineshapes will match experimental spectra.

For powder patterns of the types shown in figure 2.2, the turning points give the values of the principal components (but not the orientation of the PAS in the molecular/crystallographic frame). To obtain reasonably accurate data, a bandshape-fitting procedure is required, as discussed in section 8.2.1. However, in situations of nearly axial shielding, differentiation of σ_{xx} from σ_{yy} is difficult, that is, values of η below 0.2 are not determined accurately.

Shielding anisotropies are rather small for ^{1}H, but are up to ~200 ppm for ^{13}C and can be massive for some metallic nuclides.[2] Their values, together with those for shielding asymmetries, can be used to assign resonances to chemical sites. They also give important crystallographic information and frequently distinguish signals far better than do isotropic shifts. They help to understand electronic structure, but this is properly obtained only when the experimental measurements are combined with quantum mechanical calculations of shielding. One could say that measurement of the full shielding tensor (including its orientation in $\boldsymbol{B}_0$) provides an order of magnitude more information than the isotropic value alone.

---

[2] Metal samples have so-called *Knight shifts* (which can be very large) in their resonances. These will not be dealt with in this book. Paramagnetic samples are also subject to special shift effects (see section 7.5).

2.4 INDIRECT COUPLING[3]

This type of spin–spin coupling operates via electrons and so is classed as "through bonds." It therefore gives information on molecular-level connectivity and questions of molecular conformation. The coupling parameter is given the symbol J and is highly important for solution-state NMR, but somewhat less prominent in solid-state NMR studies. This is mainly because it is generally the smallest interaction in magnitude, being normally less than 1 kHz for light elements (though for some heavy metal nuclides, it can be in the region of 10 kHz). Solution-state NMR depends only on its isotropic value (and hence it is often referred to as "scalar" coupling), but in fact it is actually a tensor property, $\boldsymbol{J}$. Any asymmetry in $\boldsymbol{J}$ is usually neglected since it is expected to be generally small and difficult to determine, but anisotropies ΔJ need to be taken into account for solid-state spectra in some (relatively unusual, at least for 13C NMR) cases. The contribution of indirect coupling to NMR energies for a pair of nuclei j and k in a sample is:

$$h^{-1}E_{\mathrm{J}} = m_{\mathrm{j}}m_{\mathrm{k}}\left\{ J_{\mathrm{iso}}^{\mathrm{jk}} + \frac{1}{3}\left(3\cos^2\theta-1\right)\Delta J^{\mathrm{jk}} \right\}$$

2.21

For the full coupling contribution to the energy, this term must be summed over all pairs of spins. For molecular solids, indirect coupling is generally confined to intramolecular interactions, but it can extend across hydrogen bonds between molecules and therefore it can be used to recognize the existence of such bonding. Its relationship to connectivity is particularly valuable for solid-state NMR.

2.5 DIPOLAR COUPLING

Dipole–dipole coupling between spins corresponds to the classical interaction between two magnets, with dipole moments $\boldsymbol{\mu}_{\mathrm{j}}$ and $\boldsymbol{\mu}_{\mathrm{k}}$, and thus operates "through space." The dipolar energy for such a pair is given by:

$$E_{\mathrm{D}} = \left\{ \frac{\boldsymbol{\mu}_{\mathrm{j}}\cdot\boldsymbol{\mu}_{\mathrm{k}}}{r^3} - \frac{3(\boldsymbol{\mu}_{\mathrm{j}}\cdot\mathbf{r})(\boldsymbol{\mu}_{\mathrm{k}}\cdot\mathbf{r})}{r^5} \right\} \frac{\mu_0}{4\pi}$$

2.22

where r is the distance between the two dipoles (see figure 2.3).

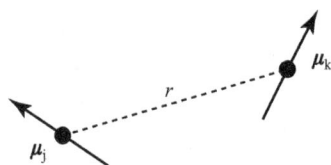

Figure 2.3. Interaction between two dipoles.

---

[3] This coupling phenomenon is frequently referred to by the clumsy designation "J coupling." This is so common in the literature that the terminology is sometimes used in this book, though on other occasions the more meaningful expression "indirect coupling" is employed.

For a heteronuclear spin pair in the high magnetic field approximation, this expression becomes (to first order in energy):

$$h^{-1}E_D = -D_{jk}m_jm_k\left(3\cos^2\theta - 1\right) \qquad 2.23$$

where θ is the angle between the internuclear vector and the magnetic field, while D_{jk} is the dipolar coupling constant between spins j and k. The latter is given (in frequency units) by:[4]

$$D_{jk} = \gamma_j\gamma_k\left(\frac{h}{4\pi^2}\right)\left(\frac{\mu_0}{4\pi}\right)r_{jk}^{-3} \qquad 2.24$$

Typical values of dipolar coupling constants are given in table 2.2. The tensor $\boldsymbol{D}$ has no anti-symmetric component and, unless there is anisotropic motion involving the two nuclear spin sites involved, it is also axially symmetric. The isotropic component of $\boldsymbol{D}$ is zero, so dipolar coupling does not affect resonance frequencies for solutions, although dipolar interactions are a primary cause of relaxation.

For a single crystal of a substance with an isolated pair of spin-$\frac{1}{2}$ heteronuclei, A and X, and a unique direction r_{AX}, the A spectrum will be a doublet of spacing $|D_{AX}(3\cos^2\theta - 1)|$, which varies in position and splitting magnitude with the crystal orientation (and is zero if θ is 54°44′).

Table 2.2. Some typical bond distances and corresponding dipolar coupling constants

Nuclei	Type[a]	r/nm	D/kHz	Comment
$^{13}C,^{13}C$	$C_{sp^3}-C_{sp^3}$	0.153	2.12	Hydrocarbons
	$C_{sp^2}-C_{sp^2}$	0.132	3.29	Unconjugated hydrocarbons
	$C_{ar}-C_{ar}$	0.140	2.79	Hydrocarbons without further conjugation
	$C_{sp}-C_{sp}$	0.118	4.61	Unconjugated hydrocarbons
$^{13}C,^1H$	$(C)-CH_3$	0.106	25.44	
	$C_{ar}-H$	0.108	23.78	
$^{13}C,^{15}N$	$C_{sp^3}-N(3)$	0.147	0.97	
$^{13}C,^{29}Si$	$C_{sp^3}-Si(4)$	0.186	0.93	
$^{13}C,^{31}P$	$C-P(=O)-C$	0.181	2.05	
$^{31}P,^{31}P$	$P(3)-P(3)$	0.221	1.82	

[a]ar = aromatic.

---

[4] Some texts define D_{jk} include a negative sign in its definition. The key point is that the energy expressed in equation 2.23 is lower when the nuclear magnetic dipoles are parallel (if γ values for both nuclei are positive).

The X spectrum behaves identically to the A spectrum. The angle θ can be evaluated by varying the orientation of the single crystal in B_0. Hence, D can be measured and r_{AX} determined. For a typical C–H bond distance (say, $r_{AX} = 0.11$ nm), the dipolar coupling constant is 22.7 kHz. Homonuclear spin pairs have more complicated spin energies (see section 4.5.1) so that equation 2.23 is no longer adequate. However, a similarly simple result is obtained, the only difference being that a single doublet is seen, with a spacing of $|\frac{3}{2}D(3\cos^2\theta - 1)|$, that is, there is an extra factor of $\frac{3}{2}$.

For polycrystalline powders, the distribution of orientations must be taken into account, as shown above for shielding. In fact for a pair of isolated heteronuclear spins (AX), the A spectrum will look like two superimposed patterns of the type shown in figure 2.2(a), one for each of the doublet resonances, with reversal of direction for one of them. This is illustrated in figure 2.4. Such a powder pattern is known as a *Pake Doublet*,[5] the outer limits of which correspond to $\theta = 0°$ and the intense "horns" to $\theta = 90°$. The separation of the horns is D, so the distance r_{AX} can be obtained even in the case of a powder. The X spectrum will be identical. The powder pattern for an isolated pair of homonuclear spins will be a single similar shape, but with a scaling factor of $\frac{3}{2}$.

In fact, the classic case of a dipolar coupling is that of the proton spectrum of gypsum, $CaSO_4 \cdot 2H_2O$, for which the protons in the water molecules form relatively isolated spin pairs. Since this case involves homonuclear interactions, the factor of $\frac{3}{2}$, mentioned above, is required for the spectrum. The unit cell contains two water molecules with different orientations, so two doublets are generally observed for a single crystal in a fixed setting of the crystal in B_0, though for some special settings only one doublet is observed. For one setting, a single line is observed, showing that r_{HH} for

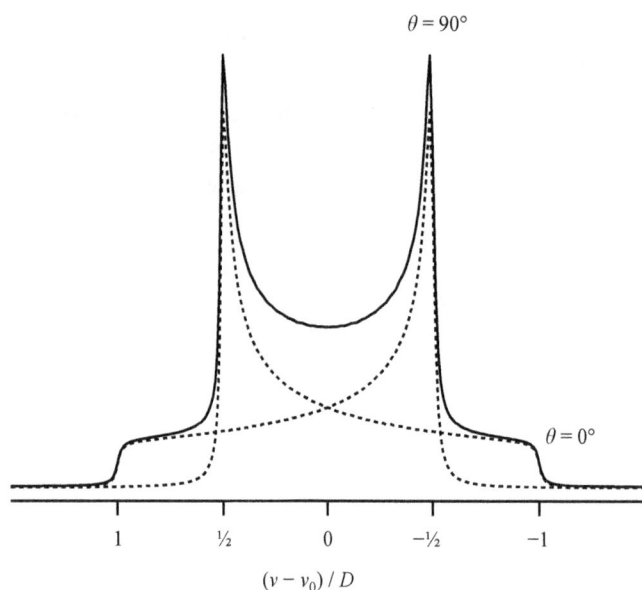

Figure 2.4. A schematic heteronuclear Pake Doublet for the A spin of an AX spin system. The dotted lines show the two subspectra (for different values of m_I for the X spin). ν_0 refers to the transition frequency under the influence of shielding.

[5] The term "Pake Doublet" originally referred to the homonuclear case.

the two different water molecules can be nearly at the magic angle $\theta = 54°44'$ to $\boldsymbol{B}_0$ simultaneously. A full analysis shows that the dipolar coupling constant is 30.7 kHz, so r_{HH} is determined to be 0.158 nm. Given the traditional difficulties of determining hydrogen positions by X-ray diffraction, such NMR experiments are potentially valuable.

In general, spectra involving dipolar interactions are much more complex because they are affected by the influence of all the spin pairs in the sample, with a huge variety of internuclear distances and orientations in the applied magnetic field. It is difficult to adequately take this into account, but suffice it to say that dipolar coupling causes extensive broadening of spectra, as is especially true for *proton* NMR of solids, where a single broad featureless line (over 50 kHz in width at half-height) is generally seen for rigid organic systems in static samples (figure 2.5). It is also typically the case for proton-coupled ^{13}C spectra, as illustrated in the upper spectrum in figure 1.1. A common aim of many attempts to determine interatomic distances by NMR is the production of samples with isolated spin pairs, for example by selective isotopic enrichment.

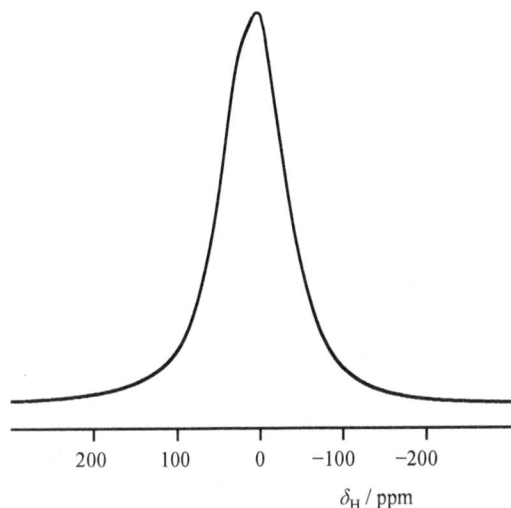

Figure 2.5. Proton spectrum of a static sample of microcrystalline alanine.

$$\delta_H \,/\, ppm$$

As already stated, molecular mobility in the liquid phase averages dipolar interactions to zero (except if the motion is anisotropic, as for liquid crystals). Any motion at the molecular level in solids will cause some averaging also, which will narrow resonances, as shown in figure 2.6 for the proton spectrum of amorphous indomethacin (which is an active pharmaceutical ingredient). In this case, motional narrowing occurs as the sample passes through its glass transition temperature (see section 1.3.4). (The motion will also simultaneously partially average shielding tensors.)

Obviously, measurement of dipolar coupling constants supplies values of internuclear distances and so is an important feature of NMR crystallography. It is even more valuable for amorphous systems, where such information is otherwise difficult to obtain. However, the effects of molecular-level mobility must be taken into account. If the crystal structure is known from diffraction experiments, details of mobility may be inferred.

90 °C

70 °C

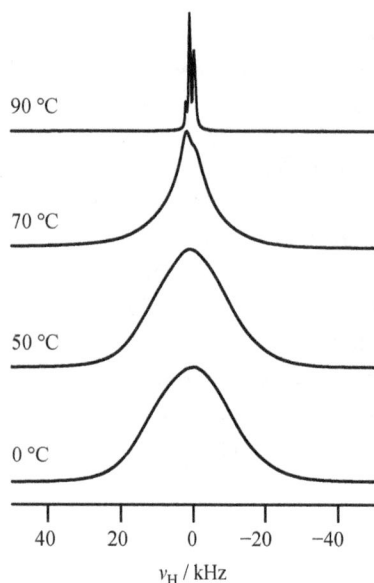

Figure 2.6. Proton spectrum of indomethacin at various temperatures, showing motional narrowing, which becomes significant above the glass transition temperature.

50 °C

0 °C

40   20   0   −20   −40

ν_H / kHz

2.6   QUADRUPOLAR COUPLING

For nuclides with spin quantum numbers greater than $\frac{1}{2}$ (only), another form of coupling exists. This is an electric interaction, not magnetic, and involves the *nuclear electric quadrupole moment*, eQ, of the nuclide. An introduction to this topic will be given here, but details are presented in chapter 6. Values of Q are listed in tabulations of nuclear spin properties (see table 6.1 and appendix B). This property can be positive (as for ^{27}Al) or negative (as for ^{17}O), and its value depends on the nature of the charge distribution within the nucleus. The quadrupole moment couples with the *electric field gradient* (EFG), eq, at the nucleus, which originates from the surrounding electrons. The EFG is a tensor property and therefore so is quadrupolar coupling. While this effect is electric in nature, it does affect the spin energy and so influences NMR spectra. Now q is a traceless tensor,[6] so the isotropic average, $\frac{1}{3}(q_{XX} + q_{YY} + q_{ZZ})$, is zero (like that of D). Thus, resonance frequencies for solutions are basically unaffected by the quadrupolar interaction. However, it still contributes significantly to relaxation and therefore to line broadening—so much so that for nuclei with large values of Q, it may be very difficult to observe the signal for the solution state.

For solids, the quadrupolar effect can be very important. Indeed, it often dominates the appearance of spectra, since the quadrupole coupling constant may be hundreds of megahertz. On the other hand, when the nucleus in question is at a site of cubic symmetry, the EFG is in principle zero. An example is the EFG at ^{33}S in a sulfate ion; however, it will only be negligibly small if the ion retains its tetrahedral symmetry in the solid state, whereas frequently the local environment lowers the symmetry.

---

[6] The symbol V is often used in the literature instead of eq for the EFG tensor.

Only two parameters are in principle needed to characterize the magnitude of q in any given situation, namely its anisotropy and asymmetry (see section 2.2). Since q is traceless, the former is simply given by the largest component in magnitude. This may be either positive or negative, the principal axes being denoted by the order $|q_{ZZ}| \geq |q_{YY}| \geq |q_{XX}|$.[7] However, it is more normal to use, instead of the EFG anisotropy itself, a factor proportional to it, known as the *nuclear quadrupole coupling constant*,[8] χ. This parameter and the quadrupolar asymmetry, η, are defined as:

$$\chi = \frac{e^2 Q q_{ZZ}}{h}$$
2.25

$$\eta = \frac{q_{XX} - q_{YY}}{q_{ZZ}}$$
2.26

where χ is expressed in frequency units but η is dimensionless ($0 \leq \eta \leq 1$).

If the quadrupolar energy is much smaller than the Zeeman energy, it can be treated as a first-order perturbation (see chapter 6 for details). In such a case, the energy, $E_Q^{(1)}$, for nuclear spin I is given by:

$$h^{-1} E_Q^{(1)} = \left\{ \frac{3m_I^2 - I(I+1)}{8I(2I-1)} \right\} [(3\cos^2\theta - 1) + \eta \cos 2\phi \sin^2\theta] \chi$$
2.27

where m_I is the spin-component quantum number and the parameters θ and ϕ are the polar angles of B_0 in the principal axis system of the tensor.

The full quadrupolar energy term must be summed over all the types of quadrupolar nuclei present. However, unlike the situation for dipolar interactions (which occur for spins pairwise), the quadrupolar effect involves single spins acting independently, which simplifies theoretical treatments.

If the asymmetry is zero, the first-order quadrupolar contribution to the resonance frequency for a transition from m_I to $m_I - 1$ is:

$$v_Q^{(1)} = \frac{3(2m_I - 1)}{8I(2I-1)} (1 - 3\cos^2\theta) \chi$$
2.28

Since there are $2I + 1$ energy levels (figure 2.7 shows the case for a spin-$\frac{3}{2}$ nucleus), there will be $2I$ nondegenerate lines in the spectrum for each value of θ other than $54°44'$.

Most quadrupolar nuclides have spin quantum numbers that are odd multiples of $\frac{1}{2}$ (i.e., $\frac{3}{2}$, $\frac{5}{2}, \frac{7}{2}$,...). In such cases, the *central transition* ($\frac{1}{2} \leftrightarrow -\frac{1}{2}$) is unique because the first-order quadrupolar effects on the two energy levels are equal. This is true even including cases of nonzero asymmetry, since the quadrupolar energy depends on m_I^2, not m_I (see equation 2.27). Therefore, the quadrupolar

---

[7] This (traditional) definition puts the Y component between the X and Z components, contrary to the usage for the other NMR tensors.

[8] The symbol C_Q is often used for this quantity, though χ is the IUPAC-approved symbol.

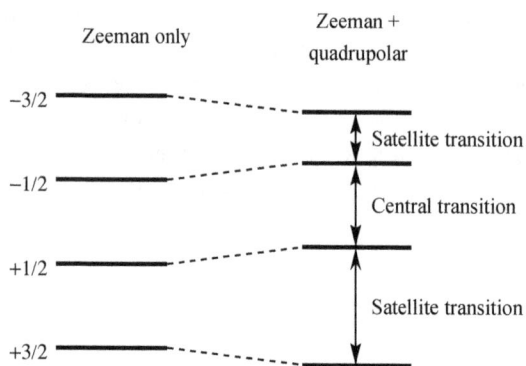

Figure 2.7. Energy levels for a quadrupolar nucleus. Left: zero quadrupole coupling constant; right: schematic introduction of a small quadrupole coupling constant (first-order effect).

contribution to the transition frequency is zero and is thus independent of the orientation. The other NMR transitions (*satellite transitions*) are, however, strongly dependent on θ, so the intensity for a polycrystalline sample will be spread over a wide frequency range. The central transition resonance will thus stand out like a sore thumb. Figure 2.8 shows a powder pattern for an $I = \frac{5}{2}$ hypothetical case. The spacing of the prominent horns (indicated by the arrow in figure 2.8) depends on the spin

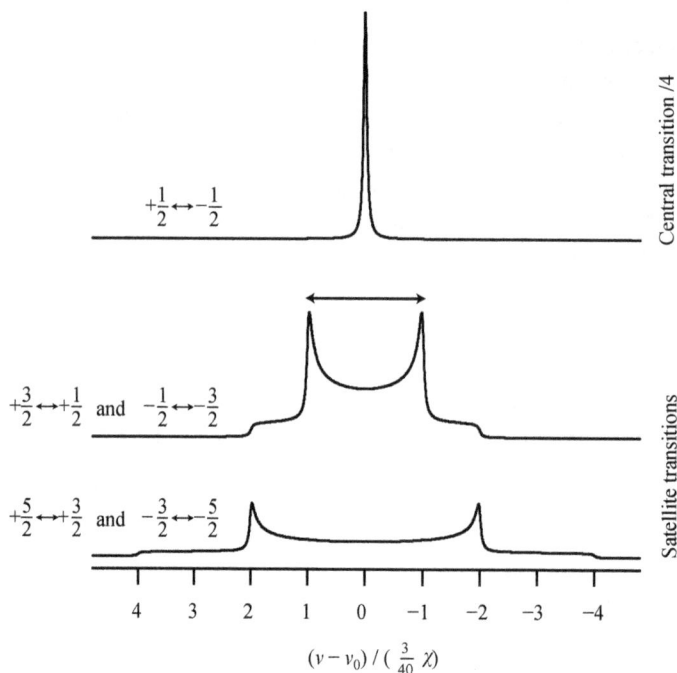

Figure 2.8. Simulated quadrupolar powder pattern for an $I = \frac{5}{2}$ case. The three types of transition (subspectra) are shown separately.

quantum number I; in the $I = \frac{5}{2}$ case, it is $\frac{3}{20} \chi$. The satellite transitions form subspectra as indicated. Figure 2.9 provides an experimental example for an $I = \frac{3}{2}$ system.

If χ is large, the satellite transitions may not be readily visible, in which case only 40% of the intensity (for $I = \frac{3}{2}$) will be detected (see inset 6.1). On the other hand, when χ is small, only a single line (though usually rather broad) may be seen, comprising all transitions. This is often the case for ^{133}Cs, which has a small value of Q (-0.343 fm^2) and is usually ionic (q_{ZZ} small). Contrast this with the cases of ^{23}Na ($Q = 10.4$ fm^2) and ^{35}Cl ($Q = -8.165$ fm^2).

However, when χ is large, as is often the case, the simple theory given above is not adequate and second-order effects become prominent, as will be described in chapter 6. The case of nuclides with integral spin is also different since there is no central transition (see inset 2.3). The only nuclides in this category usually considered to be important are ^2H, ^6Li, ^{10}B, and ^{14}N. Of these, the first two do not generally give broad spectra, since they have small quadrupole moments, and ^{11}B is usually preferred to ^{10}B, so only ^{14}N causes special difficulties.

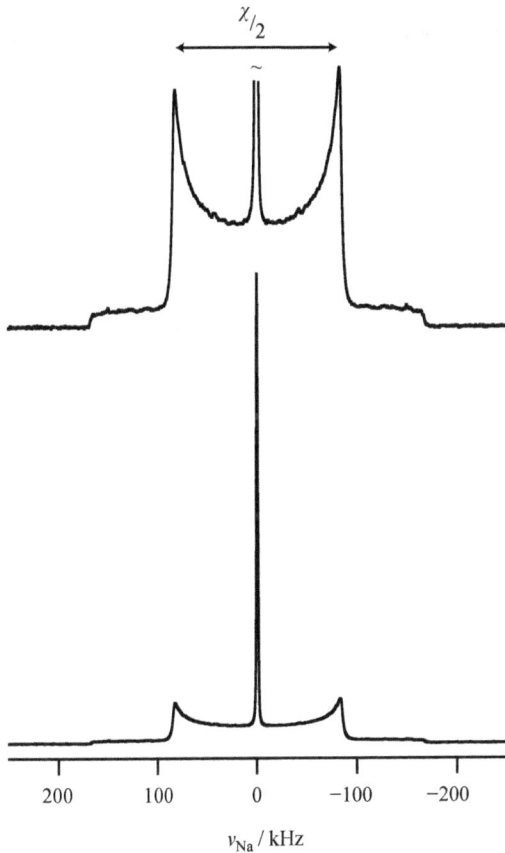

Figure 2.9. Sodium-23 spectrum of a static microcrystalline sample of sodium nitrate, which has a relatively small quadrupole coupling constant, χ, of ~335 kHz and zero asymmetry. The upper spectrum is a vertical expansion to show the turning points more clearly.

Inset 2.3. Spin-1 nuclei

Equation 2.27 shows that, for a spin-1 nucleus in an axially symmetric site, the energy levels $m_I = +1$ and $m_I = -1$ are affected in the opposite direction to the $m = 0$ level by the quadrupolar interaction. Therefore, one of the two possible $\Delta m_I = \pm 1$ transitions increases in frequency, while the other one decreases. There can be no "central transition." For a powdered sample, the transitions form subspectra that are influenced by $(3\cos^2\theta - 1)$ in opposite ways so as to give a total bandshape, which is identical to that of the homonuclear Pake Doublet, but with a scale governed by $\chi/2$ instead of D, that is with the separation of the "horns" of $3\chi/4$. Figure 2.10 shows schematically a typical deuterium spectrum for a zero-asymmetry case with a quadrupole coupling constant of 170 kHz.

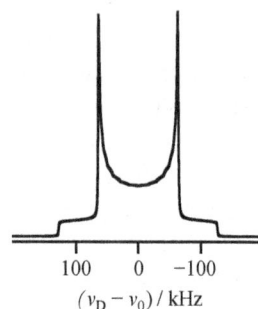

100   0   −100
$(\nu_D - \nu_0)\,/\,kHz$

Figure 2.10.

Quadrupolar data are useful in a number of ways, including acting as resonance assignment aids and as enhancements for understanding electronic structure. Accounting for quadrupolar effects is necessary to obtain chemical shifts of the relevant nuclei, with implications for structure determination, as is described in chapter 6.

2.7   MAGIC-ANGLE SPINNING

All the internal NMR interactions have a common mathematical form. In particular, there is a universal dependence on orientation of the molecular/crystallographic frame in the applied magnetic field, which, at least to first order as given in the above sections, is of the form:

$$\tfrac{1}{2}(3\cos^2\theta - 1) \qquad\qquad 2.29$$

The average value of this function for isotropic motion is zero. Thus molecular motion in solutions means that the anisotropies in σ and J do not affect the resonance frequencies. Since the isotropic values of D and q are zero, these interactions also have no significant influence on solution-state resonance frequencies.

For solids, the anisotropies cannot be ignored, as mentioned above, and they give rise to substantial line broadening. However, it was realized in the 1950s that physical rotation of macroscopic samples about an axis inclined at an angle β to the applied magnetic field would average equation 2.29 to some extent and therefore diminish the broadening effects of anisotropic interactions on solid-state spectra. In fact, it can be readily shown, by simple geometric arguments,[9] or in terms of quantum mechanics (see section 4.6.1), that:

[9] See appendix 5 of the book by Harris given in Further reading.

$$\left\langle \tfrac{1}{2}(3\cos^2\theta - 1)\right\rangle = \tfrac{1}{2}(3\cos^2\beta - 1) \times \tfrac{1}{2}(3\cos^2\Theta - 1) \qquad\qquad 2.30$$

where Θ is the angle between a relevant molecular-level direction (e.g., an internuclear distance in the case of a dipolar interaction) and the axis of rotation (see figure 2.11).

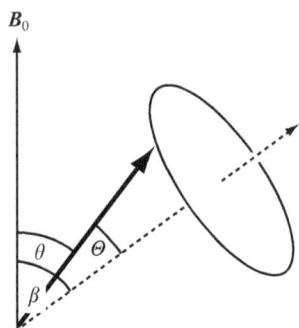

B_0

Figure 2.11. This shows the angles θ, β, and Θ, relevant to magic-angle spinning (see the text). The heavy black arrow represents a vector such as an interatomic distance, while the dashed arrow is the axis of rotation of the sample.

For a polycrystalline sample, the angle Θ can take all possible angles, just as θ can. However, the angle β is fixed by the experimentalist. Therefore, the factor $\tfrac{1}{2}(3\cos^2\beta - 1)$ acts to scale the powder pattern produced by the range of values of θ. If $\beta = 0°$ (rotation about the z axis), there is no scaling. At the other extreme, $\beta = 90°$, the spread of the powder pattern is reduced by a factor of 2, and if $\beta = 54°44'$, the effects of anisotropies vanish since the scaling factor is zero ($\cos 54°44' = 1/\sqrt{3}$). The width of the signal is reduced as if by magic, so the term "*magic-angle spinning*" (MAS) was devised to describe the phenomenon. The MAS technique is now ubiquitous in solid-state NMR.

The right-hand side of figure 2.12 shows an example for ^{13}C. The spin rate should be significantly greater than the static linewidth in order to provide a fully narrowed single signal. Normally, spin rates in the region 3–30 kHz are used, though faster rates are now feasible. At the lower end, these suffice to average ^{13}C shielding anisotropies, but higher speeds are necessary to remove the effects of H,H dipolar interactions, and it is impossible to completely average most quadrupolar effects.

At this point, it is important to appreciate that interactions can be classified as either *homogeneous* or *inhomogeneous* and that these have distinctly different responses to MAS.[10] In the latter case (shielding anisotropy, for example), a single crystal would give sharp lines and the broadening found for polycrystalline samples occurs because of the variety of orientations that are present. Therefore, MAS rates only need to exceed the linewidths which would be found for a single crystallite, so averaging by sample rotation is easy. MAS at rates substantially less than the static bandwidth for a polycrystalline sample still gives full line sharpening. However, in such a case, *spinning sidebands* are seen—resonances spaced at intervals of the spin rate from the centerband at

---

[10] The quantum mechanical origin of this distinction is discussed in inset 4.5.

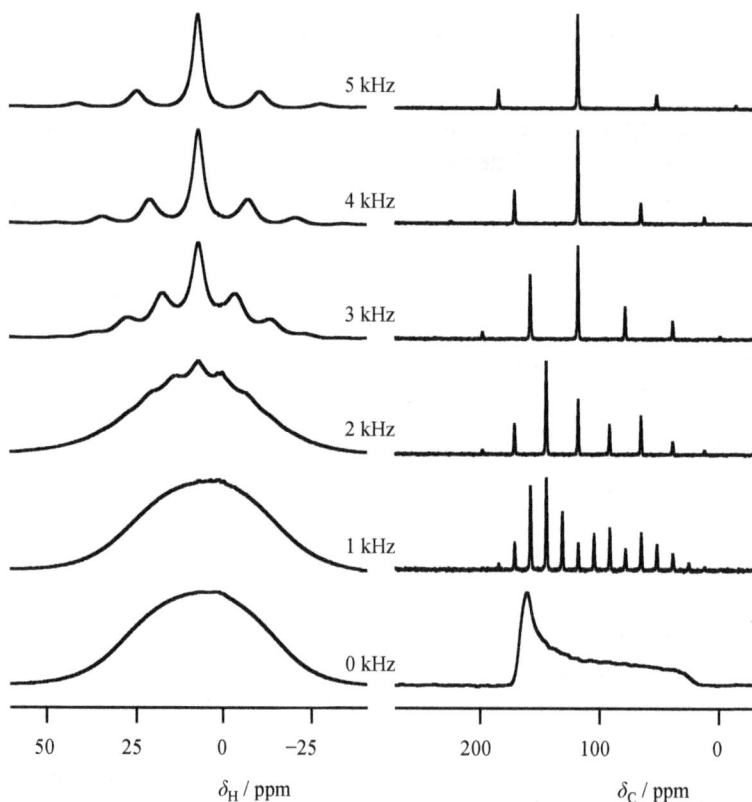

Figure 2.12. Spectra of solid dicyclopentadienyl zirconium dichloride, $(cp)_2ZrCl_2$, at various magic-angle spinning rates. Left: proton spectra; right: carbon-13 spectra (with high-power proton decoupling).

the isotropic resonance frequency (see the right-hand side of figure 2.12). Spinning sideband manifolds can be both a bane (complicating the appearance of spectra) and a blessing (allowing determination of shielding tensor parameters, for instance). The distribution of intensities for a shielding case depends on the relevant anisotropy and asymmetry. Computer programs exist to extract these parameters from the spinning sideband intensities (see figure 2.13 and section 8.2).

This procedure is often an improvement on using static spectra because the MAS spectra allow resolution of sideband manifolds of chemically different nuclei whereas the broad resonances for static samples would contain severe overlapping. Spin rates in excess of the static bandwidth of the spectrum are needed to eliminate spinning sidebands for spectra subject to inhomogeneous broadening. This is achievable when shielding anisotropies are low, as is generally the case for ^{13}C spectra but not usually the situation for heavy-metal spin-$\frac{1}{2}$ nuclei such as ^{119}Sn.

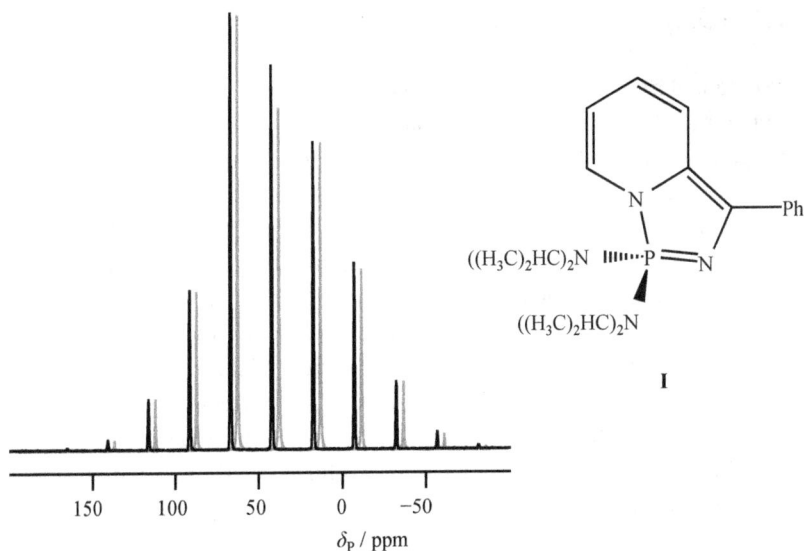

Figure 2.13. Fitting the ^{31}P spinning sideband manifold for compound **I**. The simulated spectrum is shown in black slightly shifted to the left from the experimental spectrum (in grey) for clarity of comparison. The computation shows that the shielding anisotropy, ζ, is 77.5 ppm and the asymmetry is 0.77. A 65 Hz line broadening has been applied to the simulated spectrum. The isotropic chemical shift is 42.0 ppm.

On the other hand, for homogeneous interactions (typically homonuclear H,H dipolar coupling), single crystals would still give broad lines because of the great variety in the magnitude and orientations of internuclear distances. Slow spinning would give no sharpening at all and high resolution can be obtained only at spin rates greatly exceeding the static linewidth (see the ^{1}H spectra in figure 2.12).

There is a close relation between the case of homogeneous line broadening and the phenomenon of *spin diffusion* (see inset 7.3). This involves spatial dispersion of spin orientation (without any physical diffusion of molecules). It is caused by the flip-flop of the orientations of neighboring spins under the influence of homonuclear dipole–dipole interactions. Successive flip-flops can diffuse a spin orientation in a random-walk fashion. Such spatial dispersion takes time, so the phenomenon can be used to assess the size of domains in heterogeneous materials such as semicrystalline polymers (see inset 7.3).

For most spectra of quadrupolar nuclei, the total resonance bandwidth is so large that no feasible MAS rate will be sufficient to give a single peak; therefore, numerous spinning sidebands will usually be seen. As is the case for sidebands arising from shielding anisotropy, it is feasible, by the use of suitable computer programs, to extract (quadrupolar) anisotropies and asymmetries from the intensity distributions in spinning sideband manifolds.

2.8   RELAXATION

NMR spectra are obtained from samples in strong magnetic fields by recording the response of bulk spin magnetization to radiofrequency (RF) radiation. Since the 1970s, pulses of radiation have been used, which excite a range of frequencies (to cover the relevant spectral range). The resulting signal is a time response, referred to as a *free-induction decay* (FID).

By convention, the strong applied magnetic field is said to be applied in the z direction. The RF tilts the magnetization towards the xy plane, with detection also occurring in the xy plane. Generally, signals are summed over the FIDs following many pulses in order to increase the signal-to-noise ratio (S/N) to an acceptable level (see chapter 3). During the whole NMR experiment (or whenever the system is perturbed), the net magnetization of the ensemble of spins, M, will experience a tendency to revert to its equilibrium situation. The process of regaining equilibrium is known as *relaxation*. However, there are a number of different relaxation processes, as discussed in chapter 7, depending on the experimental situation. Solid-state NMR is particularly concerned with three cases briefly described as follows:

(a) *Spin–lattice relaxation* (also sometimes known as longitudinal relaxation). This describes the process of regaining equilibrium of the z component of M (i.e., along B_0) following a perturbation (or immediately after the sample is placed in B_0). For homogeneous solid samples, this process is generally exponential, that is, describable as single exponential, with a relaxation time constant, T_1, ranging from seconds to kiloseconds, as given in equation 2.31:

$$M_z(t) - M_0 = [M_z(0) - M_0] \exp(-t/T_1) \qquad\qquad 2.31$$

where $M_z(0)$ is the magnetization in the z direction after the perturbation. The time T_1 is the inverse of a first-order rate constant for the relaxation. In multipulse experiments, T_1 is of particular importance because ample allowance for spin–lattice relaxation is normally vital between pulses (see section 3.3.1).

(b) *Spin–spin relaxation* (better described as transverse relaxation). This relates to the xy component of M (i.e., perpendicular to B_0), which is zero at equilibrium, but which can become nonzero as a result of applying RF pulses (see chapter 3). Since M_z (parallel to B_0) and M_{xy} (perpendicular to B_0) are orthogonal, transverse relaxation is entirely distinct from spin–lattice relaxation, though the mechanisms may be the same. In general, for solids, transverse relaxation is unlikely to be single exponential. The typical relaxation time constant is given the symbol T_2, which, for protons in a rigid solid, will be very short (tens of microseconds)—a markedly different situation from mobile solutions, for which $T_1 = T_2$. The value of T_2 is directly related to the linewidth or lineshape of resonances. Further information on lineshapes and on spin–spin relaxation may be found in sections 5.2 and 7.2.3.

(c) *Spin–lattice relaxation in the rotating frame.* This describes the return to equilibrium of transverse magnetization in the presence of an RF magnetic field, B_1, in the same direction. In this situation, the magnetization is said to be *spin-locked*, because the relaxation time constant, $T_{1\rho}$, is greatly extended beyond transverse relaxation. It may be likened to T_1 but as appropriate for a low magnetic field (B_1) rather than a high field (B_0). Typically, values of $T_{1\rho}$

for ^1H in solids are 1–20 ms (whereas for mobile solutions they are equal to T_1). Values of $T_{1\rho}$ are of particular importance in cross-polarization experiments (see section 3.4).

Several mechanisms contribute to relaxation. They all require a time-dependent interaction involving the nuclear spins, so the interactions (shielding, dipolar, and quadrupolar) discussed above can be involved. Naturally, the larger the energy of interaction, the more efficient is the relaxation mechanism. Thus, when present, the quadrupolar interaction will often dominate. However, both dipolar and shielding anisotropy interactions can be very important. The requisite time dependence arises from motion at the molecular level. Thus, measurement of relaxation times gives information about dynamics at the molecular level, especially if variable-temperature (or variable-field) experiments are carried out.

For solids, motions can be very complex and are often poorly understood. In simple cases, for example molecular motion which is either isotropic or involves rotation about a unique axis, T_1 and $T_{1\rho}$ pass through a well-defined minimum as the motional rate increases, with a tendency to increase toward infinity for very fast motions (as in mobile liquids) or very slow motions (as for rigid solids). For most solids, mobility at the molecular level is complex and the relaxation behavior as a function of temperature is poorly understood. However, internal rotation about $C–CH_3$ bonds tends to be relatively facile and so its effects can be well separated from those of other motions. In such situations, relaxation times may show several minima as a function of temperature. Figure 2.14 shows the case for trimethylphosphine sulfide, which has two T_1 minima corresponding to distinct motional processes, namely rotation of the methyl groups and rotation of the whole molecule about the symmetry axis.

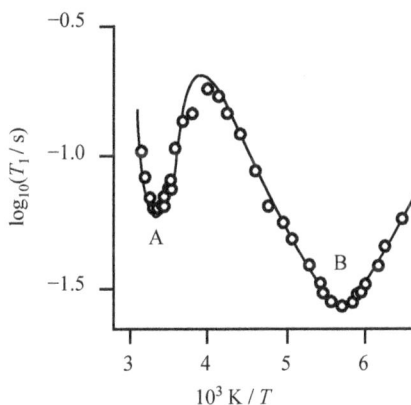

Figure 2.14. Proton spin–lattice relaxation time of solid trimethylphosphine sulfide as a function of temperature, showing minima for two types of molecular motion. Minimum A occurs because of overall rotation about the molecular axis, while minimum B corresponds to internal rotation of the methyl groups about the C–P bonds.

Minima in T_1 versus T plots relate to motions at the Larmor frequency, that is, tens or hundreds of megahertz, whereas $T_{1\rho}$ responds to motions at frequencies related to the RF power, that is, $\gamma B_1 / 2\pi$, which are generally tens of kilohertz. Transverse relaxation, on the other hand, is caused by very low-frequency motions. Thus the three relaxation types give complementary information. However, even together, they cover only three frequency regions, though of course B_1 can be varied to some extent and different spectrometers may provide several Larmor frequencies. A better strategy is to vary B_0 over a large range. This requires special equipment, now commercially available (but not common), and constitutes the subject of *relaxometry*.

Any molecular-level mobility at rates comparable to the inverse linewidth of a spectrum will also lead to partial averaging of anisotropic interactions and hence to sharpening of spectra, as already illustrated in figure 2.6.

FURTHER READING

GENERAL REVIEWS OF THE PRINCIPLES OF SOLID-STATE NMR

"Transient techniques in NMR of solids", B.C. Gerstein & C.R. Dybowski, Academic Press Inc. (1985), ISBN 0 12 281180 1.

"Nuclear magnetic resonance spectroscopy: A physicochemical view", R.K. Harris, Pearson Education Ltd. (1987), ISBN 0 582 44653 8.

"High resolution NMR in the solid state", E.O. Stejskal & J.D. Memory, Oxford University Press (1994), ISBN 0 19 507380 0.

"Introduction to solid-state NMR spectroscopy", M.J. Duer, Blackwell Publishing Ltd. (2004), ISBN 1 4051 0914 9.

CHAPTER 3

SPIN-$\frac{1}{2}$ NUCLEI: A PRACTICAL GUIDE

3.1 INTRODUCTION

This chapter gives a beginner's guide to the experimental solid-state NMR of spin-$\frac{1}{2}$ nuclei. It is a break from consideration of the underlying principles of solid-state NMR, concentrating instead on what it is necessary to know in order to obtain one-dimensional spectra, such as those shown in figure 3.1. These spectra are often valuable on their own and they are also the starting point for any of the more complex experiments described later on. Quadrupolar nuclei are treated separately in chapter 6.

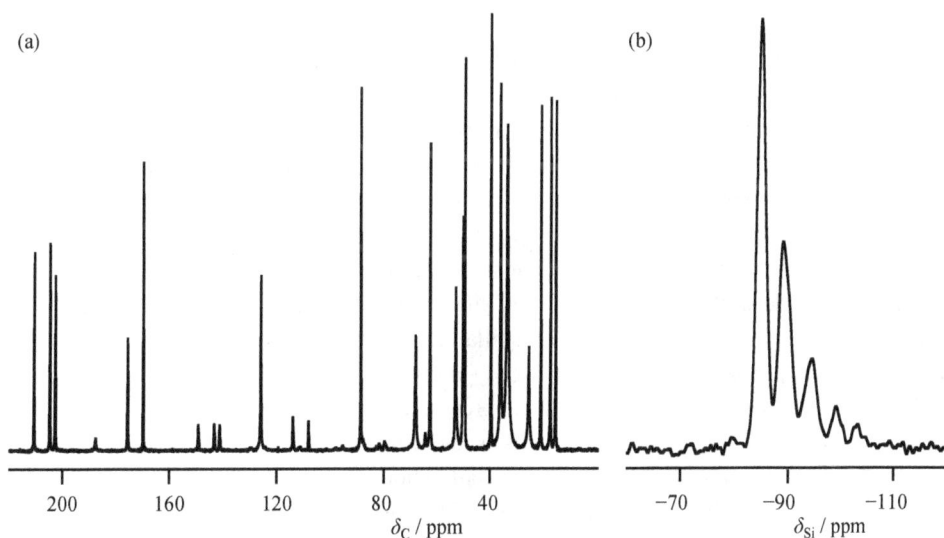

Figure 3.1. Examples of high-resolution solid-state NMR spectra: (a) a carbon-13 spectrum from a steroid; (b) a silicon-29 spectrum from a zeolite.

There are 24 elements (31 isotopes) with non-radioactive spin-$\frac{1}{2}$ nuclides (and some of these elements have quadrupolar nuclei as well); see inset 3.1. Carbon, nitrogen-15, silicon and phosphorus are the "bread and butter" nuclei for solid-state NMR and, under the right experimental conditions, their spectra are relatively easy to obtain. ^{77}Se, ^{89}Y, $^{111/113}$Cd, $^{117/119}$Sn, ^{125}Te, ^{195}Pt, ^{199}Hg and ^{207}Pb present few technical challenges, other than those associated with extensive spinning sideband manifolds arising from large shielding anisotropies. ^{57}Fe, ^{103}Rh, ^{183}W, ^{187}Os and $^{107/109}$Ag have low resonance frequencies and slow relaxation also can be a great handicap. This situation poses severe instrumental difficulties at low magnetic fields (≤ 11.7 T), and their nuclei are really accessible only to high-field instruments. In the solid state, ^{1}H and, to some extent, ^{19}F are special cases and they are dealt with separately in section 3.5. Tritium has special problems because of its radioactivity and helium-3 is not important chemically. This leaves ^{129}Xe, which has some rather specialist applications, and $^{203/205}$Tl, which is an awkward element as its resonance frequencies fall between ^{31}P and the *high-band* nuclei (^{19}F, ^{1}H, and ^{3}H) and are out of the range of most solid-state probes.

Inset 3.1. Elements with spin-$\frac{1}{2}$ isotopes
H He C
N F Si
P Fe Se
Y Rh Ag
Cd Sn Te
Xe Tm Yb
W Os Pt
Hg Tl Pb

3.2 THE VECTOR MODEL AND THE ROTATING FRAME OF REFERENCE

A vector model of the magnetization is useful for the description of the NMR experiments introduced in this and subsequent chapters. One aspect of this, the bulk magnetization vector, was introduced in chapter 2 and a second, the rotating frame of reference, is described briefly here. The text by Keeler (see Further reading) goes into more detail on this and the material introduced in section 3.2.1.

Precession is the movement of the axis of a spinning body about a second axis (see figure 3.2). If the bulk magnetization is displaced from the laboratory z axis by a *resonant* radiofrequency pulse, it will precess about $\boldsymbol{B}_0$ with a frequency equal to v_{NMR} given by equation 2.4. The effect of the pulse is easiest to describe in a coordinate frame rotating at the RF transmitter frequency (v_{RF})—the *rotating frame of reference* (figure 3.3). In such a frame, the oscillating magnetic field associated with the pulse will appear to be static (in effect v_{RF} is subtracted out of the response to the pulse) and the precession of the bulk magnetization about $\boldsymbol{B}_0$ is "eliminated". This is the way the NMR experiment is treated in chapter 4 and more details about the rotating frame can be found there. The elimination of the precession is equivalent to removing the magnetic field $\boldsymbol{B}_0$ and that makes it easier to describe the interaction of the bulk magnetization with the much weaker radiofrequency field $\boldsymbol{B}_1$.

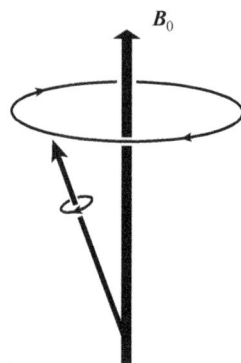

Figure 3.2. The axis of a spinning body precesses around another axis (the motion of a gyroscope).

In a coordinate frame rotating at the same rate as a rotating point, the position of the point is constant.

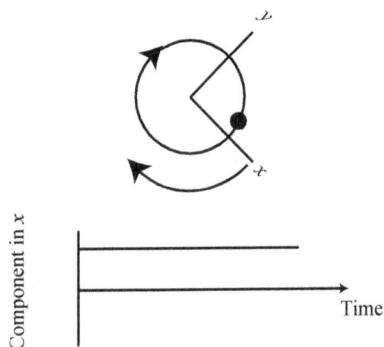

Figure 3.3. The rotating frame of reference.

Inset 3.2. A note on Fourier transform NMR spectroscopy

Precessing magnetization (see main text), generated in response to a pulse of electro-magnetic radiation, induces a voltage in a surrounding coil. It is this voltage that is the NMR signal, *interferogram*, or *free induction decay* (FID), recorded over the *acquisition time*.

This *time domain* signal is generally complicated because it contains many different precession frequencies and is rarely analyzed directly. The NMR signal is more usually presented as a frequency spectrum—the *frequency domain*. The usual mathematical method for interconverting the time and frequency domains is *Fourier trans-formation* (FT).

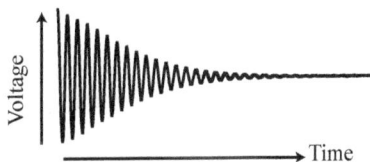

3.2.1 PULSE ANGLE

Pulses are usually described as being applied relative to the axes of the rotating frame.[1] In this frame, a pulse applied with phase x rotates the magnetization vector, at a rate γB_1 (the nutation rate;

---

[1] The axes of the rotating frame should be labeled differently to those of the fixed laboratory frame (say, x', y', and z' instead of x, y, and z). However, it is customary to omit the distinction. Phases in this sense are relative labels and have no absolute physical meaning.

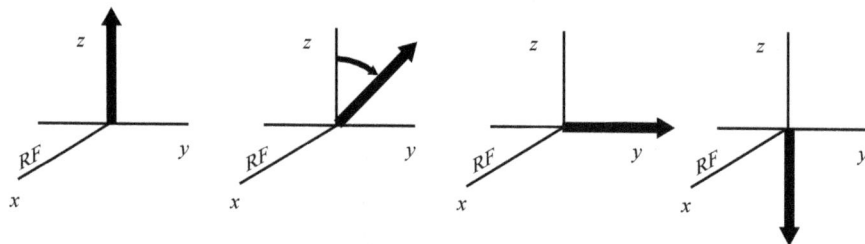

Figure 3.4. A pulse applied along x rotates the bulk magnetization vector about x.

see inset 3.3), from z, about x, toward the y axis as shown in figure 3.4.[2] The angle through which it is rotated is called the *pulse* (or tip) *angle* (θ) and is proportional to the *duration* of the pulse, τ_p:

$$\theta = \gamma B_1 \tau_p \qquad\qquad 3.1$$

where θ is in radians. Only the component of magnetization that is perpendicular to B_0 is detectable, so the greater the magnetization in the xy plane, the more intense will be the NMR signal. The maximum signal intensity will thus be obtained for a pulse angle of 90° (a *90° pulse*). Figure 3.5 shows a typical *nutation curve* (the signal as a function of pulse duration).

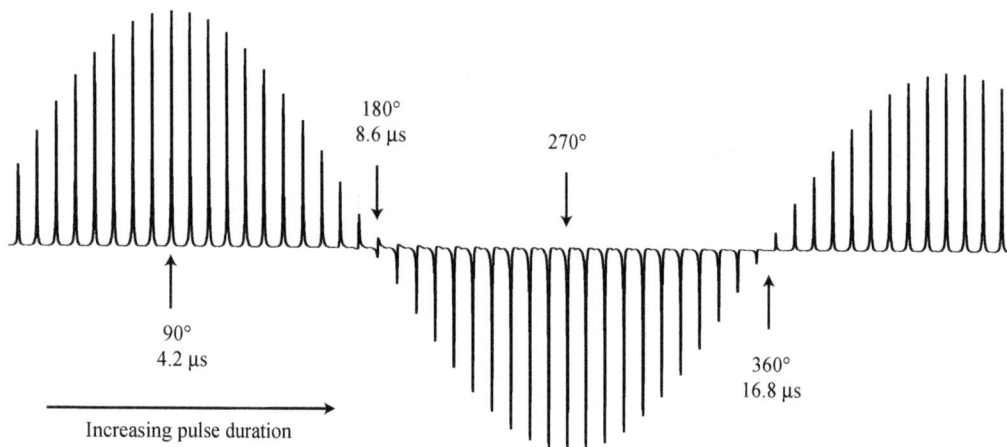

Figure 3.5. Experimental ^{31}P nutation curve. The first spectrum was recorded with a pulse of 1 μs duration, the pulse duration then increases successively in steps of 0.4 μs to the right. The overall loss of signal with increasing pulse duration arises from RF inhomogeneity.

---

[2] This rotation is what is expected from the classical physics of two interacting orthogonal magnetic fields (the bulk magnetization and B_1; in the rotating frame and on-resonance, B_0 disappears). Conventions for the direction of rotation vary—consistency is the important factor in describing an experiment.

A nutation curve is routinely recorded to determine the optimum 90° pulse duration and to calibrate the RF field strength. Because of experimental features such as a variation in RF field strength across the sample (RF *inhomogeneity*), it is not unusual to find that the observed 90° pulse is slightly different from (often less than) half of the 180° pulse duration and it may be difficult to pinpoint exactly the times for which the signal is zero. Nevertheless, it is easier to visually determine or estimate the zero-signal points rather than the signal maxima, and half the difference between the 180° and 360° pulse durations often gives a better indication of the true magnitude of the RF field strength (see inset 3.3).

Inset 3.3. Pulse angle and field strength

The RF field strength associated with a pulse is a magnetic flux density and has the unit tesla (T). The field associated with a pulse, B_1, typically has a magnitude of several milliteslas. If, say, this is 10 mT, a pulse duration, τ_p, of 3 µs would produce a pulse angle of 90° for silicon-29 (from equation 3.1).

The strength of the resonant RF field is usually discussed in terms of an equivalent rotation rate (the *nutation rate* of the affected spins):

$$\nu_1 = \gamma B_1/2\pi = 1/(4\tau_{90})$$ 3.2

where τ_{90} is the 90° pulse duration. So a 3 µs 90° pulse corresponds to an 83.3 kHz RF field. Note the distinction between ν_1 (the nutation rate produced *by* the on-resonance pulse) and ν_{RF} (the frequency *at which* it is applied).

In FT-NMR, the pulse of RF irradiation excites a range of frequencies as illustrated in figure 3.6 (this is the whole point of using a pulse). Exciting a large *spectral width* is often important in solid-state NMR, so short (often termed *hard*) pulses are generally required as these give a wide

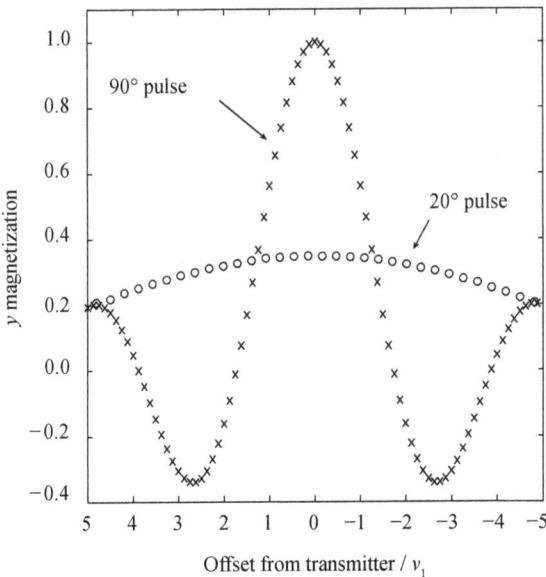

Figure 3.6. The y component of the magnetization resulting from an RF pulse along x as a function of transmitter offset (expressed relative to the nutation frequency). A 90° pulse excites more efficiently but does so over a small range: a range of approximately $\pm 0.5\nu_1$ (the nutation frequency) is within 10% of the maximum excitation. A shorter pulse, of the same amplitude, with a smaller tip angle, generates less signal but excites over a wider range ($\pm 2\nu_1$ is within 10% of the maximum excitation for this pulse).

frequency coverage; it is commonplace to work with 90° pulses with a duration of 4 μs or less. A 90° pulse can be expected to excite a frequency range roughly equivalent to $1/(4\tau_{90})$ with a reasonable degree of uniformity (so the 3 μs 90° pulse described in inset 3.3 would excite a range of about 85 kHz reasonably uniformly).

Off-resonance

So far it has been assumed that the pulse has been applied on-resonance and $v_{RF} = v_{NMR}$. In the general case, $v_{RF} \neq v_{NMR}$ and the frequency of the signal of interest, v, depends on interactions in addition to the Zeeman one, for example, shielding (see equation 2.18), so $v \neq v_{NMR}$. When $\Delta v = v - v_{RF} \neq 0$ (it has to be so for a spectrum containing more than one resonance), instead of the magnetization being rotated about $\boldsymbol{B}_1$, it rotates about an effective field along a *tilted axis*, $\boldsymbol{B}_{eff}$, which is the resultant from $\boldsymbol{B}_1$ and a residual static field $\boldsymbol{B}_r = 2\pi\Delta v/\gamma$ ($\boldsymbol{B}_0$ is no longer effectively zero in the rotating frame):

$$\boldsymbol{B}_{eff} = \boldsymbol{B}_1 + \boldsymbol{B}_r \qquad\qquad 3.3$$

This *off-resonance* situation has two consequences. Firstly, relative to the on-resonance case, it takes a longer pulse to rotate the magnetization to the xy plane and, secondly, when it arrives in that plane, it will have a phase error with respect to on-resonance magnetization. Thus, for a ^{13}C spectrum, an RF field equivalent to 80 kHz giving a 90° pulse at 100 ppm on a 9.4 T spectrometer (100 MHz for ^{13}C) would give an 83° pulse at 0 and 200 ppm. This may have consequences for the more advanced experiments described later, where precision in the pulse duration is important. The phase error in the spectrum can be removed with a first-order (frequency-dependent) correction.

3.3   THE COMPONENTS OF AN NMR EXPERIMENT

There are two main methods for exciting nuclei and generating spectra in solid-state FT-NMR. The simplest experiment consists of a pulse followed by the detection of an FID (an *acquisition*). It is variously known as pulse-acquire, single-pulse excitation (SPE), direct polarization, or direct excitation. Direct excitation is the term used here. Alternatively, magnetization can be transferred from an abundant spin such as ^{1}H to a dilute one. This *cross-polarization* technique is described in section 3.4. The previous section dealt with the excitation part of the direct-excitation experiment. The other components and some of the parameters under the control of the spectroscopist are introduced next.

3.3.1   RECYCLE DELAY

It was noted earlier that NMR is an insensitive technique. In experimental terms this means that the amount of signal generated by each radiofrequency pulse, or sequence of pulses, is usually small (of the order of microvolts) relative to the level of the electronic noise inherent in the spectrometer.

To improve sensitivity it is usual to repeat an experiment many times, adding each FID to the sum of the previous ones. The magnitude of the NMR signal is proportional to the number of repetitions (n). The noise, which is random, increases only as $\sqrt{n}$, so the higher the number of repetitions, the higher the *signal-to-noise ratio* (see inset 3.4).

Inset 3.4. Signal-to-noise ratio

The signal-to-noise ratio (S/N) is a useful measure of the quality of a spectrum and the performance of the spectrometer. The noise is usually characterized by its standard deviation and that, relative to the height of the signal in question, gives the signal-to-noise ratio. Invariably, the spectroscopist seeks the highest possible signal-to-noise ratio.

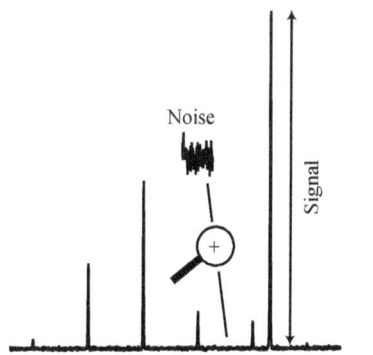

Repeating an experiment raises the issue of how long to wait between repetitions. The physical characteristic that relates to this is the spin–lattice relaxation time, T_1 (which was introduced in section 2.8 and is discussed further in chapter 7). In solution-state NMR, there is often no need to wait between finishing one acquisition and starting the next because $T_1 \approx T_2$ (and so is of the order of the acquisition time; see the next section): as soon as the FID has decayed to zero, the pulse sequence can be repeated. In the solid state, however, T_1 is usually several orders of magnitude longer than the acquisition time, so even after the FID has decayed to zero, it is necessary to wait until the magnetization has returned to equilibrium before applying another pulse.

The time between successive repetitions of the NMR experiment (the experiment is defined here as the excitation period plus the acquisition time) is called the *recycle delay* or *pulse delay*. As already noted, the optimum recycle delay depends on T_1, so is sample-, nucleus- and environment-dependent and it cannot be readily predicted (see chapter 7). Using an inappropriate delay can lead to signals being missed or their intensities being wrongly represented (see figure 3.7) or it can waste spectrometer time.

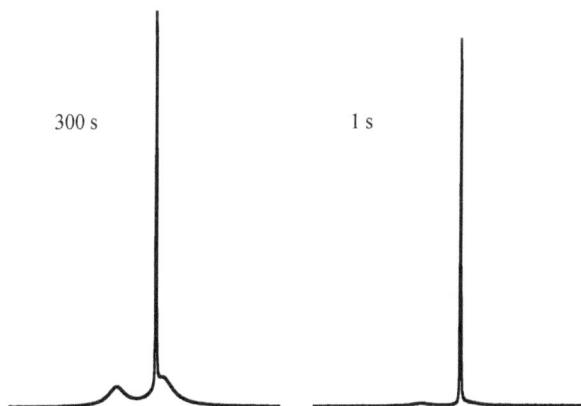

300 s   1 s

Figure 3.7. Two ^{31}P spectra acquired from the same sample but with different recycle delays. At the shorter recycle, the broad, low-frequency shoulder on the narrow line is barely detectable and the high-frequency signal is significantly underrepresented.

So what is the best recycle delay to choose? That depends on whether the experiment needs to be quantitative or whether a high signal-to-noise ratio is of most importance. It also depends on the type of experiment: as will be discussed in section 3.4.3, cross-polarization experiments tend not to be quantitative, so here the key issue is generating the highest signal-to-noise ratio.

Suppose a direct-excitation experiment is being carried out and the sample has been in the magnet long enough for equilibrium (a Boltzmann distribution of populations) to have been reached. At the beginning of the experiment (time $t = 0$), the magnetization along the z axis, M_z, is the equilibrium magnetization, M_0, and the magnetization in the xy plane, M_{xy}, is zero. After a 90° pulse, $M_z = 0$ and $M_{xy} = M_0$. (It is the magnetization M_{xy} that is detected in the FID.) After the pulse, spin–lattice relaxation begins to return magnetization to the equilibrium state (i.e., to the z axis), as described by equation 3.4 and shown graphically in figure 3.8.

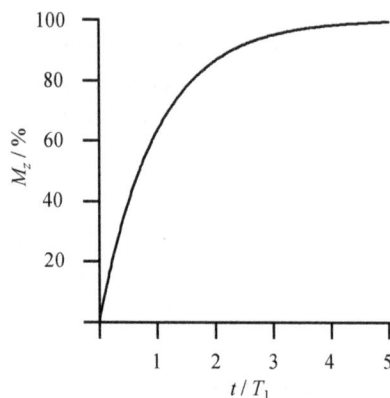

Figure 3.8. The amount of z magnetization at time t (relative to T_1) after a 90° pulse.

$$M_z(t) = M_0(1 - \exp(-t/T_1)) \qquad\qquad 3.4$$

Now suppose the experiment is repeated and a second 90° pulse is applied. If the recycle delay is short compared with T_1 so that the second pulse is applied before M_z recovers to the value M_0, the pulse will produce a reduced amount of signal. Conversely, if the recycle delay is very long, the condition $M_z = M_0$ might have been reached some time before the pulse is applied and spectrometer time has been wasted. To observe the fully relaxed signal after each pulse, the optimum recycle delay is $5 \times T_1$ (which recovers 99.3% of the signal).

Such a recycle delay is appropriate when quantitative intensity information is required, so for any sample that gives multiple signals (with differing T_1 values), setting the recycle delay to five times the *longest* T_1 will ensure that the signal intensities will be quantitative. (This was the recycle delay used for the spectrum shown in figure 3.1(b).) However, if accurate intensity information is not an issue, then a trade-off can be made; reducing the recycle delay reduces the amount of signal obtained per repetition, but allows more repetitions in a given time, potentially increasing the signal-to-noise ratio. As figure 3.9 shows, the highest signal-to-noise ratio is obtained when the recycle delay is $1.26 \times T_1$ (~72% of the full signal is obtained on each repetition). It would appear, then, that the choice of recycle

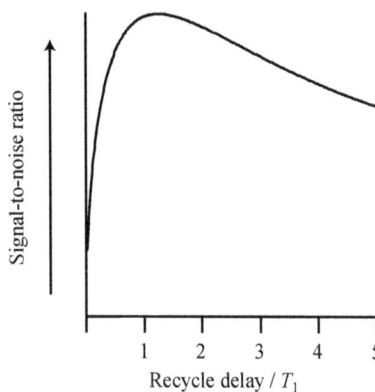

Figure 3.9. The signal-to-noise ratio, for a given total experiment time and 90° pulses, as a function of the recycle delay (expressed as a multiple of T_1).

delay requires prior knowledge of T_1. However, if a signal can be observed within a few minutes, a short sequence of experiments with differing recycle delays is usually sufficient to determine an appropriate recycle delay (see figure 3.10).[3]

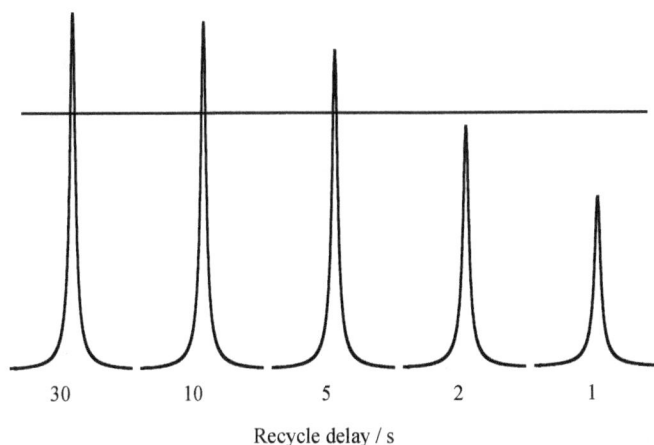

Figure 3.10. Spectra recorded with different recycle delays. The horizontal line is drawn at 72% of full intensity. The maximum intensity is obtained with a 30 s recycle, although little signal is sacrificed with a 10 s delay. In terms of the signal-to-noise ratio in any given total time, the optimum recycle is around 3 s.

Recycle delay / s

3.3.2 ACQUISITION TIME

The acquisition time is the time during which the signal (the FID) is detected. Getting this right is important to the appearance, and information content, of the spectrum. If the acquisition time is too long, the unnecessarily acquired noise in the time domain results in additional noise in the spectrum. If it is too short the FID is *truncated*. Fourier transformation of a truncated signal results in "wiggles" at the base of the lines in the spectrum (figure 3.11).[4] Although it is possible to compensate for a truncated FID with additional mathematical processing (as is commonly done for two-dimensional data sets) and an overlong acquisition can be multiplied by an *apodization function* (see Further reading), it is far better to optimize the acquisition time to start with, if the signal is sufficiently visible.

3.3.3 RECEIVER GAIN

The *receiver gain* is the amount of amplification applied to the raw signal. The receiver (the part of the spectrometer that digitizes the signal detected in the NMR coil) has a finite range. Very-high-amplitude signals will overflow the receiver and so will not be properly digitized. Significant

---

[3] There is also a trade-off between pulse angle and recycle delay. If T_1 is known, then the pulse angle, θ, that results in the maximum amount of magnetization in the xy plane for a chosen recycle delay is given by $\cos\theta = \exp(-recycle\ delay/T_1)$. This angle is known as the Ernst angle.

[4] These *sinc* wiggles appear only if the signal is *zero-filled* (as is usually the case, see inset 3.10) prior to Fourier transformation.

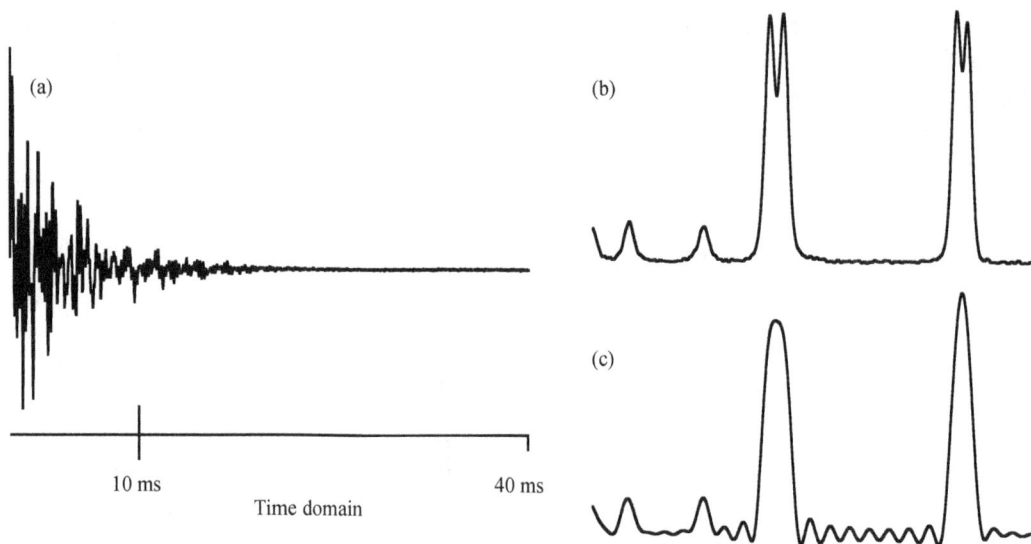

Figure 3.11. (a) A free induction decay (FID) acquired for 40 ms so that the signal decays to the level of the noise. (b) Part of the spectrum from the Fourier transformation of the full FID shown in (a). (c) The same part of the spectrum from the Fourier transformation of an FID obtained with a 10 ms acquisition time (but zero-filled to the same total number of data points as (b)). This truncation leads to characteristic baseline wiggles on either side of the affected peaks and a general loss of *resolution*.

clipping of the signal leads to baseline artifacts in the spectrum (figure 3.12). Usually, the receiver gain can be adjusted to avoid such problems, but in extreme cases (e.g., ^{1}H observation) it may be necessary to add some attenuation to the signal pathway. Too low a gain can result in artifacts when digitizing very weak signals (*digitization noise*).

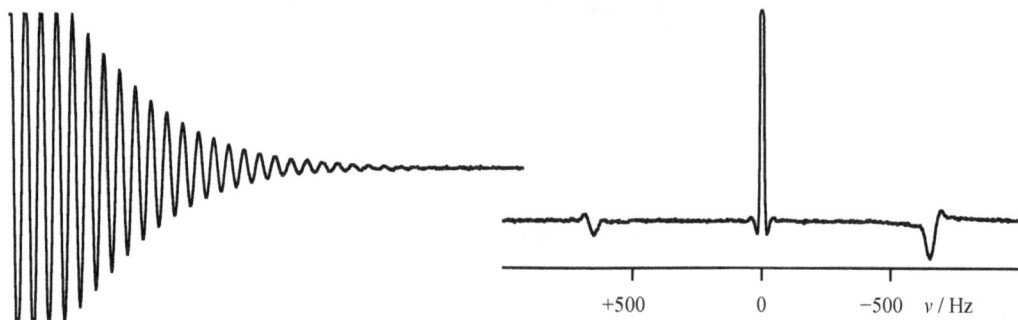

Figure 3.12. The "clipped" top and bottom of this free induction decay (FID) are characteristic of too much signal prior to digitization. The subsequent Fourier transform results in the dips at the foot of the line and the artifacts in the baseline.

3.3.4 SPECTRAL WIDTH

This is the frequency range under observation. The spectral width is determined by the rate at which the data points of the FID are *sampled* and the time between sampling points is called the *dwell time*. The maximum spectral width is determined by the fastest available sampling rate. Too high a spectral width is not usually a problem given that the pulse has a sufficient excitation range (see section 3.2.1) and that a large number of data points can be handled.[5] Too small a spectral width, however, can result in signals that fall just outside the spectral width being *folded back* into the spectrum (which is a potential trap for the unwary since the apparent chemical shift values for folded peaks will be incorrect). This situation is most likely to be encountered, for spin-$\frac{1}{2}$ nuclei, with the heavy metals (cadmium, tin, lead, mercury, and platinum) when large spinning sideband manifolds are recorded. An example is shown in figure 3.13. The only remedy is to repeat the experiment with a more appropriate spectral width.

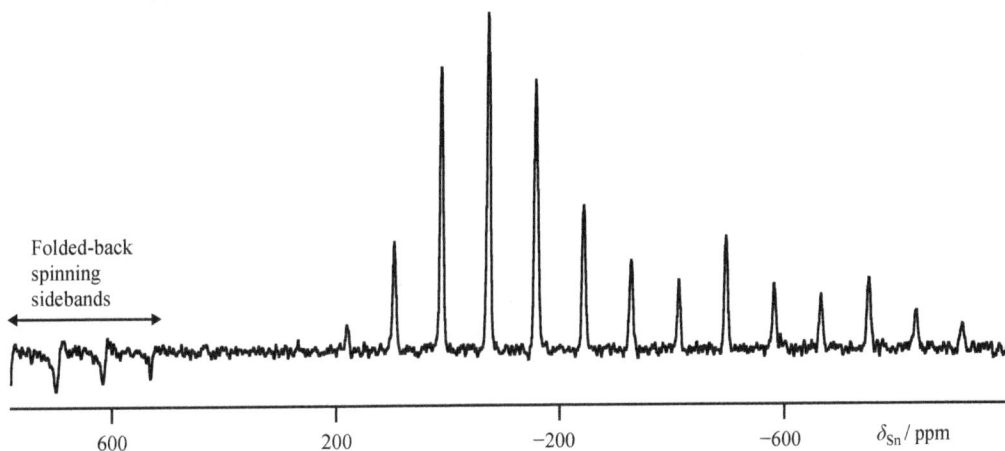

Figure 3.13. Too small a spectral width can result in signals being folded back into the spectrum. The extent of this ^{119}Sn spinning sideband manifold has caught out the spectroscopist. The way folded-back signals appear in the spectrum tends to be instrument specific.

3.3.5 DEAD-TIME

Signals extending over large spectral widths can be a problem in solid-state NMR—that is, if the spectrum is to have a flat baseline.[6] In any NMR experiment, excitation by relatively

---

[5] In fact, modern spectrometers automatically *oversample* (which is equivalent to using a very high spectral width) and then digitally filter and *downsample* to give a more manageable number of data points (all usually hidden from the spectrometer operator). This process is unlikely to result in folded back signals.

[6] Sometimes this may simply be a case of aesthetics, but if intensity information is required a flat baseline is usually essential.

high-voltage RF pulses (typically hundreds of volts in solid-state NMR) is followed by the detection of a low-voltage (typically microvolt) response. Ideally, acquisition needs to start immediately after the pulse. However, if it is started too soon, the beginning of the FID can be distorted by *pulse breakthrough* (real pulses are not perfectly rectangular and the decaying "tail" of the pulse can be detected by the receiver). The same spectrometer circuitry also has to cope with both high- and low-voltage conditions and it tends to *ring* from the shock of the high-voltage pulses. The time that it takes for the pulse to decay and the circuitry to settle down is called the *dead-time* (see figure 3.14). No meaningful data points can be obtained during the dead-time so it is not possible to follow the evolution of the signal from its start. In practical terms, this results in a distortion to the phasing of the spectrum (spectral phasing is determined by the way the real and imaginary parts of a complex Fourier transformation are combined, but that is something that is covered in other texts; see Further reading). This is particularly noticeable when the spectrum extends over a large frequency range as it does in the case shown in figure 3.15. A first-order phase correction can be applied to the spectrum to produce signals that are in-phase, but that introduces a *roll* to the baseline. Matching the dead-time to an integer number of dwell periods, together with numerical prediction of the early part of the signal decay, can minimize the baseline roll. Alternatively, the baseline correction routines in the spectrometer software can be used to produce a flat baseline. Another technique for cases involving large spinning sideband manifolds is to carry out the Fourier transform from the top of the first *rotary echo*. Under magic-angle spinning (MAS), orientation-dependent interactions give rise to rotary echoes. Echoes form because the resonance frequency from a given crystallite changes as the rotor position changes during spinning and each time the rotor returns to its original position the resonance frequency returns to its original value. Rotary echoes are illustrated in figure 3.15(d). An echo can also be used (in fact it is usual to do so, see inset 3.5) to record the broad lines from non-spinning samples (the refocused signal at the top of an echo is a good approximation to observing the signal from zero time).

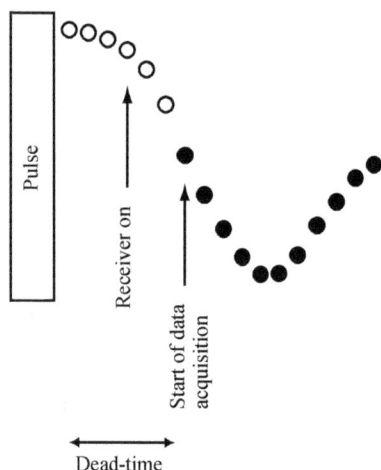

Figure 3.14. After a pulse there is a dead-time, while the spectrometer circuitry settles down, when no data points can be recorded. The missed points are represented here by the open circles. Typically, the receiver is switched on a few microseconds before it is instructed to acquire data (so avoiding any artifacts associated with the switch on).

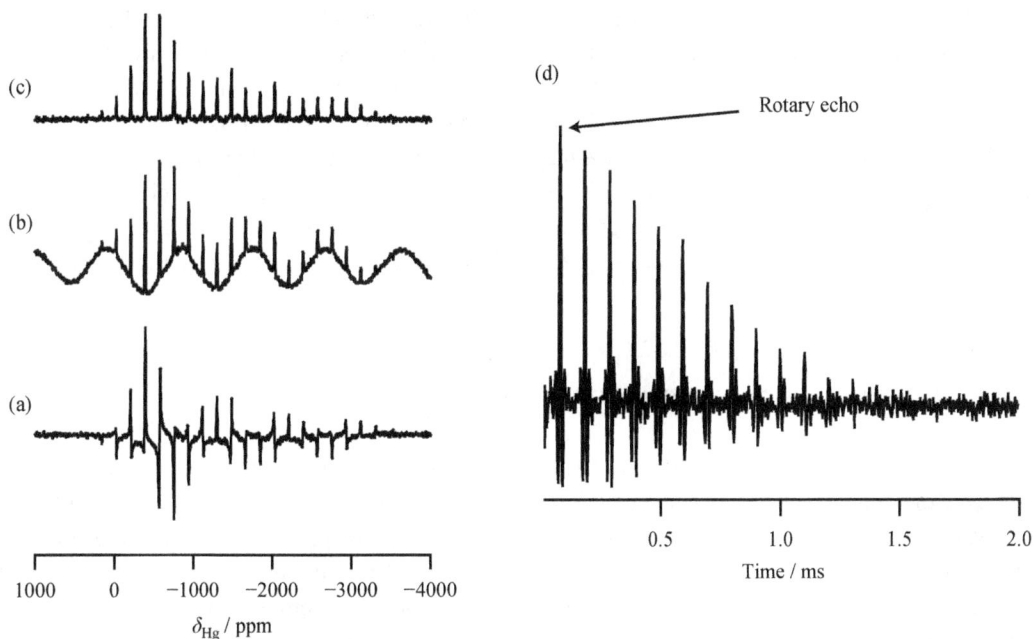

Figure 3.15. The ^{199}Hg magic-angle spinning spectrum from Hg(SCN)$_2$ acquired with a 500 kHz spectral width and at a spin rate of 9.78 kHz. There was a 15 μs dead-time delay between the end of the pulse and the first data point. (a) With no first-order phase correction. Note the flat baseline but extensive distortion to the phases of the spinning sidebands. The centerband is the in-phase signal at −1300 ppm. (b) The same data set but with a large first-order phase correction (−3529°). The sidebands now all have the same phase but the baseline rolls. (c) The same data set but transformed from the top of the first rotary echo shown in (d). Only a small (96°) first-order phase correction was necessary to produce this spectrum.

Inset 3.5. The solid (or quadrupolar) echo

A significant amount of signal can be lost in the spectrometer dead-time for the broad lines encountered with non-spinning samples. When shielding anisotropy is negligible most of the missing signal can be recovered using a solid (or quadrupolar) echo pulse sequence (see also section 4.4.2):

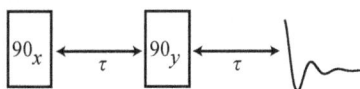

The ^{1}H spectra on the right were obtained (a) with and (b) without a solid echo. The integrated intensity for the band in (b) is 23% of that shown in (a) from the same sample. In (b) the dead-time between the pulse and the first point of the FID was 10 μs.

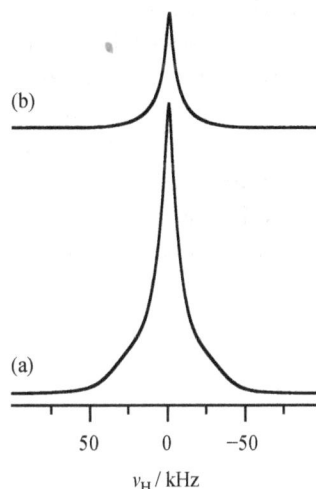

The dead-time is usually a small number of microseconds—but it is probe and frequency dependent. Acquiring data while the probe is still ringing results in the corruption of the NMR signal and that leads to an irregular distortion of the spectral baseline. Probe ringing (and hence the dead-time) increases at low observation frequencies and is a particular problem for low-gamma nuclei (loosely defined as anything with a magnetogyric ratio less than that of ^{15}N), although it does depend on the strength of the magnetic field and the study of low-gamma nuclei is one reason for going to high field strengths. Some probes, for experiments where pulses and data acquisition at the same frequency are interleaved, are designed to give very short dead-times (at the expense of sensitivity).

3.3.6 SPINNING SIDEBAND SUPPRESSION

The spectrum shown in figure 3.15(c) consists of a centerband at the isotropic chemical shift and a *manifold* of spinning sidebands. But which line is the centerband? As the position of the centerband is invariant to the spin rate, the easiest way to find out is to change the spin rate and rerecord the spectrum. Any line that changes position must be a sideband. The centerband is at −1300 ppm relative to $Hg(CH_3)_2$ in this spectrum. In this case the sideband manifold does not obscure a second centerband (this was evident from a spectrum obtained at a second spin rate) so, other than effectively diluting the signal, it does not pose a problem. Indeed, computer fitting of the sideband intensities gives information on the shielding tensor and hence on the electronic environment of the nucleus (see section 8.2.3). It would take a spin rate in excess of 100 kHz to produce a sideband-free spectrum for this sample (in the 7 T magnet used for this measurement).

Sidebands are not always so extensive. They are usually barely detectable in ^{29}Si spectra such as the one shown at the beginning of this chapter (figure 3.1) and they account for the lowest intensity signals in the carbon spectrum in that figure. In the latter case, their identity is easily confirmed by a change in spin rate and they do not interfere significantly with any of the centerbands in the spectrum. However, sometimes sidebands can be a problem and it may not always be desirable or possible to spin the sample fast enough to remove them from the spectrum. In such cases, spinning sideband suppression techniques can be useful. Two common ones are TOtal Sideband Suppression (TOSS), although the word total is something of a misnomer, and Sideband ELimination by Temporary Interruption of Chemical Shift (SELTICS). Both techniques suppress sidebands rather than refocus them so their intensity is lost from the spectrum (which might be important if intensity information is required). Both techniques are reasonably good at removing first-order sidebands (centerband $\pm 1 \times v_r$ where v_r is the spin rate) but neither is suitable for removing large manifolds of sidebands such as those shown in figure 3.15.

3.3.7 DECOUPLING

As was mentioned in chapter 1, a hydrogen-rich sample requires ^{1}H decoupling if a high-resolution spectrum of any other nuclide is to be obtained. Effective decoupling, at least at modest magnetic fields and sample spin rates ($\leq$ 10 T and 20 kHz, say), can be obtained by turning on a *continuous-wave* (CW)[7] decoupling pulse that lasts for the duration of the acquisition.

Given that the acquisition time is often many milliseconds, a decoupling pulse will be orders of magnitude longer than a typical excitation pulse. Experimentally, then, some care is required when decoupling: the RF field strength (decoupling power) and duration should not exceed the specification for the probe and the total RF on-time also should not be too high a proportion of the total experiment time. It is often useful to specify a *duty-cycle* limit of approximately 10%.

$$duty\text{-}cycle = \frac{total\ RF\ on\text{-}time}{total\ repetition\ time} \qquad\qquad 3.5$$

It is sometimes necessary to use a very long acquisition time (> 100 ms) to avoid truncating an FID. This situation may occur for samples with high molecular mobility such as gels or rubbers. As these also tend to have short relaxation times (allowing short recycle delays), it is easy to reach the duty-cycle limit. However, such samples also tend not to require very high decoupling power (the molecular motion partially averages the dipolar interactions involving ^{1}H) so the duty-cycle limit can be exceeded *providing* the decoupling power is reduced.

For organic materials, it is well worth using as much decoupling power as is available, up to the limit the probe can withstand, as figure 3.16 illustrates. However, setting too high a decoupling power can result in a spectacular electrical breakdown in the probe (known as *arcing*). Minor arcing can result in an increased and nonuniform noise pattern in the spectrum. Serious arcing can obliterate the signal that is being detected (see figure 3.17).

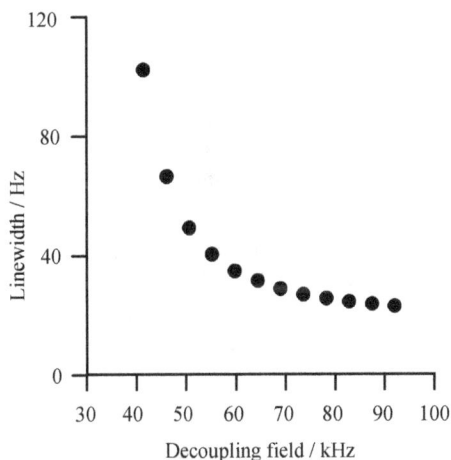

Figure 3.16. The line-narrowing effect of increasing the continuous-wave (CW) decoupling power, expressed as the ^{1}H nutation rate, is illustrated for the backbone carbon in crystalline polyethylene.

---

[7] "Continuous wave" is a term in widespread use, but it should be noted that for solid-state NMR the decoupling is *not* actually continuous. Here, "continuous wave" is merely a label for decoupling with no amplitude or phase modulation.

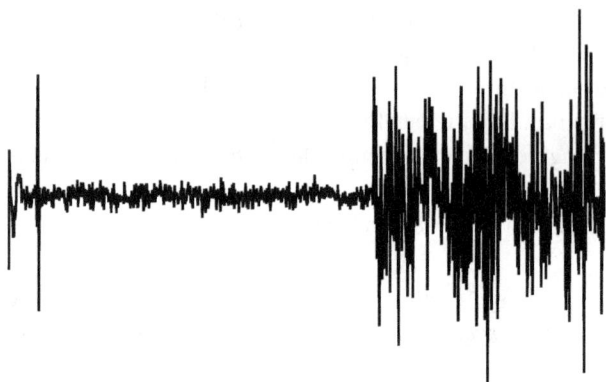

Figure 3.17. Catastrophic arcing such as that shown here is the result of an electrical breakdown in the probe, for example, between the turns of the coil, and destroys the spectrum. Apart from using too high a decoupling power, arcing can occur when there is an electrical problem in a probe (such as a badly soldered contact), when the air supply is dirty (or wet), or if there is sample leakage.

Continuous-wave decoupling has the advantage of simplicity, but more elaborate decoupling schemes are becoming commonplace, particularly for the carbon-13 NMR of organic materials. *Two-pulse phase-modulated* (TPPM) decoupling is one such scheme that can be used to increase decoupling efficiency, especially at high magnetic field, high sample spin rate, or when decoupling power is limited. The subject of decoupling is discussed in more detail in section 5.1.2. The factors governing linewidth are discussed in section 5.2.

3.3.8 SPECTRAL REFERENCING

Most of the high-resolution experimental spectra in this book are presented with a scale that has 0 ppm properly defined. That is, 0 ppm is the chemical shift of a recognized standard compound. The primary reference compounds are liquids or solutions, and while they can be put into a solid-state probe, there are good reasons why this is not always desirable: (i) they cannot be used to calibrate the class of experiments introduced in the next section (which work only on a solid sample) and (ii) some of them are volatile, poisonous, or reactive (and sealing them, intact, into a glass tube small enough to fit into a solids probe is not easy).

In solid-state NMR, referencing is usually *external*—that is, a separate reference spectrum is obtained from a standard sample. Solid-state spectra are not affected by solvent susceptibility as solutions are, so external referencing is reliable—although it should be carried out on a regular basis to counter drift in the applied magnetic field. If a liquid sample is impractical, a secondary solid standard is required. Ideally, this will give an easily detectable, narrow line (so that its position can be defined accurately) and will have a known chemical shift with respect to the appropriate recognized standard compound. However, because the solid-state spectrum from a quadrupolar nucleus may be magnetic field dependent in a way a solution is not (as explained in chapter 6), solutions are used routinely to calibrate and reference experiments on such nuclei in solids. Table 3.1 lists commonly used chemical shift reference materials for solid-state NMR.

Table 3.1. Common chemical shift reference compounds (including some for quadrupolar nuclei) relative to recognized standard compounds (see the reference in footnote f).

Nuclide	Reference for solid-state NMR (chemical shift / ppm)
^{1}H	Adamantane (1.9)[a] Glycine (8.5)[b,c]
^{13}C	Adamantane (38.5, CH$_2$)[c,d] Glycine (176.5, COO$^-$)[c]
^{15}N	NH$_4$^{15}NO$_3$ (−5.1, NO$_3$)[c] Glycine (−346.8)
^{19}F	C$_6$F$_6$ (−164.9)[e]
^{29}Si	tetrakis(trimethylsilyl)silane (−9.8,−135.4)
^{31}P	CaHPO$_4$·2H$_2$O (1.0)
^{77}Se	(NH$_4$)$_2$SeO$_4$ (1040)
^{119}Sn	Sn(C$_6$H$_{12}$) (−97.4)
^{199}Hg	[Hg(dmso)$_6$][O$_3$SCF$_3$]$_2$ (−2313)
^{7}Li	LiCl (0)[f]
^{11}B	BF$_3$/O(CH$_2$CH$_3$)$_2$ (0)
^{17}O	H$_2$O (0)
^{23}Na	NaCl (0)[f]
^{27}Al	Al(NO$_3$)$_3$ (0)[f]
^{45}Sc	Sc(NO$_3$)$_3$ (0)[g]
^{51}V	β-NaVO$_3$ (−519) (solid)
^{59}Co	K$_3$Co(CN)$_6$ (0)[f]

[a] At high (>10 kHz) spin rates (when the centerband is well separated from the spinning sidebands).

[b] When used under high-resolution, multipulse conditions. Literature values vary considerably so this value should be used with caution.

[c] This is the highest frequency centerband in the spectrum.

[d] Ξ = 25.145 970 and 25.145 743% for the CH$_2$ and CH, respectively (see text).

[e] There is some debate about this value. CFCl$_3$ is no longer freely available so resolving this issue is a problem.

[f] In aqueous solution; details given in Pure Appl. Chem. **80** (2008) 59.

[g] 0.11 M in 0.05 M HCl.

Referencing methods that use documented absolute frequencies for different nuclei, defined through the parameter Ξ, relative to the ^1H resonance from tetramethylsilane in dilute solution in chloroform ($\Xi = 100\%$)[8] can be used in the solid state, although this may require some ingenuity in the absence of the deuterium lock used to do this in solution-state NMR. (With the possible exception of biochemical solid-state applications, solid-state experiment times are generally too short and lines are too broad for any drift in magnetic field to be noticed, so a deuterium lock is not usually used in solid-state measurements.)

3.3.9 TEMPERATURE CALIBRATION

Carrying out a solid-state NMR measurement at any temperature leads to the question: What is the true sample temperature?

Due to the mechanics of MAS, it is not feasible to insert a temperature-monitoring device directly into the sample. Instead, it is usual to have a thermocouple in the heated/cooled gas stream close to the sample. But the temperature of the nearby gas is not necessarily an accurate indicator of the sample temperature because, even in the air bearing used for MAS, there is frictional heating of the rotor (and so the sample). Some sort of temperature calibration is therefore necessary if the true sample temperature is to be known. Two of the various methods for doing this are mentioned here.

In solution-state NMR, neat methanol is a well-known "thermometer" and the chemical shift difference in parts per million, Δ, between the CH_3 and OH proton signals as a function of temperature is well documented:

$$T/K = 403.0 - 29.5\,\Delta - 23.8\,\Delta^2 \qquad\qquad 3.6$$

For solid-state NMR it can be used either neat (if it can be contained) or on a solid support. It is useful for calibrating temperature as a function of spin rate (see figure 3.18) under "ambient" conditions and for giving an absolute temperature. Because most probes can operate over a wide temperature range, other temperature-sensitive substances have to be used for temperatures at which methanol is not useful (below −90 °C or above +60 °C). One such is solid lead nitrate (see figure 3.18). The chemical shift of the ^{207}Pb signal is very sensitive to temperature (0.70 ppm / °C) and can be used to determine the temperature *difference* from a known point (there is no accepted absolute relationship between temperature and the chemical shift of the single lead resonance), for temperatures up to about 250 °C (it decomposes at 290 °C). The shape of the line (see figure 3.19) also gives an indication of the temperature variation in the sample (this increases with extremes of temperature and can be several degrees depending on the shape and size of the sample space). The only major drawback with using lead nitrate is its toxicity.

---

[8] For details see Pure Appl. Chem. **80** (2008) 59.

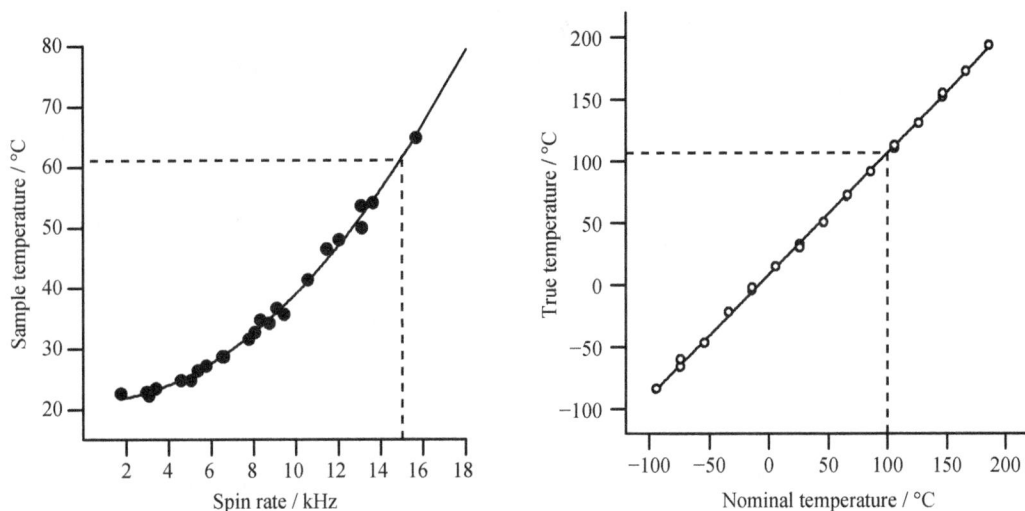

Figure 3.18. (Left) Sample temperature as a function of spin rate for a 4 mm (rotor outside diameter) magic-angle spinning (MAS) probe determined using neat methanol and with no temperature control. At 15 kHz the sample temperature exceeds 60 °C. (Right) Temperature calibration graph for an MAS probe at constant spin rate. Under the conditions used for this calibration, the true sample temperature is about 8 °C above the temperature set on the spectrometer.

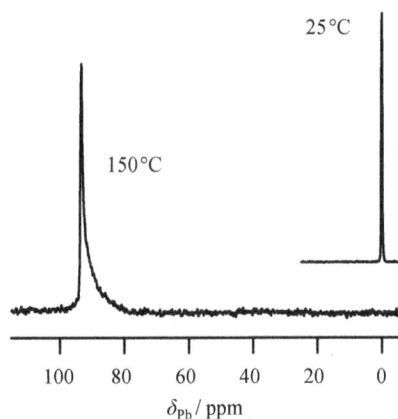

Figure 3.19. ^{207}Pb spectrum from lead nitrate at nominal temperatures 25 and 150 °C. In the high-temperature spectrum, the parts per million spread of the signal equates to a temperature range from around 133 to 158 °C.

    With all measurements at non-ambient temperature, it is important to wait (usually at least 10 minutes) after changing the temperature (or spin rate) for a proper thermal equilibrium to be established. Two consecutive measurements will determine whether the temperature is stable.

    The effects of heating from the RF irradiation cannot always be neglected. High duty cycles and long, high-power pulses can cause sample heating. The extent of this tends to be more sample dependent than spinning-induced heating and it is, therefore, more difficult to calibrate the effect on the sample temperature. Samples with high dielectric constants (those that are highly ionic or have a high water content) may need special care. Measurements on delicate (e.g., biomolecular) systems should be carried out at reduced temperatures to decrease the risk of sample degradation.

3.4 CROSS POLARIZATION

Cross polarization (CP) involves the transfer of magnetization (or polarization) from the nuclei of one element to those of another. The most commonly encountered cross polarization experiment involves transfer of magnetization from *abundant* ^{1}H (or, occasionally, ^{19}F) spins to *dilute* X ones, where X is any other spin-$\frac{1}{2}$ nucleus. It is worth noting, however, that cross polarization is not limited to this combination.

The cross-polarization experiment is more complicated than that for direct excitation (see section 3.4.1), so what are its advantages? There are two main ones:

* Because the magnetization originates from ^{1}H, the recycle delay is limited by the recovery of the ^{1}H magnetization and not that of the X spins. That is, the recycle delay depends on T_1^H and not T_1^X. Usually $T_1^H \ll T_1^X$ (see inset 3.6), so this means that the pulse sequence can be repeated much more rapidly than in a direct-excitation experiment, so significantly increasing the signal-to-noise ratio in the spectrum.

* In the limit of 100% magnetization transfer, there is a signal enhancement by a factor equal to γ^H/γ^X due to the difference in equilibrium populations (see equation 2.6). For ^{13}C the maximum enhancement is ~4 and for ^{15}N it is ~10, which represent 16- or 100-fold reductions, respectively, in experiment time. The enhancement actually achieved depends on the efficiency of the cross-polarization process which varies with the nature of the sample and the experimental conditions.

The impact that these two factors might have on experiment time is exemplified in inset 3.6 and illustrated in figure 3.20.

Inset 3.6. Cross polarization vs. direct excitation

For 3-methoxybenzoic acid, the contrast between T_1^H and T_1^C is striking. Due to rapid spin diffusion between the protons, there is a single value for T_1^H for the molecule of 1.7 s. Although T_1^C for the methyl group is short (~1 s) because of the mobility of that group, the average value for the other carbons is extremely long (190 s). To record the same number of repetitions takes 112 times longer by direct excitation than by cross polarization (given that an appropriate recycle delay of 1.2 times T_1^H or T_1^C for cross-polarization and direct-excitation experiments, respectively, is used). Suppose also that cross polarization gives the maximum signal enhancement of a factor of 4 over that obtained by direct excitation. It therefore takes 1792 ($4^2 \times 112$) times as long for the direct-excitation experiment to reach the same signal-to-noise ratio as the cross-polarization one. Put another way, to match a cross-polarization experiment lasting 1 h, a direct-excitation one would take over 10 weeks! Cross-polarization and direct-excitation spectra from a complex organic molecule are compared in figure 3.20.

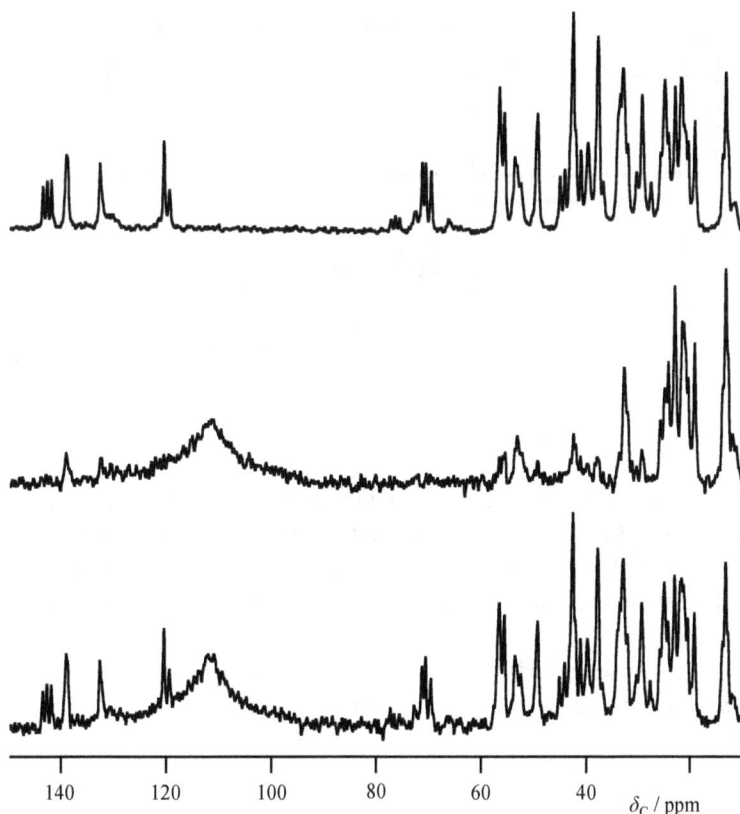

Figure 3.20. Three carbon-13 magic-angle spinning spectra from the plant sterol stigmasterol. The top spectrum was obtained using a cross-polarization experiment in 22 minutes (448 repetitions with 3 s recycle delay). The middle spectrum was obtained using a direct-excitation experiment also in 22 minutes (448 repetitions with 3 s recycle delay). Note the lower signal-to-noise ratio and selectivity of the experiment—only the signals from fast-relaxing carbons are obtained. The signals from the other carbons in the molecule are observed only when the recycle delay is increased. The bottom spectrum was obtained using a direct-excitation experiment with 120 s recycle delay (448 repetitions, total experiment time nearly 15 h). The broad line near 110 ppm in the direct-excitation spectra arises from the CF_2 carbon in the Teflon® used as rotor caps. Because it contains no protons, the Teflon® signal is not detected in the cross-polarization experiment.

3.4.1 THE CROSS-POLARIZATION EXPERIMENT

The cross-polarization pulse sequence is shown in figure 3.21. The initial ^1H 90° pulse directed along x, say, rotates the ^1H magnetization onto the y axis. The second, 90°-phase-shifted ^1H pulse *spin-locks* the magnetization (see section 2.8). During this spin-lock period, a pulse is applied simultaneously at the X frequency. The time for which these two pulses are applied is called the *contact time* and the X pulse will be referred to as the contact pulse. Under the right conditions (see below), magnetization will transfer from ^1H to X, and X magnetization will build up during the contact time.

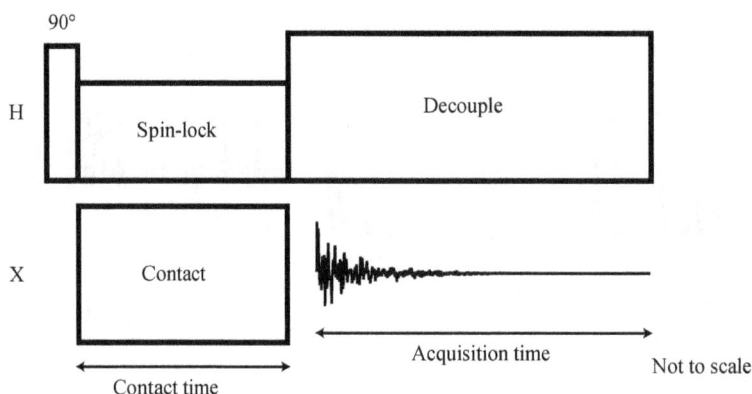

Figure 3.21. The cross-polarization pulse sequence.

The NMR signal of the X spins is then measured during the acquisition time, which, as the sample contains ^{1}H, is carried out under ^{1}H high-power decoupling. The contact time might last for up to tens of milliseconds (if the probe can take it) for the polarization transfer to reach its peak. This is in contrast to the pulses in a direct-excitation experiment which, typically, have durations of a few microseconds.

The right conditions for polarization transfer occur when the ^{1}H and X RF fields fulfill the *Hartmann–Hahn match condition* (the fields are then said to be *matched*):

$$\gamma^H B_1^H = \gamma^X B_1^X \text{ or, from equation 3.2, } \nu_1^H = \nu_1^X \qquad 3.7$$

In other words, cross polarization occurs when the H and X nutation frequencies are equal. It is also a prerequisite of cross polarization that there is an interaction (generally dipolar coupling) between the two types of nucleus.

The match condition is typically determined experimentally by observing the X signal as a function of the amplitude (or power) of either the ^{1}H spin-lock pulse or the X contact pulse (while keeping the other RF

Inset 3.7. Volts, watts and decibels

$V_{pp} = 2\sqrt{2P \times 50\ \Omega}$ or $V_{pp} = 20\sqrt{P},$ where P is the power in watts and V_{pp} is the peak-to-peak voltage (at an impedance of 50 Ω).

A decibel value is the ratio of two powers: $P(\text{dB}) = 10\log_{10}\left(P_1/P_2\right)$. When quoted as dBm P_2 has a reference value of 1 mW.

So 1 mW ≡ 632 mV (0 dBm), 100 W ≡ 200 V (50 dBm), and 1 kW ≡ 632 V (60 dBm).

---

field constant). It is usually safer if the former is varied while the contact pulse is kept constant as this is less likely to result in too much RF power being applied to the probe, since $\gamma^H > \gamma^X$, $B_1^H < B_1^X$ at the matching condition (inset 3.7 relates to the measurement of RF power). Figure 3.22 shows a typical ^{13}C match profile from hexamethylbenzene (HMB).[9] At a modest spin rate (3 kHz), a well-defined

---

[9] Hexamethylbenzene (HMB) is often used as a set-up compound for ^{13}C because it gives a well-defined match profile irrespective of other instrument variables such as decoupling efficiency or shimming. Adamantane (see inset 7.8) is a highly mobile solid that gives very narrow lines and is a good check for lineshape.

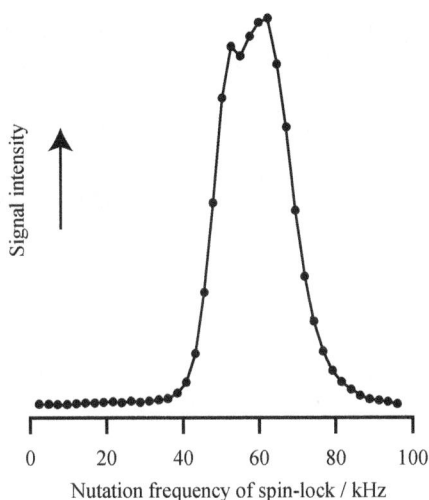

Figure 3.22. The match profile from the methyl signal for hexamethylbenzene at a spin rate of 3 kHz. The solid line is a visual guide only. The amplitude of the ^{1}H spin-lock pulse was varied while the ^{13}C contact pulse had a fixed amplitude.

maximum is observed (this is relatively uncommon behavior—the match profile is broader and flat-topped for more typical materials).

For HMB the match behavior becomes more complicated at higher spin rates (see section 5.1.3) and in other materials similar complexity can arise at low spin rates when hydrogen is dilute in the sample or when there is a high degree of molecular motion (as there is for adamantane, which is commonly used in setting up experimental conditions for a spectrometer).

Any X pulse, including a contact pulse, will result in a signal being generated by direct excitation as well as, and independent of, that originating from cross polarization. This is most noticeable for nuclei with high natural abundance such as ^{31}P or for isotopically labeled samples where the direct-excitation signal has a high intensity. This unwanted signal is suppressed using a *phase cycle*.

Phase Cycling

The phase cycle is a key part of most solid-state NMR experiments. As has already been discussed, the pulses in a pulse sequence are labeled with a phase. The receiver has a phase too (see inset 3.8). Over a cycle of successive repetitions of the pulse sequence, it is usually the case that these phases change—the phase cycle. In the simplest of cases, the phase cycling uses the quadrature phases x, y, $-x$, and $-y$ (or 0, 90, 180, 270) to compensate for

---

Inset 3.8. Phase-sensitive (quadrature) detection

The "raw" NMR signal is split into two to give "real" and "imaginary" components that have a phase difference of 90°. This is a necessary precursor to the complex Fourier transform that converts the time-domain signal into the frequency spectrum. The details of the way the signal is split are not important here–and in any case tend to change with developments in the electronics within the spectrometer. The important point is that a receiver nominally with phase x will detect both the x and y components of the signal.

any missetting of the phases (exactly how this works is not important here). The receiver phase simply needs to follow the pulse. So the pulse-acquire experiment might be phase cycled as:

Repetition	1	2	3	4
Pulse phase	0	180	90	270
Receiver phase	0	180	90	270

The phase cycle in a cross-polarization pulse sequence is designed to do a further job—to select only signal that is generated by cross polarization (i.e., to suppress the signal that is generated directly from the contact pulse on the X-channel). A typical phase cycle might be:

Repetition	1	2	3	4
^{1}H 90° pulse	0	180	0	180
^{1}H spin-lock pulse	90	90	90	90
X contact pulse	0	0	90	90
Receiver	0	180	90	270
Phase of cross-polarization signal[a]	0	0	0	0
Phase of direct-excitation signal[a]	90	−90	90	−90

[a] Real component, relative to the receiver.

Over the four repetitions of the pulse sequence in this phase cycle, the cross-polarization signal is always detected as positive (relative to the receiver phase) but the sum of the direct-excitation signal is zero over the phase cycle. The alternating reversal of the phase of the 90° pulse is sometimes referred to as *spin-temperature inversion*.

Designing complex phase cycles (like the one illustrated in inset 3.9) is an advanced skill and takes some practice (and understand). An error in the phase cycle is a common cause of unexpected behavior from a new pulse sequence. See the texts in the Further reading for detailed information on devising phase cycles.

3.4.2 CONTACT TIME

During the contact time, the X magnetization builds up toward a steady state corresponding to an equilibrium between the ^{1}H and X magnetizations. At the same time the spin-locked ^{1}H and X magnetizations decay through spin–lattice relaxation in the rotating frame (section 2.8) at rates of $1/T_{1\rho}^H$ and $1/T_{1\rho}^X$, respectively. These processes are summarized in figure 3.23. If $T_{1\rho}^X \gg T_{1\rho}^H$ then $T_{1\rho}^X$

Inset 3.9. Phase cycles for coherence selection

The phase cycle is also a key part of the coherence selection in multiple quantum experiments. Here the quadrature phases are not the only ones that can be used; 45, 90, 135 … are found in experiments where double quantum coherences are selected and 30, 60, 90 … can be used for triple quantum coherence selection. Figure 6.16(a) shows a two-pulse, triple-quantum/single-quantum correlation experiment that has the following 24-repetition phase cycle:

Repetition	1	2	3	4	5	6	7	8	9	10	11	12
Pulse 1	0	60	120	180	240	300	90	150	210	270	330	30
Pulse 2	90	90	90	90	90	90	180	180	180	180	180	180
Receiver	0	180	0	180	0	180	90	270	90	270	90	270

Repetition	13	14	15	16	17	18	19	20	21	22	23	24
Pulse 1	180	240	300	0	60	120	270	330	30	90	150	210
Pulse 2	270	270	270	270	270	270	0	0	0	0	0	0
Receiver	180	0	180	0	180	0	270	90	270	90	270	90

As this case illustrates, phase cycles can be many repetitions long. So that the phase cycle can do the job it is designed for, it is always a good idea to acquire an integer multiple of the number of repetitions in a phase cycle. That may not be as trivial as it sounds for a complex two-dimensional experiment with a many-repetition phase cycle—particularly if a long recycle delay is needed and the total time required becomes prohibitive.

can be safely ignored and the amount of signal (S) obtained with a contact time t_c is given by equation 3.8

$$S(t_c) = S_0 \left(1 - \exp(\frac{-t_c}{T_{XH}})\right) \exp(\frac{-t_c}{T_{1\rho}^H})$$ 3.8

where S_0 is the maximum obtainable intensity in the absence of relaxation and T_{XH} is loosely defined as the time constant for cross polarization.

Strongly dipolar-coupled hydrogen nuclei in rigid solids tend to have a common $T_{1\rho}^H$ value because spin diffusion is fast on the NMR timescale. However, T_{XH} is determined by the local strength of the dipolar coupling between X and H, and so is specific to the X environment.

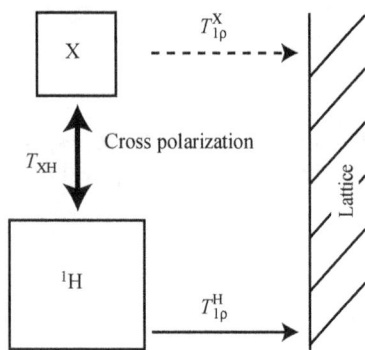

Figure 3.23. The relationship of the parameters influencing a cross polarization experiment. $T_{1\rho}^{X}$ is usually much longer than the other time constants and is often ignored.

For example, T_{SiH} for $(SiO)_3SiOH$ and $(SiO)_4O$ environments in silicates are very different as they have different relationships to the hydrogen. Similarly, T_{CH} for a CH_2 carbon will be different from that for a quaternary carbon (e.g., around 30 and 440 µs, respectively, in isoleucine). For a complex sample, there is no guarantee that the signal intensities at a particular contact time will be proportional to the number of nuclei giving rise to that signal. This is a disadvantage of cross polarization, but is analogous to the situation encountered in solution-state NMR where the relative intensities of slowly relaxing quaternary carbons are often not quantitative. In the solid state, quantitivity can be restored at the expense of recording multiple spectra as a function of contact time and fitting the result to equation 3.8 to obtain S_0 (see section 7.4). Experience suggests, however, that equation 3.8 is often too simplistic (see figure 3.24).

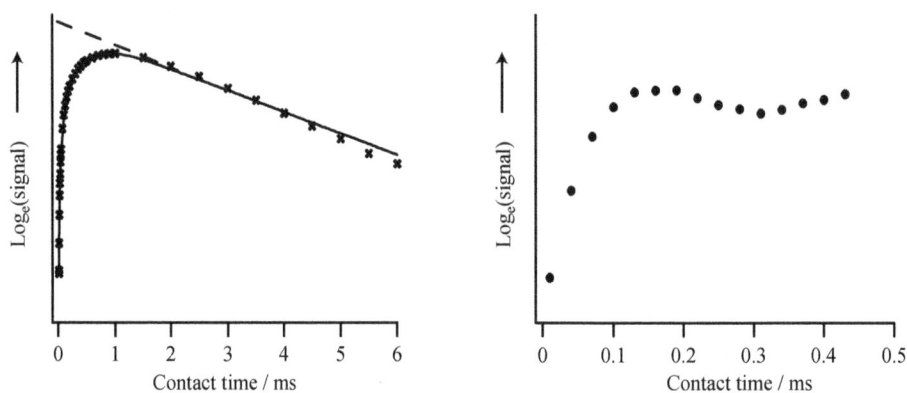

Figure 3.24. Contact time behavior for the carboxylic acid signal from L-isoleucine (left). The observed signal (crosses) can be fitted (the solid line) using equation 3.8 with $T_{XH} = 0.44$ ms and $T_{1\rho}^{H} = 2.7$ ms. S_0 (from the intercept on the signal axis indicated by the dashed line) is 160% of the maximum *observed* intensity. The short-contact time behavior for the nitrogen in ^{15}N labeled proline is shown on the right. Equation 3.8 is no longer useful for simulating the observed behavior, although the *transient oscillation* can be used to measure the $^{15}N,^{1}H$ dipolar coupling (see section 8.3.1).

Occasionally, $T_{1\rho}^{H}$ is long (tens of milliseconds) and, providing the ^{1}H spin-lock is maintained to the end of the acquisition time,[10] a ^{1}H 90° *flip-back* pulse (of opposite phase to the initial one) can return ^{1}H magnetization to the z axis and circumvent the need for a long recycle delay, hence speeding up experiments. This is illustrated in figure 3.25.

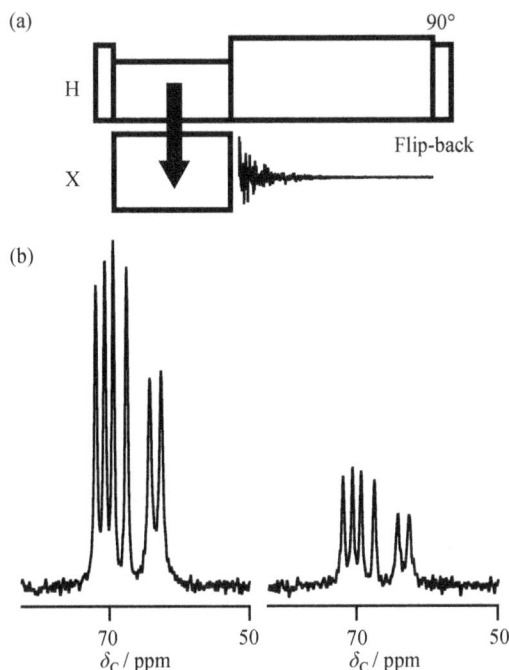

Figure 3.25. (a) The flip-back modification to the cross-polarization pulse sequence and (b) ^{13}C spectra from mannitol obtained with a 60 s recycle delay with (on the left) and without (right) flip-back (all other acquisition conditions being equal). Without flip-back, a 300 s recycle delay is required to produce the equivalent result—increasing the experiment time by a factor of five.

Despite the potential loss of quantitation and the added complexity of the experiment, cross polarization is an essential tool for the study of dilute spins at natural abundance. It makes ^{13}C and ^{15}N spectroscopy feasible in the solid state. It provides an extra tool for studies involving ^{29}Si and is a useful aid to speeding up experiments on the heavy metals ^{77}Se, $^{113/111}$Cd, $^{117/119}$Sn, ^{195}Pt and ^{199}Hg. But does cross polarization always work effectively? The short answer is no. Cross polarization relies on dipolar coupling between ^{1}H and X and, to a lesser extent, between ^{1}H and ^{1}H,[11] so anything that reduces or removes this or gives rise to a very long T_{XH} or very short $T_{1\rho}^{H}$ can interfere with it. Molecular motion in a gel or rubbery material can sufficiently interrupt the dipolar coupling to render the sample invisible to a cross-polarization experiment. This is a case where a direct-excitation experiment

---

[10] In figure 3.21 the spin-lock and decoupling pulses have the same phase so the spin-lock is maintained even when there is an increase in RF field for decoupling. The spin-lock is lost if the ^{1}H irradiation is turned off, as it would be for a dipolar dephasing experiment (see section 7.2.4), or if more complex decoupling schemes than simple CW irradiation are used.

[11] ^{1}H,^{1}H coupling enables spin diffusion to be efficient among the protons and this leads to increased polarization transfer, making it more likely that a signal enhancement closer to the theoretical γ^H/γ^X maximum will be achieved.

may be the only viable one. There is also a problem if $T_{1\rho}^{H}$ is very short (in the microsecond range) since there is not enough time for a cross-polarization signal to build up before the ^{1}H magnetization is lost. This situation often arises in materials that are paramagnetic or contain a paramagnetic impurity.

3.4.3 DIRECT EXCITATION OR CROSS POLARIZATION? A SUMMARY

Does the sample contain hydrogen? With no hydrogen, ^{1}H–X cross-polarization is not possible, so direct excitation is the only option (this is likely to be the case with a very low hydrogen content too). This may mean very long accumulation times as T_1^{X} (and hence the recycle delay) is likely to be long. It may prove difficult to obtain a high signal-to-noise ratio for low abundance nuclei.

Is the sample a rigid solid? If it is, cross polarization is the only realistic choice for obtaining a spectrum with a high signal-to-noise ratio for natural abundance ^{13}C or ^{15}N. For ^{31}P and, often, ^{29}Si, direct excitation can also yield useful information even though long recycle delays may be required (2–5 minutes are typical). For the heavy metals, the situation is complicated by potentially large spinning sideband manifolds (which disperse the signal; see section 3.3.4) or low resonance frequencies (which present some instrumental challenges) and it is more difficult to generalize.

Is the sample a soft or mobile solid (rubber or gel like)? Under these conditions cross polarization may be ineffective, so direct excitation may be the only feasible experimental method. However, in such materials, T_1^{X} is likely to be relatively short so long recycle delays may not be required.

Will intensity information be used quantitatively? If so, a direct-excitation experiment can be made quantitative easily (by setting the recycle delay to $5 \times T_1^{X}$). If this is not feasible for dilute nuclei, then a series of cross-polarization spectra, as a function of contact time, followed by numerical analysis, will be required (see section 7.4).

Are low-intensity signals of particular interest? If they are, then cross polarization (if it is effective) is likely to give a spectrum with the highest signal-to-noise ratio.

Is the sample heterogeneous? Depending on the nature of the components (rigid/mobile, ^{1}H rich/poor), heterogeneous samples may present some special challenges and a combination of cross-polarization and direct-excitation experiments may be required to characterize the sample. Given an appropriate recycle delay, the direct-excitation measurement will give the most quantitative picture of the composition of the sample.

Is the sample paramagnetic? Wholly paramagnetic samples cause special problems. Nuclei close to a paramagnetic center may be undetectable (their resonance lines being too broad to distinguish them from the baseline). More distant ones may be observable but probably only with a direct-excitation experiment ($T_{1\rho}^{H}$ is likely to be too short to allow efficient cross polarization). Often a small amount of a paramagnetic impurity can be tolerated and in some cases can cause a helpful shortening of the T_1^{X} relaxation time (see section 7.5).

3.5 HIGH-RESOLUTION SPECTRA FROM ^{1}H (& ^{19}F)

It was noted at the beginning of this chapter that ^{1}H and, to some extent, ^{19}F are special cases. This is because, despite the most advanced experimental techniques available at the time of writing,

^{1}H and ^{19}F linewidths, for true solids, remain stubbornly high (often several 100 Hz) and this fact limits the chemical information available from a spectrum. This is in stark contrast to the solution state, where a proton NMR spectrum is usually the starting point for the characterization for any soluble, proton-containing material. Solution-state linewidths are typically of the order of 1 Hz. In the solid state, strong homonuclear dipolar coupling is the main contributor to high linewidths. For ^{1}H, the problem is compounded by its small chemical shift range (~20 ppm) and that results in poorly resolved spectra (see inset 3.10). The chemical information from a ^{1}H spectrum may be limited to the general identification of signals as aliphatic, aromatic or, when present, strongly hydrogen bonded (which usually have a distinct high-frequency shift). The situation is slightly better for ^{19}F because the chemical shift range is over 400 ppm and useful chemical information can be usually obtained. Examples of solid-state ^{1}H and ^{19}F spectra are given in figure 3.26.

Inset 3.10. Spectral resolution

Spectral resolution is a measure of the degree to which two closely spaced signals in a spectrum can be separately identified. The resolution depends on linewidth and line separation as illustrated in the figure (right), which shows two lines with Gaussian shape and half-height width Δv, (a) at the same chemical shift, (b) $0.8\Delta v$ apart—unresolved (the flat top is characteristic of this situation), (c) Δv apart—just resolved (the limit of resolution), and (d) $2\Delta v$ apart—resolved.

A related issue is digital resolution. This is the frequency difference of successive data points in the spectrum. It is inversely proportional to the number of points in the Fourier transform. Fourier transformation of the acquired data points only (i) may result in limited resolution. Adding zeros to the FID (zero-filling) so that the number of points transformed is twice that acquired (ii) gives the maximum resolution. Further zero-filling simply improves the aesthetic appearance of the spectrum (iii).

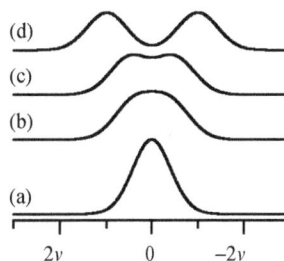

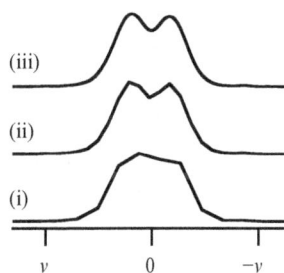

To obtain "high-resolution" proton spectra, it is necessary to overcome the homonuclear dipolar coupling. High sample spin rates can achieve this to a certain extent—the higher the better (usually above 25 kHz before a significant impact is observed). For crystalline solids, high magnetic fields are also a distinct advantage. However, even with the highest spin rates achievable with traditional MAS technology, resolution is still disappointing from the point of view of routine sample characterization.

The alternative to mechanically rotating the sample is to manipulate the sample with a carefully designed sequence of RF pulses and interleave these with data acquisition. The background to this

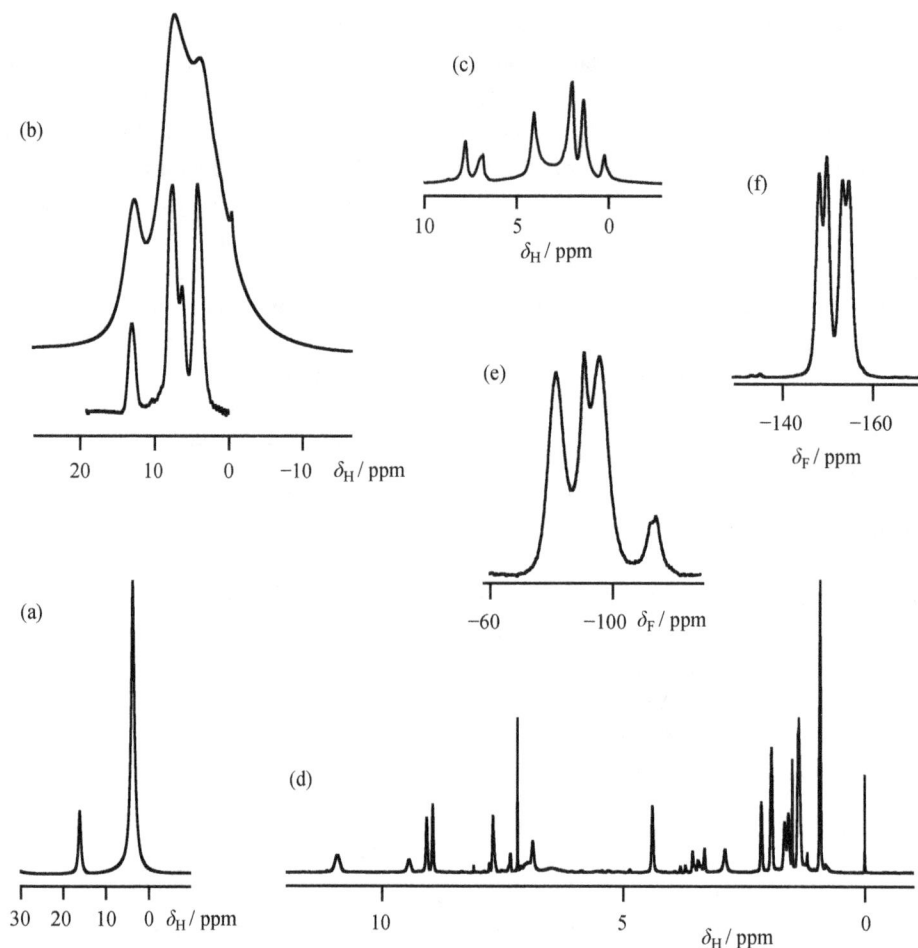

Figure 3.26. Examples of solid-state ^{1}H and ^{19}F spectra. (a) Naturally resolved ^{1}H spectrum from the mineral octosilicate (spin rate 5 kHz). (b) ^{1}H spectra from 3-methoxybenzoic acid, magic-angle spinning (MAS) at 15 kHz (upper trace) and with homonuclear decoupling (a windowed phase-modulated Lee-Goldburg, wPMLG, spectrum) and spinning at 10 kHz. (c) ^{1}H spectrum from an organic compound in a silica matrix recorded at a spin rate of 60 kHz at a field of 21.1 T. (d) ^{1}H high-resolution (HR) MAS spectrum from an organic compound bound to a swollen polystyrene bead. (e) ^{19}F spectrum from polyvinylidenedifluoride. (f) ^{19}F spectrum from octafluoronaphthalene. Note the different scale ranges. Figure 3.26(c) supplied courtesy of Dr Anne Lesage.

homonuclear decoupling is discussed in chapter 4 and is visited again in section 5.5. Combining multiple-pulse homonuclear decoupling with two-dimensional methods, often detecting on a nucleus such as carbon and obtaining a heteronuclear correlation spectrum (see section 5.4.3), is generally a more promising approach.

These multipulse techniques can also be applied to fluorine, although such extreme measures are probably needed only in systems with strong homonuclear dipolar coupling such as inorganic fluorides or perfluorinated organics. Sample spin rates of the order of 10–20 kHz are often adequate for obtaining useful fluorine spectra. In organofluorine systems, fluorine atoms may be dilute within a molecule and in such cases homonuclear coupling is not a significant issue; instead it is efficient decoupling of the protons that becomes important. However, decoupling ^{1}H while observing ^{19}F is not trivial because of the similarity in their resonance frequencies ($v_F = 0.94v_H$) and a specially designed probe and RF filters are needed to do this successfully. An added complication at low field (less than 11.7 T) is the Bloch–Siegert effect,[12] which needs to be taken into account when quoting chemical shift values and it may also have an impact on linewidths under ^{1}H decoupling.

Cross polarization from ^{1}H to ^{19}F is possible. Because of the similarity in their magnetogyric ratios there is no significant signal enhancement; however, if the relaxation behaviors are different (proton significantly faster), then there is a potential gain in signal-to-noise ratio by carrying out the cross-polarization experiment. The potential selectivity of the experiment can also be exploited. The dynamics of the cross-polarization process may not have the relative simplicity of equation 3.8 in systems where both ^{1}H and ^{19}F are strongly coupled (equation 3.8 is based on assumption that the X spins are dilute). Cross polarization from ^{19}F to ^{1}H is also feasible and may be useful for spectral editing.

Hydrogen- and fluorine-containing polymers are often used in the construction of NMR probes and in the materials used to cap rotors (the rotors themselves are usually made from a ceramic such a zirconia or silicon nitride and, when clean, contain no hydrogen or fluorine). These components will contribute a signal to the spectrum and that can become a problem if the signal from the sample under study has a low intensity. It is usual to choose materials so that this *background* signal is minimized. For example, Teflon® (tetrafluoroethylene) or Kel-F® (trifluorochloroethylene) contain only residual traces of hydrogen so are ideal for use when recording proton spectra but should be avoided when observing fluorine.[13] Instead, a hard-wearing polyimide such as Vespel® is ideal for fluorine. In practice, some compromises may have to be made and it is not always possible to fully avoid background signals. In such cases it is advisable to record a spectrum from the empty rotor and, if necessary, subtract it from the spectrum obtained from the sample under study. Pulse sequences are also available for *reducing* the intensity of background signals (although it is still advisable to record a spectrum from the empty rotor).

^{1}H (or ^{19}F) do not always experience strong homonuclear coupling and combinations of molecular mobility and relative isolation may mean that well-resolved spectra can be obtained without any special experimental methods. Soft solids, such as gels or tissue samples, fall into this category. Resolved ^{1}H spectra can be obtained from these with MAS probes designed for work with true solids. However, these are often not optimized for the highest resolution in the same way that solution-state probes are and resolution can be improved by using a specifically designed high-resolution MAS (HRMAS) probe (see figure 3.26(d)).

---

[12] When $\gamma B_1^H / 2\pi < |v_H - v_X|$ ^{1}H decoupling has no detectable impact on the chemical shift of the observed X nuclei. If this condition does not hold (as it might not for ^{19}F observation with ^{1}H decoupling), then a small shift in the X resonance frequency may be observed. This is known as the *Bloch–Siegert* effect.

[13] Because these materials contain no ^{1}H, they are also ideal for use when observing ^{13}C using ^{1}H to ^{13}C cross polarization (in effect they are invisible). They will, however, contribute signal to a direct-excitation spectrum.

At the other extreme, the broad lines obtained from a non-spinning, hydrogen-rich material yield no chemical information directly but can give useful insights into the physical characteristics of the sample. This is a topic that is examined in chapter 7.

In this chapter, basic experimental NMR methods for the study of solids have been introduced. Although there is a vast range of possible solid-state NMR measurements, one or other of these methods is central to most of them. The components of the experiment that control the appearance of the spectrum it produces have also been introduced.

In the next chapter the theme changes from practical issues to the quantum mechanics of solid-state NMR. This chapter should give the reader an insight into the theory of solid-state NMR. It is not essential to the understanding of the subsequent chapters, so if this insight is not required yet, go on to the rest of the book and come back to it later.

FURTHER READING

DESCRIPTION OF THE NMR EXPERIMENT

"Understanding NMR spectroscopy", J. Keeler, John Wiley & Sons (2005), ISBN 978 0 470 01787 6.

PULSE NMR AND PRACTICAL ISSUES RELATING TO SIGNAL PROCESSING

"Modern NMR techniques for chemistry research", A.E. Derome, Pergamon Press (1987), ISBN 0 08 032513 0.
"High-resolution NMR techniques in organic chemistry", T.D.W. Claridge, Elsevier (2009), ISBN 978 0 08 054818 0.

CHAPTER 4

QUANTUM MECHANICS OF SOLID-STATE NMR

4.1 INTRODUCTION

The previous chapters have deliberately avoided delving deeply into the theory underlying solid-state NMR. Indeed many solid-state NMR experiments can be understood in terms of the vector model familiar from the solution state (and discussed in section 3.2), so that most experiments can be interpreted using straightforward qualitative pictures of the different NMR interactions and their effects on the spectrum. As the subject has progressed, however, the complexity of experiments and the associated theoretical background have considerably increased. This chapter aims to provide an introduction to the various theoretical tools used in solid-state NMR. The Further reading lists some more detailed and thorough texts, but this concise overview should give the reader sufficient insight to be able to explore further and to tackle the NMR literature with confidence. Those only interested in using the results of solid-state NMR will probably prefer to skip over this chapter and go to the applications of solid-state NMR, ignoring the references to theory when some of the more complex experiments are discussed.

The theory of solid-state NMR necessarily involves quantum mechanics which often appears to bear little relationship to the quantum mechanics covered in undergraduate chemistry courses. However, as will be seen below, the quantum mechanics of NMR is in many ways simpler and more elegant than typical quantum chemistry.

Ultimately, all quantum mechanics reduces to solving the Schrödinger equation $\hat{H}\psi = E\psi$, where ψ is a wavefunction for the system, E is an allowed energy corresponding to the wavefunction ψ, and $\hat{H}$ is the *Hamilton operator*, or *Hamiltonian*, which describes the energy of the system. We can write $\hat{H}$ schematically for a system of nuclear spins and the surrounding "system":

$$\hat{H} = \hat{H}_{nuc} + \hat{H}_{nuc,surr} + \hat{H}_{surr}$$

where $\hat{H}_{nuc}$ contains terms that only depend on the nuclear spins, $\hat{H}_{surr}$ refers to terms with no dependence on the nuclear spin state (such as molecular vibration and rotation), while $\hat{H}_{nuc,surr}$ contains terms that involve both the nuclear spins and the surrounding environment, for example, the J coupling, which is mediated by the electronic wavefunction.

This total Hamiltonian is tremendously complex, with degrees of freedom (independent variables) for the nuclear and electron spins, the positions of the atoms, etc. Fortunately the "coupling" (degree of interaction) between the nuclear spins and their environment is extremely small, which allows the problem to be vastly simplified. The largest of the NMR interactions introduced in chapter 2, the Zeeman interaction, is only of the order of 100s of MHz[1] and all other energy terms in $\hat{H}$ (vibration, rotation, electronic, etc.) are associated with much higher frequencies. The nuclear spin energy has negligible impact on the other degrees of freedom of the total wavefunction and these other components evolve on much faster timescales. The external degrees of freedom can therefore be integrated out, leaving a Hamiltonian that only depends on the nuclear spins. In this *spin Hamiltonian*, $\hat{H}_{surr}$ is a constant and so can be ignored, while the terms involving both nuclear spin and external coordinates, $\hat{H}_{nuc,surr}$, can be simplified to terms involving nuclear spin operators and constants that depend on the external system. For instance, the (isotropic) J-coupling Hamiltonian between two nuclear spins, I and S, over the NMR timescale, becomes

$$\langle \hat{H}_J \rangle = J_{IS}\, \hat{\boldsymbol{I}} \cdot \hat{\boldsymbol{S}}$$

So while the full J-coupling Hamiltonian involves both the spin coordinates and the entire electronic wavefunction, the corresponding spin Hamiltonian just involves the states of the two relevant nuclear spins and a coupling constant, J_{IS}, which results from integrating out the other degrees of freedom.

This "decoupling" of spin and space coordinates greatly simplifies the analysis of NMR. While the wavefunction for a multi-electron system has no analytical solution, so that deriving increasingly accurate solutions is the main challenge of quantum chemistry, the Schrödinger equation for a number of interacting nuclear spins is generally embarrassingly easy to solve. The drawback of splitting off the nuclear spin Hamiltonian from the complete system is that terms involving the external degrees of freedom are reduced to empirical constants, such as coupling constants and chemical shifts. *Calculating* these constants from first principles (*ab initio* quantum chemistry) is difficult since it requires the full multi-electron Hamiltonian. Historically, NMR parameters have generally been treated as empirical quantities that are compiled, and empirical correlations are then derived between their values and chemical/structural features. Increasingly, however, quantum chemistry is able to provide robust estimates of NMR parameters, particularly shielding and quadrupole coupling constants, allowing the NMR parameters to be tied in a more direct and meaningful way to the underlying molecular structure (see section 8.8.2).

4.2   THE HAMILTONIANS OF NMR

NMR theory is further simplified by working at sufficiently high magnetic field that the Zeeman interaction dominates all the other interactions. As discussed in chapter 6 on quadrupolar NMR,

---

[1] Note that the nuclear spin Hamiltonian is typically expressed in terms of frequency units (Hz, MHz, etc.), which are the natural units for NMR interactions. A slightly different convention, common in theoretical work, uses *angular* frequencies (rad s^{-1}). The two conventions are usually distinguished by using ν for frequencies and ω for angular frequencies.

this is not always achievable for nuclei with large quadrupole moments and low NMR frequencies, but, even here, the quadrupole interaction can usually be treated as a perturbation on the dominant Zeeman term. As there is little chemical interest in performing NMR at low magnetic fields, we will work exclusively within this *high field approximation*.

If we consider a single spin with nuclear spin quantum number I, the different spin (angular momentum) states can be conveniently represented in the Dirac bracket notation, $|I, m_I\rangle$, where m_I is the magnetic component quantum number (see section 2.1). The Hamiltonian for the Zeeman interaction is

$$\hat{H}_Z = -\gamma B_0 \hat{I}_z \qquad\qquad 4.1$$

where the magnetic field, B_0, has its conventional alignment along the z axis, and $\hat{I}_z$ is the operator for the z component of the nuclear spin angular momentum. The Schrödinger equation is trivial to solve for this $\hat{H}$ since the $2I + 1$ spin states are already eigenfunctions of $\hat{I}_z$

$$\hat{I}_z |I, m_I\rangle = m_I \hbar |I, m_I\rangle \qquad\qquad 4.2$$

Hence $\qquad\qquad \hat{H}|I, m_I\rangle = -\gamma \hbar B_0 m_I |I, m_I\rangle \qquad\qquad 4.3$

This recovers the simple expression for the energy of a nuclear spin in a magnetic field presented in chapter 2:

$$E(m_I) = -m_I h v_{\mathrm{NMR}} \quad \text{where} \quad v_{\mathrm{NMR}} = \frac{\gamma B_0}{2\pi} \qquad\qquad 4.4$$

4.2.1 SPIN OPERATORS

The term $\hat{I}_z$ used to represent the z magnetization operator in equation 4.1 deserves further comment. This is one of a set of operators that extract information about the nuclear spin angular momentum from the nuclear spin wavefunction (see equation 4.2). Ultimately all Hamiltonians involving nuclear spins are defined in terms of such operators.

The $\hat{I}_x$ and $\hat{I}_y$ operators correspond to x and y components of the angular momentum respectively. It is often simpler, however, to express the x and y operators in terms of the so-called *raising* and *lowering* operators

$$\hat{I}_+ = \hat{I}_x + \mathrm{i}\hat{I}_y \qquad \hat{I}_- = \hat{I}_x - \mathrm{i}\hat{I}_y \qquad\qquad 4.5$$

These names are derived from their effect on the Zeeman eigenstates:

$$\hat{I}_+ |I, m_I\rangle = \hbar \sqrt{I(I+1) - m_I(m_I+1)} \, |I, m_I+1\rangle \qquad\qquad 4.6$$

$$\hat{I}_- |I, m_I\rangle = \hbar \sqrt{I(I+1) - m_I(m_I-1)} \, |I, m_I-1\rangle \qquad\qquad 4.7$$

If we consider, for example, a spin-$\frac{1}{2}$ ($I = \frac{1}{2}$), then $\hat{I}_+ \mid \frac{1}{2}, -\frac{1}{2}\rangle = \hbar \mid \frac{1}{2}, \frac{1}{2}\rangle$ or, using the conventional shorthand $\mid \frac{1}{2}, \frac{1}{2}\rangle \equiv \mid \alpha\rangle$ and $\mid \frac{1}{2}, -\frac{1}{2}\rangle \equiv \mid \beta\rangle$, then $\hat{I}_+ \mid \beta\rangle = \hbar \mid \alpha\rangle$, that is, $\hat{I}_+$ "raises" the $m_I = -\frac{1}{2}$ (β) state to $m_I = \frac{1}{2}$ (α). Note from equations 4.6 and 4.7 how trying to raise a state of maximum m_I or lowering a state of minimum m_I fails: $\hat{I}_+ \mid I, I\rangle = \hat{I}_- \mid I, -I\rangle = 0$.

Many NMR problems can be expressed simply using such symbolic representations for spin operators. For example, the quantum mechanical description of the NMR experiment in section 4.4.1 uses spin operators to express the state of the system. Particularly when dealing with multi-spin systems with strong couplings between them (as commonly encountered in solid-state NMR), however, operator treatments can become unwieldy. In these cases it is often simpler to work with *matrix representations* of operators. Numerical simulations as discussed in appendix D, for example, invariably make use of matrix representations.

The matrix representation of an operator, $\hat{O}$, is obtained by evaluating all possible combinations of bras and kets, $\langle I, m_I' \mid \hat{O} \mid I, m_I \rangle$. Thus for spin-$\frac{1}{2}$

$$
\hat{I}_z = \begin{array}{c} \\ \langle \alpha \mid \\ \langle \beta \mid \end{array} \begin{pmatrix} +\frac{1}{2} & 0 \\ 0 & -\frac{1}{2} \end{pmatrix}
$$

with column labels $\mid \alpha\rangle$, $\mid \beta\rangle$ above. (4.8)

$$
\hat{I}_+ = \begin{array}{c} \langle \alpha \mid \\ \langle \beta \mid \end{array} \begin{pmatrix} 0 & 1 \\ 0 & 0 \end{pmatrix} \qquad \hat{I}_- = \begin{array}{c} \langle \alpha \mid \\ \langle \beta \mid \end{array} \begin{pmatrix} 0 & 0 \\ 1 & 0 \end{pmatrix}
$$

with column labels $\mid \alpha\rangle$, $\mid \beta\rangle$ above each. (4.9)

Note that the factor of $\hbar$ has been dropped from the spin operators. This corresponds to the normal NMR convention of expressing the spin Hamiltonian and its components (such as coupling constants) in terms of frequency rather than energy.

It follows from equation 4.5 that $\hat{I}_x = (\hat{I}_+ + \hat{I}_-)/2$ and $\hat{I}_y = -\mathrm{i}(\hat{I}_+ - \hat{I}_-)/2$, hence

$$
\hat{I}_x = \begin{pmatrix} 0 & \frac{1}{2} \\ \frac{1}{2} & 0 \end{pmatrix} \qquad \hat{I}_y = \begin{pmatrix} 0 & -\frac{\mathrm{i}}{2} \\ \frac{\mathrm{i}}{2} & 0 \end{pmatrix}
$$

(4.10)

Inset 4.1 shows an example of deriving matrix representations for a spin with a higher value of I.

4.2.2 SECULAR & NON-SECULAR TERMS

Because the other interactions are usually small in comparison with the Zeeman interaction, the eigenbasis of the total nuclear spin Hamiltonian is the same as the Zeeman eigenbasis $\mid I, m_I \rangle$

Inset 4.1. Matrix representations of spin operators

Matrices for general spin quantum numbers can be generated using equations 4.2–4.7. For example, a spin-$\frac{3}{2}$ has four spin states, $m_I = -\frac{3}{2}, -\frac{1}{2}, +\frac{1}{2}$, and $+\frac{3}{2}$, and the associated representations are 4×4 matrices:

$$\hat{I}_z = \begin{array}{c} \\ \langle+\tfrac{3}{2}| \\ \langle+\tfrac{1}{2}| \\ \langle-\tfrac{1}{2}| \\ \langle-\tfrac{3}{2}| \end{array} \begin{pmatrix} \tfrac{3}{2} & 0 & 0 & 0 \\ 0 & \tfrac{1}{2} & 0 & 0 \\ 0 & 0 & -\tfrac{1}{2} & 0 \\ 0 & 0 & 0 & -\tfrac{3}{2} \end{pmatrix}$$

with column headings $|+\tfrac{3}{2}\rangle \ |+\tfrac{1}{2}\rangle \ |-\tfrac{1}{2}\rangle \ |-\tfrac{3}{2}\rangle$

The non-zero matrix elements for $\hat{I}_+$ will be $\langle \frac{3}{2} | \hat{I}_+ | \frac{1}{2} \rangle$, $\langle \frac{1}{2} | \hat{I}_+ | -\frac{1}{2} \rangle$, and $\langle -\frac{1}{2} | \hat{I}_+ | -\frac{3}{2} \rangle$, that is, $\langle m_I + 1 | \hat{I}_+ | m_I \rangle$. Hence

$$\hat{I}_+ = \begin{array}{c} \\ \langle+\tfrac{3}{2}| \\ \langle+\tfrac{1}{2}| \\ \langle-\tfrac{1}{2}| \\ \langle-\tfrac{3}{2}| \end{array} \begin{pmatrix} 0 & \sqrt{3} & 0 & 0 \\ 0 & 0 & 2 & 0 \\ 0 & 0 & 0 & \sqrt{3} \\ 0 & 0 & 0 & 0 \end{pmatrix}$$

with column headings $|+\tfrac{3}{2}\rangle \ |+\tfrac{1}{2}\rangle \ |-\tfrac{1}{2}\rangle \ |-\tfrac{3}{2}\rangle$

(except when large quadrupole interactions are present, as discussed in chapter 6). Moreover, *components of the interaction Hamiltonian that do not commute with the Zeeman interaction can be neglected.*

This is a subtle but recurring theme in NMR quantum mechanics and is worth considering in detail. In classic perturbation theory, we consider the effect of a perturbing Hamiltonian, $\hat{H}'$, on the eigenvalues and eigenstates of a dominant Hamiltonian, $\hat{H}_0$. Working in the eigenbasis of $\hat{H}_0$, with the set of states, $|n\rangle$, the first-order correction to the energy of level n is just $\langle n | \hat{H}' | n \rangle$. In other words, the off-diagonal elements of $\hat{H}'$ (expressed in the $\hat{H}_0$ eigenbasis) have no effect to first order on the energy. In more formal terms, the perturbing Hamiltonian can always be divided into a component that commutes with $\hat{H}_0$ (see inset 4.2), $\hat{H}'_{\text{secular}}$, and a component that does not, $\hat{H}'_{\text{non-secular}}$. The *secular approximation* involves discarding $\hat{H}'_{\text{non-secular}}$ since it has no effect on the NMR frequencies to first order. We will see an alternative, but equivalent, way of looking at this problem when considering the rotating frame of reference in section 4.2.4.

The shielding Hamiltonian, responsible for the chemical shift, provides a concrete example of the distinction between "secular" and "non-secular" terms. The full shielding Hamiltonian is

$$\hat{H}_S = \frac{\gamma}{2\pi} \hat{I} \cdot \boldsymbol{\sigma} \cdot \boldsymbol{B}_0 \qquad \qquad 4.11$$

where $\boldsymbol{B}$ is the magnetic field vector, $\hat{\boldsymbol{I}}$ is the vector $(\hat{I}_x, \hat{I}_y, \hat{I}_z)$, and σ is the shielding tensor. If the field is along z, that is, $\boldsymbol{B}_0 = (0, 0, B_0)$, this simplifies to

$$\hat{H}_S = \frac{\gamma}{2\pi} \left(\hat{I}_x, \hat{I}_y, \hat{I}_z \right) \begin{pmatrix} \sigma_{xx} & \sigma_{xy} & \sigma_{xz} \\ \sigma_{yx} & \sigma_{yy} & \sigma_{yz} \\ \sigma_{zx} & \sigma_{zy} & \sigma_{zz} \end{pmatrix} \begin{pmatrix} 0 \\ 0 \\ B_0 \end{pmatrix} \qquad 4.12$$

$$= \nu_{NMR} \left(\hat{I}_x\, \sigma_{xz} + \hat{I}_y\, \sigma_{yz} + \hat{I}_z\, \sigma_{zz} \right) \qquad 4.13$$

However $\hat{I}_x$ and $\hat{I}_y$ do not commute with $\hat{I}_z$ (a basic principle of quantum angular momentum), and so these terms are non-secular with respect to the Zeeman interaction and can be neglected. Only the

---

Inset 4.2. Reminder of some basic matrix terminology

Commuting and non-commuting matrices   The order in which two matrices, A and B, are multiplied is important since frequently $AB \neq BA$. Such matrices are said to be *non-commuting*. On the other hand, if $AB - BA = 0$, then A and B are said to *commute*, and the order of matrix multiplication can be freely interchanged. The *commutator* between A and B, $AB - BA$, is often written in the shorthand form $[A, B]$.

Matrix trace   The trace of a matrix is the sum of its diagonal elements

$$\text{tr}\left(A\right) = \sum_i A_{ii} \qquad 4.14$$

For instance, the expectation value of an operator $\hat{O}$ given the density operator $\hat{\rho}$ is $\langle \hat{O} \rangle = \text{tr}\left(\hat{\rho}\hat{O}\right)$.

Handy relation: The trace of a product is invariant to cyclic permutation, $\text{tr}(ABC) = \text{tr}(CAB)$.

Matrix inverse   The *inverse* of a square matrix A is denoted by A^{-1} and has the property

$$A^{-1}A = 1 \qquad 4.15$$

where 1 is an identity matrix.
Handy relation: $(AB)^{-1} = B^{-1}A^{-1}$

Transpose and conjugate transpose   *Transposing* a matrix A (common symbol A^T) involves swapping matrix elements across the diagonal. The *conjugate transpose*, $A^\dagger$, involves taking the conjugate of the matrix before or after transposition:

$$A^T_{ij} = A_{ji} \qquad A^\dagger_{ij} = \left[\left(A^*\right)^T \right]_{ij} = \left(A^T\right)^*_{ij} = A^*_{ji} \qquad 4.16$$

Handy relation: $(AB)^T = B^T A^T$

term in $\hat{I}_z$ will add to the diagonal and so contribute directly to the energies of the Zeeman eigenstates. Hence the shielding Hamiltonian can be simplified to $v_{\text{NMR}}\sigma_{zz}\hat{I}_z$, as previously assumed in section 2.3.

Alternatively, the matrix representation of the total Zeeman and shielding Hamiltonians for a spin-$\frac{1}{2}$ would be

$$\hat{H}_Z + \hat{H}_S = \begin{pmatrix} -v_{\text{NMR}}\left(1 - \sigma_{zz}\right)/2 & v_{\text{NMR}}\left(\sigma_{xz} - i\sigma_{yz}\right)/2 \\ v_{\text{NMR}}\left(\sigma_{xz} + i\sigma_{yz}\right)/2 & v_{\text{NMR}}\left(1 - \sigma_{zz}\right)/2 \end{pmatrix} \qquad 4.17$$

The diagonal terms of the shielding Hamiltonian give the first-order correction to the eigenvalues, hence the overall NMR frequency (given by the difference of the eigenvalues) is $v_0 = v_{\text{NMR}}\left(1 - \sigma_{zz}\right)$, as expected. The off-diagonal terms would contribute to a second-order correction to the energy, but this can be readily neglected; the effect of the first-order term on the NMR frequencies is measured in parts per million, and so the second-order correction due to the $\hat{I}_x$ and $\hat{I}_y$ components of the shielding Hamiltonian is of the order of $1:10^{12}$.

4.2.3 COUPLING HAMILTONIANS

The full Hamiltonians for NMR couplings have a common form:[2]

$$\hat{H} = \hat{\boldsymbol{I}}_1 \cdot \boldsymbol{R} \cdot \hat{\boldsymbol{I}}_2 \qquad 4.18$$

where $\hat{\boldsymbol{I}}_j$ indicates the vector of spin operators $(\hat{I}_x, \hat{I}_y, \hat{I}_z)$ for spin j, and $\boldsymbol{R}$ is a tensor (see section 2.2 for an introduction to tensors and section 4.6 for a more detailed treatment). As previously discussed in section 2.2, $\boldsymbol{R}$ can always be decomposed into a rank-0 term (isotropic), a rank-1 term (which can be ignored), and a rank-2 term.

The final form of the equation 4.18 depends on the nature of the interaction. For instance, the rank-2 component of the J (or indirect dipole–dipole) coupling is either eliminated by molecular tumbling in the solution state or can be merged into the much larger direct dipole–dipole interaction (see equation 8.3). Hence the J coupling between two spins I and S is normally reduced to its isotropic component

$$\hat{H}_J \text{ (general, isotropic)} = J_{\text{IS}}\,\hat{\boldsymbol{I}} \cdot \hat{\boldsymbol{S}} \qquad 4.19$$

where the scalar (dot) product between the spin operators expands to $\hat{I}_x\hat{S}_x + \hat{I}_y\hat{S}_y + \hat{I}_z\hat{S}_z$. In the homonuclear case, I and S have the same Larmor frequency, v_I, and the Zeeman Hamiltonian is

$$\hat{H}_Z = -v_I\left(\hat{I}_z + \hat{S}_z\right) \qquad 4.20$$

---

[2] The NMR shielding Hamiltonian, equation 4.11, also has this form, but with the external magnetic field instead of the second spin.

In this case, the secular approximation does not provide further simplification to $\hat{H}_J$ since it commutes with $\hat{I}_z + \hat{S}_z$ (exercise for the reader!).

If, however, the difference in NMR frequencies is large in comparison with J_{IS}, either because they form a heteronuclear pair or because their chemical shifts are very different, then further terms can be discarded. The dominant component of the Hamiltonian is:

$$\hat{H}_Z = -v_I \hat{I}_z - v_S \hat{S}_z \qquad 4.21$$

In this case, only the $\hat{I}_z \hat{S}_z$ terms of $\hat{H}_J$ commutes with $\hat{H}_Z$, and the coupling Hamiltonian can be reduced to

$$\hat{H}_J \left(\text{heteronuclear, isotropic}\right) = J_{IS} \hat{I}_z \hat{S}_z \qquad 4.22$$

This corresponds to the *weak coupling* limit familiar from solution-state NMR spectra.

The full dipole–dipole coupling Hamiltonian between I and S in the terms of equation 4.18 is

$$\hat{H}_D = -\hat{\boldsymbol{I}} \cdot 2\boldsymbol{D} \cdot \hat{\boldsymbol{S}} \qquad 4.23$$

where $\boldsymbol{D}$ is the dipolar coupling tensor. $\boldsymbol{D}$ in its principal axis system is given by[3]

$$\boldsymbol{D}^{\mathrm{P}} = \begin{pmatrix} -D_{IS}/2 & 0 & 0 \\ 0 & -D_{IS}/2 & 0 \\ 0 & 0 & D_{IS} \end{pmatrix} \qquad 4.24$$

where D_{IS} is the dipolar coupling constant between I and S (see section 2.5). Note how $\boldsymbol{D}$ has zero trace (i.e., no isotropic component) and zero asymmetry ($D_{XX}^{\mathrm{P}} = D_{YY}^{\mathrm{P}}$), as previously presented in table 2.1.

The algebra of the reduction to the secular component is considerably more involved and we will jump straight to the result (see Further reading for details):

$$\hat{H}_D \left(\text{general}\right) = -D_{IS} P_2 \left(\cos\theta\right) \left(2\hat{I}_z \hat{S}_z - \tfrac{1}{2}\left[\hat{I}_+ \hat{S}_- + \hat{I}_- \hat{S}_+\right]\right) \qquad 4.25$$

where $P_2(x)$ is the second-order Legendre polynomial, $P_2(\cos\theta) = (3\cos^2\theta - 1)/2$. The second term of equation 4.25 is often referred to as the "flip-flop" term since it has the effect of interconverting the states $\langle\alpha\beta|$ and $\langle\beta\alpha|$ (see equation 4.76).

As with the J-coupling Hamiltonian above, this Hamiltonian commutes with the Zeeman Hamiltonian for a like-spin pair, but only the *zz* term survives for a heteronuclear spin pair:

$$\hat{H}_D \left(\text{heteronuclear}\right) = -2D_{IS} P_2 \left(\cos\theta\right) \hat{I}_z \hat{S}_z \qquad 4.26$$

which is equivalent to the expression for the nuclear spin energies given in equation 2.23.

---

[3] $\boldsymbol{D}$ and equation 4.23 can be formulated in different ways, but these will expand to the same overall expression for the dipole–dipole coupling Hamiltonian.

The presence of the additional flip-flop term in the homonuclear Hamiltonian of equation 4.25 is responsible for the difference between the spectra for a homo- and a hetero- nuclear spin pair referred to in section 2.5 and discussed in detail in section 4.5.1.

As mentioned above, the secular approximation begins to break down for the quadrupolar coupling, and so discussion of the quadrupolar coupling Hamiltonian is deferred to chapter 6.

4.2.4 RADIOFREQUENCY & THE ROTATING FRAME

The final term required for this quantum treatment of the NMR experiment is the Hamiltonian for the radiation used to excite the nuclear spins. The Hamiltonian for a nuclear spin interacting with electromagnetic radiation is

$$\hat{H}_{RF}(t) = 2\frac{\gamma B_1}{2\pi}\left[\cos(2\pi\nu_{RF}t + \phi)\hat{I}_x + \sin(2\pi\nu_{RF}t + \phi)\hat{I}_y\right] \qquad 4.27$$

where B_1 is the intensity of the oscillating magnetic field and ϕ is its phase (with zero conventionally corresponding to initial x phase). ν_{RF} denotes the frequency of the radiation from the radiofrequency (RF) transmitter.[4] The reason for the leading factor of 2 will appear later.

The time dependence of $\hat{H}_{RF}(t)$ greatly complicates solving the Schrödinger equation, and the normal solution is to move into an *interaction frame* rotating about the z axis at frequency ν_{RF}, as introduced qualitatively in section 3.2. The transformation from the normal *laboratory frame* to this rotating frame corresponds to the following transformation of the axis system

$$\hat{I}_x \rightarrow \cos\left(\omega_{RF}t\right)\hat{I}'_x + \sin\left(\omega_{RF}t\right)\hat{I}'_y \qquad 4.28$$

$$\hat{I}_y \rightarrow -\sin\left(\omega_{RF}t\right)\hat{I}'_x + \cos\left(\omega_{RF}t\right)\hat{I}'_y \qquad 4.29$$

that is, the xy axes rotate around the z axis with angular frequency ω_{RF} (the use of angular frequencies avoids factors of 2π).

After multiplying out and applying some trigonometric identities, equation 4.27 reduces to

$$\hat{H}'_{RF}(t) = \frac{\gamma B_1}{2\pi}\left[\left(\cos\phi + \cos\left(2\omega_{RF}t + \phi\right)\right)\hat{I}'_x + \left(\sin\phi + \sin\left(2\omega_{RF}t + \phi\right)\right)\hat{I}'_y\right] \qquad 4.30$$

that is, one component of the RF is now independent of time in the new (rotating) frame, while the other is rotating at an angular frequency of $2\omega_{RF}$.

---

[4] The frequency of electromagnetic radiation used to observe NMR, ν_{RF}, and nutation frequency of the nuclear spin magnetization *resulting* from the resonant absorption, ν_1, must be distinguished carefully. Various notations are used in the literature, for example, ν_{RF} often refers to a nutation frequency.

The frequency of this second component of the oscillating magnetic field is very much faster than the NMR interactions (in the rotating frame) and so averages away over the NMR timescale.[5] Hence the RF Hamiltonian reduces to

$$\hat{H}_{RF} = \frac{\gamma B_1}{2\pi}\left(\cos\phi\,\hat{I}_x + \sin\phi\,\hat{I}_y\right) \qquad 4.31$$

Note how the factor of 2 present in equation 4.27 has disappeared along with the discarded component of the magnetic field. The primes used to denote the rotating frame have been dropped since the rotating frame will be used from now on.

The key feature of equation 4.31 is that the RF Hamiltonian is time independent in this frame of reference, in contrast to the laboratory frame Hamiltonian (equation 4.27). This greatly simplifies the task of determining the evolution of the nuclear spins. Even when the implicit approximation that the Zeeman interaction is much larger than the other components of the nuclear spin Hamiltonian becomes slightly questionable, for example, when the quadrupolar interactions are very large, it is normal practice to add correction terms using second-order perturbation theory rather than abandon the rotating frame altogether.

The transformation into the rotating frame also affects the Zeeman / chemical shift Hamiltonian. The Zeeman + shift Hamiltonian in the laboratory frame is

$$\hat{H}_{Z+S} = -v_0\,\hat{I}_z \qquad 4.32$$

(for a single spin), where v_0 refers to the Larmor frequency (including the shielding term). The effect of this Hamiltonian is to cause the net nuclear magnetization to precess about z at its NMR frequency. The effective precession rate in the rotating frame will be $\Delta = v_0 - v_{RF}$. Hence the effective Hamiltonian in the rotating frame is

$$\hat{H}_{Z+S} = -\Delta\hat{I}_z \qquad 4.33$$

The total Hamiltonian will then be

$$\hat{H}_{Z+S+RF} = -\Delta\hat{I}_z + v_1\left(\cos\phi\,\hat{I}_x + \sin\phi\,\hat{I}_y\right) \qquad 4.34$$

where

$$v_1 = \frac{\gamma B_1}{2\pi} \qquad 4.35$$

By definition the RF needs to be close to the Larmor frequency, v_0, for nuclear magnetic resonance to be observed, and so the offset term, $-\Delta\hat{I}_z$, will be negligible in comparison with the RF term.

---

[5] The *Bloch–Siegert* effect is a shift in the effective NMR frequency due to this neglected component. Its size is inversely proportional to the NMR frequency and so is negligible for typical high-field NMR. Analogous effects can be seen at high field if irradiating one nucleus while observing another with a very similar Larmor frequency. For example, shifts in ^{19}F resonance frequencies due to ^{1}H decoupling can be readily observed since ^{19}F and ^{1}H NMR frequencies differ only by about 12%.

Hence the net magnetization vector will rotate around the rotating-frame B_1 vector at a rate v_1. This is the *nutation frequency* discussed in chapter 3. If, however, the RF is far from resonance and/or very weak, the $\Delta \hat{I}_z$ term will dominate and the spins will precess around z, unaffected by the RF irradiation.

4.3   THE DENSITY MATRIX

Understanding spin dynamics in solution-state NMR is relatively straightforward largely because the only significant interactions directly affecting the spectrum are chemical shifts and J couplings. Moreover, these interactions commute with each other in the weak coupling approximation (that the J couplings are much smaller than the differences in Larmor frequencies between inequivalent spins). This allows the effects of the shift and coupling Hamiltonians to be considered separately, and relatively simple rules can be derived to describe the effects of the NMR interactions and an RF pulse sequence. See Further reading for books that describe such *product operator* treatments. This rarely applies in solid-state NMR, and so a more thoroughly quantum mechanical treatment is generally required. This is outlined below (again see the Further reading for more gentle introductions to the quantum mechanics of NMR).

The state of an individual system is determined by its wavefunction, Ψ. This can always be expressed as a linear combination of basis functions (generally just the Zeeman eigenbasis) that is,

$$|\Psi\rangle = \sum_n c_n |n\rangle \quad \text{and} \quad \langle\Psi| = \sum_m c_m^* \langle m| \qquad 4.36$$

and so the "value" of an operator, $\hat{O}$, (its *expectation value*) will be

$$\langle\hat{O}\rangle = \langle\Psi|\hat{O}|\Psi\rangle = \sum_{m,n} c_m^* c_n \langle m|\hat{O}|n\rangle \qquad 4.37$$

If we have an ensemble of identical systems (i.e., systems with the same Hamiltonian), then the ensemble-averaged expectation value of $\hat{O}$ is

$$\overline{\langle\hat{O}\rangle} = \overline{\sum_{m,n} c_m^* c_n \langle m|\hat{O}|n\rangle} \qquad 4.38$$

The matrix elements, $\langle m|\hat{O}|n\rangle$, are identical between the different systems, hence the measured expectation value reduces to

$$\overline{\langle\hat{O}\rangle} = \sum_{m,n} \overline{c_m^* c_n} \langle m|\hat{O}|n\rangle \qquad 4.39$$

that is, the only quantities we need to define are the ensemble-averaged coefficients $\overline{c_m^* c_n}$. Hence it is sufficient to define the state of the system using the matrix, ρ, with elements

$$\rho_{mn} = \overline{c_m^* c_n} \qquad 4.40$$

ρ is referred to as the *density matrix* and is the matrix representation of the *density operator*, $\hat{\rho}$ (although ρ and $\hat{\rho}$ are often used interchangeably). The expectation value of $\hat{O}$ is then simply

$$\langle \hat{O} \rangle = \mathrm{tr}\left(\hat{\rho} \hat{O} \right) \qquad\qquad 4.41$$

where tr is the matrix trace (see inset 4.2). We lose information about the wavefunction of an individual system, but this is not a great loss as the NMR experiment invariably involves measurements on large ensembles.

For example, the equilibrium state of the nuclear spins within a magnetic field involves net magnetization along the z axis, with net excess of spin magnetization, $M_0/2$, in, say, the a state. The density matrix[6] for the equilibrium state would therefore be

$$\rho = \begin{array}{cc} & \begin{array}{cc} |a\rangle & \quad |\beta\rangle \end{array} \\ \begin{array}{c} \langle a| \\ \langle \beta| \end{array} & \begin{pmatrix} M_0/2 & 0 \\ 0 & -M_0/2 \end{pmatrix} \end{array} \qquad\qquad 4.42$$

We can then obtain expectation values for the different angular momentum operators of equation 4.9, for example:

$$\langle \hat{I}_z \rangle = \mathrm{tr}\left(\hat{\rho}\hat{I}_z \right) = \mathrm{tr}\left[\begin{pmatrix} M_0/2 & 0 \\ 0 & -M_0/2 \end{pmatrix} \begin{pmatrix} \tfrac{1}{2} & 0 \\ 0 & -\tfrac{1}{2} \end{pmatrix} \right] = M_0/2 \qquad\qquad 4.43$$

$$\langle \hat{I}_x \rangle = \mathrm{tr}\left[\begin{pmatrix} M_0/2 & 0 \\ 0 & -M_0/2 \end{pmatrix} \begin{pmatrix} 0 & \tfrac{1}{2} \\ \tfrac{1}{2} & 0 \end{pmatrix} \right] = \mathrm{tr}\begin{pmatrix} 0 & 0 \\ 0 & 0 \end{pmatrix} = 0 \qquad\qquad 4.44$$

that is, the average state corresponds to pure magnetization along z.

Note how the diagonal elements of the density matrix, $\rho_{mm} = \overline{c_m^* c_m}$, directly correspond to the populations of the states. The physical significance of the off-diagonal elements is less immediately obvious, but consider the density matrix

$$\rho = \begin{array}{cc} & \begin{array}{cc} |a\rangle & |\beta\rangle \end{array} \\ \begin{array}{c} \langle a| \\ \langle \beta| \end{array} & \begin{pmatrix} 0 & 1 \\ 0 & 0 \end{pmatrix} \end{array} \qquad\qquad 4.45$$

Clearly the average populations of the a and β states are zero, but the system is still in a well-defined state. This is impossible classically since a two-level system must be in one of the two available states at any one time. Quantum mechanics, however, allows different states of existence, *coherences*, which involve mixtures of the eigenstates. Hence we need a matrix to hold the populations

---

[6] Strictly speaking this is the *reduced* density matrix. See later for the distinction.

of the states (diagonal elements) and the coherences (off-diagonal elements) that can be created between the different states.

Figure 4.1 illustrates different ways of representing the state of the set of nuclear spins. The vector model, figure 4.1(a), is useful when there is a single spin type or when couplings between spins can be ignored. It is difficult, however, to represent coherences other than simple x, y, and z magnetization in this way. Especially when problems involve an isolated pair of spins, it can be useful to consider an energy level diagram such as figure 4.1(b). The relative populations of the four possible states are indicated by the occupancy of each level while the arrows mark possible coherences between the states. Coherences can be illustrated with such diagrams, for example, the single-headed arrow would indicate a coherence (of order +2) between the $|\alpha\alpha\rangle$ and $|\beta\beta\rangle$ states, but this could be confused with a transition or a relationship between the *populations* of the states involved. The density

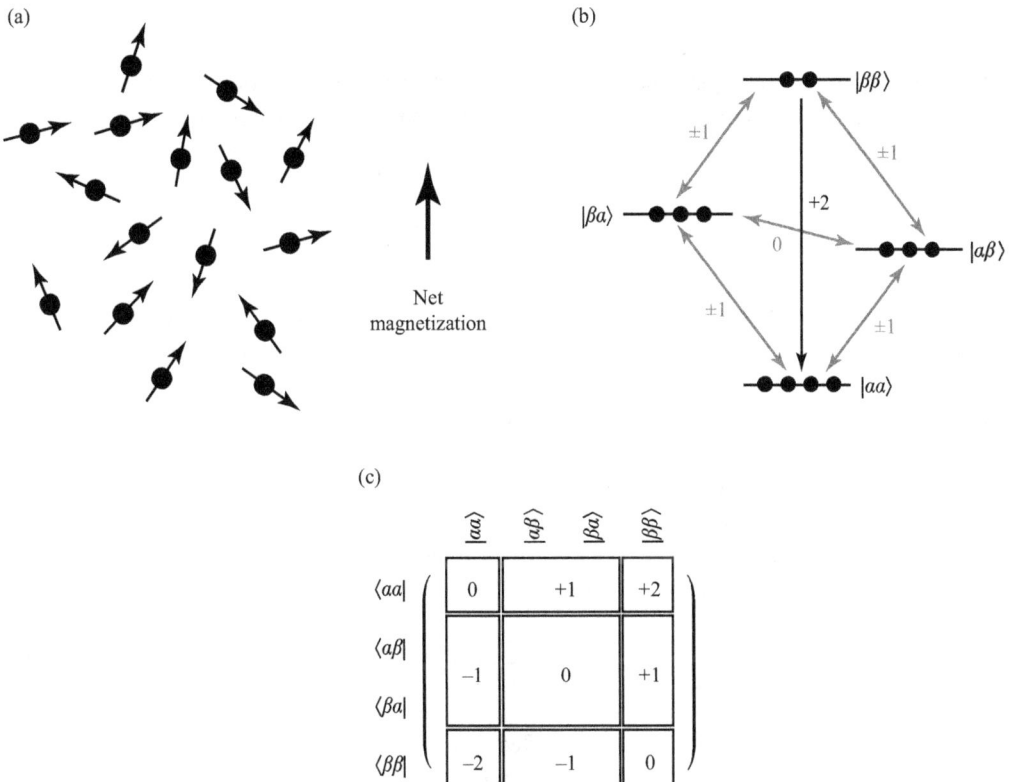

Figure 4.1. Different ways of representing coherences in NMR: (a) the vector model of a collection of spins in a magnetic field and the resulting net magnetization; (b) schematic representation of the energy levels for two spin-$\frac{1}{2}$ nuclei. Filled circles represent relative populations of the spin states and double-headed arrows correspond to potential coherences between the states. (c) The density matrix for this system. Numbers give the coherence order, p, for the blocks, see section 4.5.1.

matrix representation, figure 4.1(c), provides the most compact representation of the different possible states of the system.

For instance, the population distribution in (b) would correspond to entries of 4 ($|\alpha\alpha\rangle$), 3 ($|\alpha\beta\rangle$), 3 ($|\beta\alpha\rangle$), and 2 ($|\beta\beta\rangle$) down the diagonal of the matrix, while the +2 coherence would correspond to a non-zero value for the matrix element connecting $\langle\alpha\alpha|$ and $|\beta\beta\rangle$. Note how the density matrix can be divided into blocks of a given "*coherence order*". This important concept is explored in section 4.5.1.

The NMR experiment can now be described in this formalism, starting with the density operator that describes the initial equilibrium state of the system. At equilibrium there are no net coherences between the states, and the relative populations of the eigenstates are given by the Boltzmann distribution. The equilibrium density operator is[7]

$$\hat{\rho}_{eq} = \frac{1}{Z}e^{-h\hat{H}/kT} \qquad\qquad 4.46$$

where $\hat{H}$ is the total nuclear spin Hamiltonian and Z is a normalization factor (the partition function). Since the nuclear spin energies are so small in comparison with thermal energies, the exponential can be expanded

$$\hat{\rho}_{eq} \propto \left[\hat{1} - \frac{h}{kT}\hat{H}\right]/(2I+1) \qquad\qquad 4.47$$

where Z equals the number of spin states, $2I+1$, in this limit (see inset 4.3 for an explanation of how to take exponentials of matrices such as $\hat{H}$). $\hat{1}$ is the identity operator and so represents an equal distribution of population between the different states. This term is of no interest (since it does not evolve under $\hat{H}$) and can be dropped. The second term corresponds to the tiny changes to the populations caused by the lifting of the degeneracy of nuclear spin states by the spin Hamiltonian $\hat{H}$.

Hence we will work with the *reduced density operator*, $\hat{\rho}_{eq'}$, which gives the difference of the full density operator from the situation of equal population. The expression for the equilibrium density matrix is further simplified using the high-field approximation, that is, $\hat{H} \approx \hat{H}_{Zeeman}$. The equilibrium density operator for a single spin can then be written simply

$$\hat{\rho}_{eq'} = -\frac{h}{(2I+1)kT}\hat{H} = M\hat{I}_z \qquad\qquad 4.48$$

where M is a constant of proportionality that effectively expresses the size of the nuclear spin magnetization. For instance, $\rho_{eq'}$ for a single spin-$\frac{1}{2}$ in these terms would be

$$\rho_{eq'} = M\begin{pmatrix} 1/2 & 0 \\ 0 & -1/2 \end{pmatrix} \qquad\qquad 4.49$$

---

[7] The factor of h follows from expressing the Hamiltonian in terms of frequency rather than energy.

which corresponds to net positive population of pure α states and a net negative population of β states.

Applying the Schrödinger equation to the density operator gives the *Liouville–von Neumann equation*

$$\dot{\hat{\rho}} = \frac{d\hat{\rho}}{dt} = -2\pi \mathrm{i}\left[\hat{H}, \hat{\rho}\right] \qquad 4.50$$

for the evolution of the density operator under the Hamiltonian, $\hat{H}$ (again expressed in frequency units).

Integrating equation 4.50 gives the density operator at time t

$$\hat{\rho}(t) = \hat{U}(t,0)\,\hat{\rho}(0)\hat{U}(t,0)^{-1} \qquad 4.51$$

where $\hat{\rho}(0)$ is the density operator at time $t = 0$ and $\hat{U}(t, 0)$ is the *propagator*—the matrix that "propagates" the density operator from time zero to time t. For a time-independent Hamiltonian, the propagator is given by

$$\hat{U}(t,0) = \mathrm{e}^{-2\pi \mathrm{i}\hat{H}t} \qquad 4.52$$

4.4 DENSITY OPERATOR TREATMENTS OF SIMPLE NMR EXPERIMENTS

The density operator approach comes in to its own when multiple spins are involved, but it is useful to start with some simple experiments that can be easily visualized in vector model terms.

4.4.1 THE BASIC NMR EXPERIMENT

Consider a system of identical spins at thermal equilibrium. From equation 4.48, the density operator at time zero is

$$\hat{\rho}(0) = M\hat{I}_z \qquad 4.58$$

The system Hamiltonian (in a frame rotating at the same rate as the RF) consists of a small chemical shift offset term

$$\hat{H}_{\mathrm{sys}} = \Delta\hat{I}_z \qquad 4.59$$

Because $\hat{\rho}(0)$ and $\hat{H}_{\mathrm{sys}}$ commute (they are simply proportional to each other), the Liouville–von Neumann equation, equation 4.50, reassuringly predicts no evolution of the density operator $\left(\dot{\hat{\rho}} = 0\right)$.

If we apply RF of phase x, the Hamiltonian is now

$$\hat{H}_{\mathrm{sys+RF}} = \Delta\hat{I}_z + v_1\hat{I}_x \qquad 4.60$$

Inset 4.3. More advanced matrix algebra

Matrix diagonalization   In matrix terms, solving the Schrödinger equation, $H\psi = E\psi$, involves finding the eigenvalues of the Hamiltonian matrix H (which correspond to the allowed energies, E). The eigenvectors of H describe the transformation from the initial basis set into the basis that diagonalizes H. This is written symbolically:

$$H = V\Lambda V^{-1} \tag{4.53}$$

where V is the matrix of eigenvectors and Λ is a diagonal matrix with the eigenvalues, Λ_i, along the diagonal.

For example, considering a Hamiltonian matrix containing a dominant diagonal term and weak off-diagonal component:

$$H = \begin{pmatrix} 10 & 1 \\ 1 & 10 \end{pmatrix} \tag{4.54}$$

The eigenvalues and eigenvectors are

$$V = \begin{pmatrix} \cos 45° & \sin 45° \\ \sin 45° & \cos 45° \end{pmatrix} \quad \Lambda = \begin{pmatrix} 9 & 0 \\ 0 & 11 \end{pmatrix} \tag{4.55}$$

V is a rotation matrix, with transformation $V^{-1}HV$ generating the diagonal matrix Λ containing the eigenvalues (energies).

Handy relation: if two matrices commute, then the same matrix V will diagonalize both matrices, that is, they share a common eigenbasis.

Matrix exponentials   Exponentiation is defined for matrices as for simple scalars:

$$e^{kA} = 1 + kA + k^2 \frac{A^2}{2!} + k^3 \frac{A^3}{3!} + \cdots \tag{4.56}$$

where k is a scalar.

In practice this series converges slowly and the exponentials in equation 4.52 are typically evaluated via the matrix eigenbasis:

$$e^{kA} = V e^{k\Lambda} V^{-1} \tag{4.57}$$

where the elements of the exponential of the diagonal matrix Λ are simply $[\exp(\Lambda)]_{ii} = \exp(\Lambda_i)$.

Handy relations: (i) if matrices A and B commute, then $e^{A+B} = e^A e^B = e^B e^A$. If A and B represent Hamiltonians, for example, then this means that the propagators (see equation 4.52) for A and B can be evaluated and applied to the density operator independently; (ii) $[e^A]^{-1} = e^{-A}$.

where v_1 is the nutation rate. As discussed in section 4.2.4, the offset, Δ, can be neglected if $v_1 \gg \Delta$, hence the propagator for such a "hard" (or ideal) pulse of duration τ is (equation 4.52)

$$\hat{U}_{\text{pulse}} = e^{-i\theta\hat{I}_x} \qquad 4.61$$

where the tip angle is $\theta = 2\pi v_1 \tau$. This propagator corresponds to rotation about the x axis, and so the density operator at the end of the pulse will be[8]

$$\hat{\rho}(\tau) = \hat{U}_{\text{pulse}} \,\hat{\rho}(0)\hat{U}_{\text{pulse}}^{-1} \qquad 4.62$$

$$= e^{-i\theta\hat{I}_x}M\hat{I}_z e^{i\theta\hat{I}_x} \qquad 4.63$$

$$= M\left(\hat{I}_z \cos\theta + \hat{I}_y \sin\theta\right) \qquad 4.64$$

that is, $\hat{\rho}(\tau) = M\hat{I}_y$ after a 90°_x pulse. This corresponds to the nuclear spin magnetization now being aligned with the y axis.

When the RF is turned off, the Hamiltonian returns to equation 4.59 and the propagator for the following period of *free precession* (evolution under the system Hamiltonian in the absence of RF) is

$$\hat{U}(\tau+t, \tau) = e^{-2\pi i\Delta\hat{I}_z t} \qquad 4.65$$

This simply corresponds to rotation about the z axis at rate Δ. Hence the density operator is

$$\hat{\rho}(\tau+t) = M\hat{I}_y \cos(-2\pi\Delta t) + M\hat{I}_x \sin(-2\pi\Delta t) \qquad 4.66$$

Note how in periods of free precession, the evolution of the density operator does not change the order, p, of coherences; initial y ($p = \pm 1$) magnetization evolves into a mix of x and y (also ± 1). By contrast, evolution under RF mixes different coherence orders, for example, the interchange between z ($p = 0$) and y ($p = \pm 1$) magnetization in equation 4.64.

To determine the NMR signal, we need to evaluate the expectation values of $\hat{I}_x$ (nominally x magnetization in the "real" channel) and $\hat{I}_y$ operators (y magnetization in the "imaginary" channel). These are combined to give the complex NMR signal, $s(t) = \langle\hat{I}_x\rangle + i\langle\hat{I}_y\rangle$, allowing positive and negative precession frequencies to be distinguished (see inset 3.8).[9] This neatly corresponds to determining the expectation values of the $\hat{I}_+ = \hat{I}_x + i\hat{I}_y$ operator. Here we can read out the coefficients of $\hat{I}_x$ and $\hat{I}_y$ directly from equation 4.66 to give

$$s(t) = \langle\hat{I}_x\rangle + i\langle\hat{I}_y\rangle = M \sin(-2\pi\Delta t) + iM \cos(-2\pi\Delta t) \qquad 4.67$$

$$= iMe^{-2\pi i\Delta t} \qquad 4.68$$

---

[8] See Further reading for proof that $e^{-i\theta\hat{I}_x}\hat{I}_z e^{i\theta\hat{I}_x} = \hat{I}_z \cos\theta + \hat{I}_y \sin\theta$.
[9] This is a somewhat handwaving treatment. The relationship between the quantum mechanical evolution and the observed NMR signal is extremely subtle. A more thorough treatment can be found in Levitt's *Spin Dynamics* (Further reading).

that is, an NMR signal of frequency Δ (relative to the transmitter frequency used for the rotating frame of reference). The factor of i corresponds to a phase shift of $90°$ relative to x phase, since the magnetization at $t = 0$ is aligned with the y axis.

Note that evaluating the NMR signal as $\text{tr}(\hat{\rho}\,\hat{I}_+)$ corresponds to detecting $p = -1$ coherences of the density matrix. For example, if the density matrix is a hypothetical pure -1 coherence, $\hat{\rho} = M\hat{I}_-$, the observable NMR signal at this point is

$$\text{tr}\left(\hat{\rho}\,\hat{I}_+\right) = \begin{pmatrix} 0 & 0 \\ M & 0 \end{pmatrix}\begin{pmatrix} 0 & 1 \\ 0 & 0 \end{pmatrix} = M \qquad\qquad 4.69$$

The comparison between the vector model and density operator treatments of this problem is illustrated in figure 4.2.

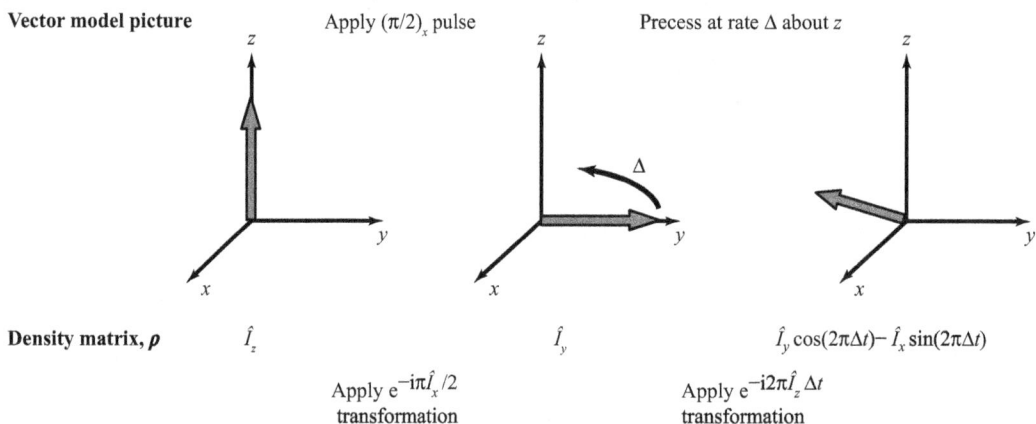

Figure 4.2. Comparison of (top) vector model and (bottom) density operator treatments of the pulse-and-acquire NMR experiment (see text for further details).

4.4.2 ECHOES & COHERENCE PATHWAY DIAGRAMS

Echoes of various kinds form the central building blocks of many NMR experiments. Two of the most important of these, the *simple* (or *Hahn*) *echo* and the *solid* (or *quadrupolar*) *echo*, are illustrated in figure 4.3.

The simple spin echo is used to "refocus" the effect of Hamiltonian terms that are "linear" in the spin of interest, such as the chemical shift and *heteronuclear* couplings. After initial excitation, in this case using a $90°_x$ pulse to create y magnetization from the equilibrium state, the xy coherences

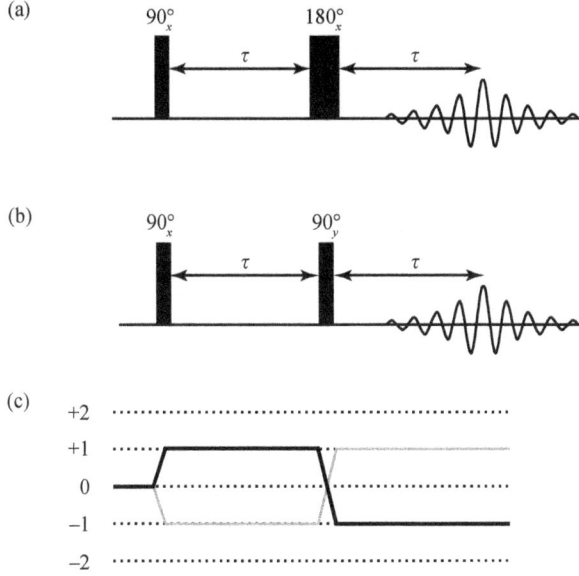

Figure 4.3. Pulse sequences for (a) a simple spin echo and (b) a solid (or quadrupolar) echo, and (c) the coherence pathway diagram that applies to (a). The bold solid line in the coherence pathway diagram marks the path taken by the coherences that are detected (−1). The lighter line traces out a mirror pathway which is not detected (see text).

(or equivalently +1 and −1 coherences) evolve during the free precession period, τ. Including just linear terms, the free precession Hamiltonian might be:

$$\hat{H} = \Delta \hat{I}_z + J \hat{I}_z \hat{S}_z \qquad 4.70$$

where Δ is the offset from the origin of the rotating frame, and J is a scalar coupling between I and S spins. The problem can, however, just be expressed in terms of the I spin, since the sub-spectra corresponding to different states of the S spin are fully independent. For example, if S is a spin-$\frac{1}{2}$, then we can consider an effective offset, $\Delta' = \Delta \pm J/2$, reducing the Hamiltonian to $\hat{H} = \Delta' \hat{I}_z$.

Using the same arguments discussed above, initial y magnetization, $\hat{\rho}(0) = \hat{I}_y$, evolves into

$$\hat{\rho}_{\text{before}}(\tau) = \hat{I}_y \cos(2\pi\Delta'\tau) - \hat{I}_x \sin(2\pi\Delta'\tau) \qquad 4.71$$

at the end of the first τ period and before the inversion pulse.

The 180°_x pulse will invert the sign of the y magnetization and leave the x component unchanged, that is, the density operator after the pulse is:

$$\hat{\rho}_{\text{after}}(\tau) = -\hat{I}_y \cos(2\pi\Delta'\tau) - \hat{I}_x \sin(2\pi\Delta'\tau) \qquad 4.72$$

The second period of free precession again gives rises to a "rotation" of $2\pi\Delta'\tau$ about z. Applying this to the $\hat{I}_y$ and $\hat{I}_x$ operators gives the final density operator

$$\hat{\rho}(2\tau) = -\left[\hat{I}_y \cos(2\pi\Delta'\tau) - \hat{I}_x \sin(2\pi\Delta'\tau)\right]\cos(2\pi\Delta'\tau)$$
$$-\left[\hat{I}_x \cos(2\pi\Delta'\tau) + \hat{I}_y \sin(2\pi\Delta'\tau)\right]\sin(2\pi\Delta'\tau) \qquad 4.73$$

$$= -\hat{I}_y\left[\cos^2(2\pi\Delta'\tau) + \sin^2(2\pi\Delta'\tau)\right] \qquad 4.74$$

Using the trigonometric identity $\cos^2\theta + \sin^2\theta = 1$, the density operator at 2τ simplifies to

$$\hat{\rho}(2\tau) = -\hat{I}_y \qquad 4.75$$

that is, the density operator corresponds to magnetization along $-y$. In other words, the effect of the chemical shift and heteronuclear couplings is refocused *irrespective of the effective offset* Δ'. In typical situations where there is a spread of offset frequencies, the xy magnetization will decay as the spins precess at different frequencies during the initial τ period, but this dephasing is reversed during the second τ to create an "echo" at 2τ when all the spin packets are aligned along $-y$.

Figure 4.3(c) shows the *coherence pathway diagram* that applies to the spin-echo (and also the quadrupole-echo) experiment. These diagrams summarize the changing state of the spin system during the experiment in terms of a "pathway" showing the order of coherences (zero quantum, single quantum, etc.) present at different steps of the experiment. The pathway always starts at zero order, corresponding to equilibrium z magnetization, and finishes with observable $p = -1$ coherence magnetization during the acquisition period.

During free precession periods (no pulses), the different coherences evolve but do not change order. For instance, initial $\hat{I}_x$ magnetization will evolve under the influence of a heteronuclear coupling to a spin S to create mixed coherences of the form $\hat{I}_x\hat{S}_z$. These are still, however, single quantum coherences, and so the coherence pathways are horizontal during such delay periods. By contrast, RF pulses mix coherences between the different orders that the spin system can support, allowing the pathway to jump between coherence orders. It is important to note that the coherence pathway only summarizes the (desired) coherences present at a particular point and does not predict the detailed state of the system. For instance, the presence of an echo at 2τ must be deduced from analytical treatment of the density operator.

Considering the coherence pathway diagram for the spin-echo experiments, figure 4.3(c), the excitation pulse converts initial z magnetization into y magnetization (a sum of +1 and −1 coherences). The 180° pulse then interchanges +1 and −1 coherence. Finally, the −1 coherence is detected during the acquisition period. Hence the pathway of interest, marked in bold, corresponds to $0 \rightarrow +1 \rightarrow -1$.

As previously referred to in section 3.4.1, the purpose of phase cycling is to ensure that unwanted pathways are suppressed. For example, an imperfect initial 90° pulse will leave some residual zero-order (population) coherence, which will translate to unwanted additional signals if these coherences are propagated through to −1 coherence by later pulses. The key concept behind any form of phase cycling is that changing the phase, ϕ, of the RF pulse has a different effect on different coherence orders. Specifically for a given change in coherence order due to a pulse, Δp, changing the phase of the pulse by $\Delta \phi$ will change the phase of the relevant coherences in the density matrix by $-\Delta \phi \Delta p$. The phase cycle is designed so that unwanted coherences acquire different phases over the course of the cycle such that the unwanted terms cancel when the overall signal is added together. The pulse phases in the spin-echo experiment of figure 4.3(a), for example, lead to magnetization being refocused on the $-y$ axis. Shifting the phase of the refocusing π pulse by $90°$ results in refocusing along the $+y$ axis. Hence the desired coherence, which has passed through the $\Delta p = -2$ pathway, has acquired a phase shift of $180°$. If the receiver is shifted by $180°$ between the two experiments, then the desired signal will add constructively, while unwanted signals with $\Delta p = 0$ (corresponding to components of the density matrix that have not been inverted) will be cancelled. Since achieving perfect inversion across the full spectral width/sample is often difficult, this phase cycling is very important in spin-echo experiments. See Further reading for more details on the construction of phase cycles.

Figure 4.3(b) shows the pulse sequence for the solid (or quadrupolar) echo experiment. This has strong similarities to the simple spin echo (indeed the coherence pathway diagram is identical), but the details are different since the goal is to refocus the evolution under "bilinear" Hamiltonians, such as the quadrupole or homonuclear coupling Hamiltonians.[10] In particular, the refocusing pulse is a $90°$ pulse; a $180°$ pulse has no effect on the evolution under bilinear Hamiltonians, see section 4.7.3.

4.5   THE DENSITY MATRIX FOR COUPLED SPINS

This may seem a great deal of effort to describe the basic NMR experiment, and, indeed, the density operator treatment for a single spin-$\frac{1}{2}$ is equivalent to the much simpler description of the NMR experiment provided by the vector model of NMR. The density operator approach comes into its own, however, for treating systems with more than two levels.

4.5.1   EXAMPLE 1: A DIPOLAR-COUPLED HOMONUCLEAR SPIN PAIR

It was noted without detailed explanation in section 2.5 that the splittings due to homonuclear couplings were 50% larger than corresponding heteronuclear couplings. This first example shows how this factor arises.

The matrix elements of equation 4.25 first need to be evaluated in the "product" eigenbasis for a pair of spin-$\frac{1}{2}$ nuclei I and S, which consists of the four different combinations of α and β

---

[10] The analytical derivation of the echo response is more involved than for the simple spin echo and can be found in the Further reading. In contrast to the simple spin echo, the solid echo cannot be pictured in simple vector model terms. Moreover, the phase relationship of the two pulses is also significant. While the relative phases of excitation and refocusing pulses only affect the phase of the simple spin echo (the direction in xy along which the magnetization is refocused), the phase difference must be $90°$ to obtain a full solid echo.

states for two spins: $|aa\rangle$, $|a\beta\rangle$, $|\beta a\rangle$, $|\beta\beta\rangle$. The first quantum state in the ket refers to I and the second to S.[11]

The matrix representation for terms such as $\hat{I}_+\hat{S}_-$ is simply the product of the matrices[12] for $\hat{I}_+$ and $\hat{S}_-$. Alternatively, the matrix elements can be determined directly by considering the effect of the individual operators on the appropriate spins. For example, $\hat{I}_+\hat{S}_-|\beta a\rangle$ involves applying the lowering operator to S in the a state and the raising operator to the I spin in the β state

$$\hat{I}_+\hat{S}_-|\beta a\rangle = \hat{I}_+|\beta\beta\rangle = |a\beta\rangle \qquad 4.76$$

The evaluation order is not significant since operators involving different spins must, by definition, commute ($[\hat{I}_+, \hat{S}_-] = 0$). Similarly

$$\hat{I}_z\hat{S}_z|\beta a\rangle = \frac{1}{2}\hat{I}_z|\beta a\rangle = -\frac{1}{4}|\beta a\rangle \qquad 4.77$$

Working through the combinations gives the matrix representation of the dipolar coupling Hamiltonian:

$$\hat{H}_D = \begin{array}{c} \\ \langle aa| \\ \langle a\beta| \\ \langle \beta a| \\ \langle \beta\beta| \end{array} \begin{array}{cccc} |aa\rangle & |a\beta\rangle & |\beta a\rangle & |\beta\beta\rangle \\ \begin{pmatrix} d/2 & 0 & 0 & 0 \\ 0 & -d/2 & -d/2 & 0 \\ 0 & -d/2 & -d/2 & 0 \\ 0 & 0 & 0 & d/2 \end{pmatrix} \end{array} \qquad 4.78$$

where $d = D_{IS}P_2(\cos\theta)$ is used as shorthand for the dipolar coupling constant between I and S for a particular orientation θ. In the heteronuclear case, where the off-diagonal terms can be dropped, the NMR frequencies can be directly read off as $\pm d$ by inspection of the Hamiltonian matrix. This corresponds to a splitting of $2d$ in the spectrum.

In the homonuclear case, we need to diagonalize equation 4.78, and the NMR frequencies are less obvious. To determine the NMR spectrum, we also need to evaluate the sum angular momentum operators $\hat{I}_x+\hat{S}_x$, $\hat{I}_z+\hat{S}_z$, etc. These sum operators are conventionally denoted $\hat{F}_x$, $\hat{F}_z$, etc., and they are readily derived using equations 4.2–4.7 (but dropping the factors of $\hbar$). The non-zero matrix elements are:

$$\langle m_I, m_S|\hat{F}_Z|m_I, m_S\rangle = m_I + m_S \qquad 4.79$$

$$\langle aX|\hat{F}_+|\beta X\rangle = \langle Xa|\hat{F}_+|X\beta\rangle = 1 \qquad 4.80$$

$$\langle \beta X|\hat{F}_-|aX\rangle = \langle X\beta|\hat{F}_-|Xa\rangle = 1 \qquad 4.81$$

---

[11] The ordering is not significant, but obviously must be used consistently!

[12] A more formal way to construct such matrices is via the "direct" or "Kronecker" product between the 2 × 2 matrix representations for the individual spins. The direct product (usual symbol ⊗) should not be confused with simple matrix multiplication.

Hence the matrices are

$$\hat{F}_z = \begin{pmatrix} 1 & 0 & 0 & 0 \\ 0 & 0 & 0 & 0 \\ 0 & 0 & 0 & 0 \\ 0 & 0 & 0 & -1 \end{pmatrix} \quad \hat{F}_+ = \begin{pmatrix} 0 & 1 & 1 & 0 \\ 0 & 0 & 0 & 1 \\ 0 & 0 & 0 & 1 \\ 0 & 0 & 0 & 0 \end{pmatrix} \quad \hat{F}_- = \begin{pmatrix} 0 & 0 & 0 & 0 \\ 1 & 0 & 0 & 0 \\ 1 & 0 & 0 & 0 \\ 0 & 1 & 1 & 0 \end{pmatrix} \qquad 4.82$$

and so

$$\hat{F}_x = \begin{pmatrix} 0 & 1/2 & 1/2 & 0 \\ 1/2 & 0 & 0 & 1/2 \\ 1/2 & 0 & 0 & 1/2 \\ 0 & 1/2 & 1/2 & 0 \end{pmatrix} \quad \hat{F}_y = \begin{pmatrix} 0 & -i/2 & -i/2 & 0 \\ i/2 & 0 & 0 & -i/2 \\ i/2 & 0 & 0 & -i/2 \\ 0 & i/2 & i/2 & 0 \end{pmatrix} \qquad 4.83$$

Note the block structure of all the matrices, where the blocks correspond to the different values of $m_I + m_S$, that is, the eigenvalues of the sum z operator, $\hat{F}_z$. The Hamiltonian matrix of equation 4.78, for example, has non-zero elements in a 1×1 block corresponding to $\langle \hat{F}_z \rangle = 1$, a 2×2 block corresponding to the two states with $\langle \hat{F}_z \rangle = 0$ ($|\alpha\beta\rangle$ and $|\beta\alpha\rangle$), and a final 1×1 block corresponding to $\langle \hat{F}_z \rangle = -1$ ($|\beta\beta\rangle$). This block structure also applies to the density matrix, with a block linking states with $\langle \hat{F}_z \rangle = k$ and $\langle \hat{F}_z \rangle = l$ corresponding to coherences of order $p = k - l$. Hence the system of two spin-$\frac{1}{2}$ nuclei, as previously illustrated in figure 4.1(c), can support coherences of order 0 (the populations of the different eigenstates plus *zero-quantum coherences* between, e.g., $|\alpha\beta\rangle$ and $\langle \beta\alpha|$), detectable NMR coherences of order ± 1, plus *multiple-quantum coherences* of order ± 2 between the $\alpha\alpha$ and $\beta\beta$ states.

Rather than model the complete NMR experiment as in the previous section, the NMR frequencies and amplitudes can be calculated directly. If we start with x magnetization and acquire data points at integer multiples, n, of the dwell time, Δt, then the NMR signal will be

$$s(n\Delta t) = \mathrm{tr}(\hat{U}^n \hat{F}_x (\hat{U}^\dagger)^n \hat{F}_+) \qquad 4.84$$

where $\hat{U}$ is the propagator over the dwell time, $\hat{U} \equiv \hat{U}(\Delta t, 0)$. Since the propagator for each dwell time is the same, we can write $\hat{U}(n\Delta t, 0) = \hat{U}(\Delta t, 0)^n$. Equation 4.84 is most easily evaluated in the eigenbasis of the Hamiltonian. The matrix of equation 4.78 is readily diagonalized thanks to its block structure—only the central 2×2 block will have nontrivial eigenvalues and eigenvectors:

$$\Lambda = \begin{pmatrix} d/2 & 0 & 0 & 0 \\ 0 & -d & 0 & 0 \\ 0 & 0 & 0 & 0 \\ 0 & 0 & 0 & d/2 \end{pmatrix} \quad V = \begin{pmatrix} 1 & 0 & 0 & 0 \\ 0 & 1/\sqrt{2} & 1/\sqrt{2} & 0 \\ 0 & 1/\sqrt{2} & -1/\sqrt{2} & 0 \\ 0 & 0 & 0 & 1 \end{pmatrix} \qquad 4.85$$

Transforming $\hat{F}_+$ and $\hat{F}_x$ into the new eigenbasis gives

$$\hat{F}'_+ = V\hat{F}_+ V^{-1} = \begin{pmatrix} 0 & \sqrt{2} & 0 & 0 \\ 0 & 0 & 0 & \sqrt{2} \\ 0 & 0 & 0 & 0 \\ 0 & 0 & 0 & 0 \end{pmatrix} \quad \hat{F}'_x = \begin{pmatrix} 0 & 1/\sqrt{2} & 0 & 0 \\ 1/\sqrt{2} & 0 & 0 & 1/\sqrt{2} \\ 0 & 0 & 0 & 0 \\ 0 & 1/\sqrt{2} & 0 & 0 \end{pmatrix} \qquad 4.86$$

Since Λ is diagonal, the propagator, equation 4.52, is easily evaluated in the Hamiltonian eigenbasis:

$$\hat{U}^n = e^{-2\pi i \Lambda n \Delta t} \qquad 4.87$$

Expanding out the matrix products of equation 4.84 gives (another exercise for the reader):

$$s(t) = \sum_{j,k} (F'_+)_{jk} \, (F'_x)_{kj} \, e^{2\pi i (\Lambda_{jj} - \Lambda_{kk})t} \qquad 4.88$$

If we create a new matrix with elements $A_{jk} = (F'_+)_{jk} \, (F'_x)_{kj}$, we can write

$$s(t) = \sum_{j,k} A_{jk} \, e^{2\pi i (\Lambda_{jj} - \Lambda_{kk})t} \qquad 4.89$$

that is, the NMR signal is the sum of oscillations with frequencies $\Lambda_{jj} - \Lambda_{kk}$ (which correspond to differences of the eigenvalues of the Hamiltonian) and amplitudes given by the matrix elements A_{jk}.

Evaluating the "amplitude matrix" gives

$$A = \begin{pmatrix} 0 & 1 & 0 & 0 \\ 0 & 0 & 0 & 1 \\ 0 & 0 & 0 & 0 \\ 0 & 0 & 0 & 0 \end{pmatrix} \qquad 4.90$$

The two non-zero elements of A correspond to two allowed transitions of unit intensity, giving

$$s(t) = e^{2\pi i (\Lambda_{11} - \Lambda_{22})t} + e^{2\pi i (\Lambda_{22} - \Lambda_{44})t} \qquad 4.91$$

$$= e^{2\pi i (3d/2)t} + e^{-2\pi i (3d/2)t} \qquad 4.92$$

Hence a pair of lines with a frequency difference of $3d$ will be seen in the spectrum. As predicted, this splitting is 50% larger than the splitting of $2d$ observed in the heteronuclear case.

4.5.2 EXAMPLE 2: CROSS-POLARIZATION

Although cross-polarization (CP) generally involves transfer of magnetization from an abundant spin species, such as ^{1}H, the basic principles can be seen in a simple two spin picture. This example also illustrates the use of different frames of reference to simplify analysis.

The Hamiltonian for a heteronuclear spin pair, subject to RF irradiation along the x axis on both spins, is

$$\hat{H} = 2d\hat{I}_z\hat{S}_z + v_I\hat{I}_x + v_S\hat{S}_x \qquad 4.93$$

where v_I and v_S are the nutation frequencies due to the RF on the I and S spins respectively. For simplicity we assume that the RF is on-resonance for both spins. The Hamiltonian matrix in the product Zeeman basis is

$$
\hat{H} = \begin{array}{cc}
 & \begin{array}{cccc} |\alpha\alpha\rangle & |\alpha\beta\rangle & |\beta\alpha\rangle & |\beta\beta\rangle \end{array} \\
\begin{array}{c} \langle\alpha\alpha| \\ \langle\alpha\beta| \\ \langle\beta\alpha| \\ \langle\beta\beta| \end{array} &
\begin{pmatrix}
d/2 & v_S & v_I & 0 \\
v_S & -d/2 & 0 & v_I \\
v_I & 0 & -d/2 & v_S \\
0 & v_I & v_S & d/2
\end{pmatrix}
\end{array}
\qquad 4.94
$$

Note how the v_I elements connect states in which the first (I) spin flips, while the v_S elements correspond to allowed NMR transitions of the S spins.

Despite its elegance, the eigenvalues and vectors of this matrix are not obvious, and there are large terms (those involving the RF) that are off the diagonal. The situation is improved using a basis that puts the dominant terms along the diagonal by rotating the axis system so that the new z axis is along the RF spin-lock axis. The required 90° rotation about y simply permutes the labels on the operators ($x \rightarrow z \rightarrow -x \rightarrow -z \rightarrow x$) to give

$$\hat{H}' = v_I\hat{I}_z + v_S\hat{S}_z + 2d\hat{I}_x\hat{S}_x \qquad 4.95$$

$$
= \begin{pmatrix}
v_I + v_S & 0 & 0 & d/2 \\
0 & v_I - v_S & d/2 & 0 \\
0 & d/2 & v_I - v_S & 0 \\
d/2 & 0 & 0 & -v_I - v_S
\end{pmatrix}
\qquad 4.96
$$

The significance of matching the nutation rates of the two nuclei, $v_{HH} = v_I = v_S$, is then easier to appreciate (HH denotes the Hartmann–Hahn matching condition). At this condition

$$
\hat{H}' = \begin{pmatrix}
2v_{HH} & 0 & 0 & d/2 \\
0 & 0 & d/2 & 0 \\
0 & d/2 & 0 & 0 \\
d/2 & 0 & 0 & -2v_{HH}
\end{pmatrix}
\qquad 4.97
$$

Since the RF field strength must significantly exceed the dipolar coupling for a good spin-lock, $v_{HH} \gg d$, the secular approximation can be invoked to drop the off-diagonal terms connecting different diagonal elements, leading to

$$\hat{H}' \approx \begin{pmatrix} 2v_{HH} & 0 & 0 & 0 \\ 0 & 0 & d/2 & 0 \\ 0 & d/2 & 0 & 0 \\ 0 & 0 & 0 & -2v_{HH} \end{pmatrix} \qquad 4.98$$

The off-diagonal terms linking the approximately degenerate middle two states cannot be neglected, and so the dipolar coupling will "mix" these states.

Considering the matrix representations for I and S magnetization along the spin-lock axes (i.e., z)

$$\hat{I}_z = \begin{pmatrix} 1/2 & 0 & 0 & 0 \\ 0 & 1/2 & 0 & 0 \\ 0 & 0 & -1/2 & 0 \\ 0 & 0 & 0 & -1/2 \end{pmatrix} \qquad \hat{S}_z = \begin{pmatrix} 1/2 & 0 & 0 & 0 \\ 0 & -1/2 & 0 & 0 \\ 0 & 0 & 1/2 & 0 \\ 0 & 0 & 0 & -1/2 \end{pmatrix} \qquad 4.99$$

the mixing of the central states can be seen to interconvert $\hat{I}_z$ and $\hat{S}_z$. It is left as an exercise to show, using the approach set out in section 4.5.1, that the evolution of the density operator *in the conventional basis* will be

$$\hat{\rho}(t) = M\left[\hat{I}_x \cos 2\pi dt + \hat{S}_x \sin 2\pi dt \right] \qquad 4.100$$

In other words, the magnetization will oscillate between I and S spin magnetization, with the frequency of oscillation set by the strength of the dipolar coupling.[13]

It is also useful to consider this problem in terms of spin operators. Starting from equation 4.95, the Hamiltonian in the tilted frame at the Hartmann–Hahn condition can be written

$$\begin{aligned} \hat{H}' &= v_I \hat{I}_z + v_S \hat{S}_z + 2d\hat{I}_x \hat{S}_x \\ &= v_{HH}\hat{F}_z + d(\hat{I}_+\hat{S}_- + \hat{I}_-\hat{S}_+) + d(\hat{I}_+\hat{S}_+ + \hat{I}_-\hat{S}_-) \end{aligned} \qquad 4.101$$

where $\hat{F}_z = \hat{I}_z + \hat{S}_z$ is the sum magnetization operator. The second term is a flip-flop term interchanging states $|\alpha\beta\rangle$ and $|\beta\alpha\rangle$, while the third interchanges $|\alpha\alpha\rangle$ and $|\beta\beta\rangle$. If the spin-lock is strong, that is, $v_{HH} \gg d$, then the secular approximation can be invoked to discard terms that do not commute with $v_{HH}\hat{F}_z$. The Hamiltonian then reduces to

$$\hat{H}' = v_{HH}\hat{F}_z + d\left(\hat{I}_+\hat{S}_- + \hat{I}_-\hat{S}_+\right) \qquad 4.102$$

---

[13] In cases where the *IS* spin pair is sufficiently well isolated, it is possible to observe such "transient oscillations" during Hartmann–Hahn cross-polarization experimentally. As discussed in section 8.1.5, this often provides a simple approach to measuring dipolar couplings.

The initial density operator (in the tilted frame) is $M\hat{I}_z$. This commutes with the RF Hamiltonian (i.e., the magnetization is spin-locked). Hence the density operator will evolve simply under the influence of $d(\hat{I}_+ \hat{S}_- + \hat{I}_- \hat{S}_+)$. As discussed above, this flip-flop term has the effect of interconverting I and S spin magnetization.

The two formulations are exactly equivalent; discarding non-commuting spin operator terms is equivalent to neglecting off-diagonal terms connecting non-degenerate eigenstates in the matrix representation. Operator representations often have the advantage of generality (they are not specific to a spin-$\frac{1}{2}$ pair), but the necessary operator algebra can be distracting. The matrix representation is not strictly general but can be a more visual way to tackle nontrivial problems.

4.6   EULER ANGLES & SPHERICAL TENSORS

So far, we have mostly treated the NMR parameters, such as dipolar couplings, as simple empirical constants. In practice, the NMR interactions are orientation dependent and we need to consider their full tensor nature. As set out in section 2.2, it is relatively straightforward to write down expressions for NMR parameters, such as the dipolar coupling, in terms of the polar angles, θ and ϕ, defining the orientation of the principal axis system for the tensor in the laboratory frame defined by the external magnetic field.

In more complex problems, however, it is necessary to introduce additional intermediate frames. Considering a collection of nuclear spins in a crystalline framework, we first need to relate the principal axis systems of the individual interactions (dipolar, CSA, etc.) to a common *molecular frame* (M) (typically defined in terms of the crystallographic lattice). This molecular frame must then be related to the laboratory frame (L) by a set of angles that specify the orientation of the crystal.

In general three angles are needed to define a transformation in three-dimensional space. These three *Euler angles* are conventionally denoted α, β, and γ, while the set of angles specifying a transformation, (α, β, γ), is often collectively indicated by Ω. For instance, the transformation[14] from the molecular frame to the laboratory frame may be represented by Ω_{LM}. Inset 4.4 explains how these angles are commonly defined. In some cases, specification of all three Euler angles is unnecessary. For example, the α angle is irrelevant when transforming a dipolar coupling *from* its principal axis system to another frame since the dipolar interaction is symmetric about its Z principal axis (the internuclear vector) in the absence of molecular motion. On the other hand, α must be specified if the interaction has a non-zero asymmetry, η. Similarly, the γ angle is unimportant when transforming *into* the final laboratory frame, since the nuclear spin Hamiltonian is unaffected by rotations around the magnetic field axis (this follows from the high-field approximation discussed in section 4.2).

As discussed in inset 4.4, the simple 3×3 matrix representation of rank-2 tensors is not well adapted to expressing the effect of rotations and usually leads to extremely unwieldy expressions when a series of frame transformations is required. The remainder of this section discusses an alternative representation of tensors that better expresses symmetry under rotation and leads to more elegant and compact descriptions of anisotropic NMR interactions.

---

[14] Subtle issues such as active *vs.* passive rotations are being glossed over here. See, for example, Levitt's *Spin Dynamics* and Mueller's *Tensors and Rotations in NMR* (Further reading) for details.

Inset 4.4. Euler angles and tensor rotation

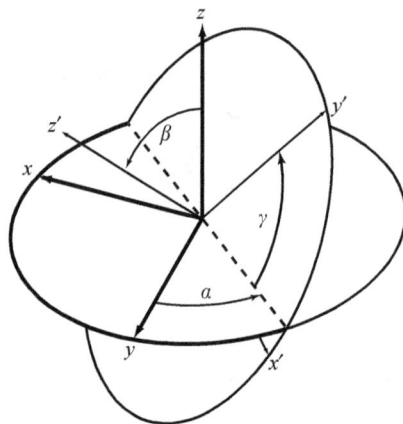

Figure 4.4. Definition of the Euler angles (Rose convention)[*] for the transformation of the original axes (x, y, z) to a new axis system (x', y', z'). The axes are first rotated by a about the z axis to bring the y axis onto the nodal line defined by xy and $x'y'$ planes. This is followed by rotation about the new y axis through β and then a final rotation through γ is applied about the final z axis (z').

Figure 4.4 shows the most common definition of the three Euler angles, a, β, and γ, specifying the rotation of one coordinate system to another. Unfortunately, the definition of the angles is not unique and different systems have been used in the literature. Care is needed if reporting Euler angles or when compiling orientational information from different sources.

If we consider a typical rank-2 tensor, such as a shielding tensor, in its conventional matrix (Cartesian) form:

$$A = \begin{pmatrix} \sigma_{xx} & \sigma_{xy} & \sigma_{xz} \\ \sigma_{yx} & \sigma_{yy} & \sigma_{yz} \\ \sigma_{zx} & \sigma_{zy} & \sigma_{zz} \end{pmatrix} \qquad 4.103$$

The tensor in the new basis is given by $A' = RAR^{-1}$ where the rotation matrix R is

$$R(a, \beta, \gamma) = \begin{pmatrix} \cos a \cos \beta \cos \gamma - \sin a \sin \gamma & -\cos a \cos \beta \sin \gamma - \sin a \cos \gamma & \cos a \sin \beta \\ \sin a \cos \beta \cos \gamma + \cos a \sin \gamma & -\sin a \cos \beta \sin \gamma + \cos a \cos \gamma & \sin a \sin \beta \\ -\sin \beta \cos \gamma & \sin \beta \sin \gamma & \cos \beta \end{pmatrix} \qquad 4.104$$

The final matrix A' is not shown as it is rather cumbersome and not particularly informative. The main text discusses an alternative representation of tensors that is better adapted to frame transformations.

---

[*] Elementary theory of angular momentum, M. E. Rose, Wiley (1957).

In this *spherical tensor* representation, the tensor is decomposed using a basis with the same properties under rotation as the spherical harmonics familiar from atomic orbital or angular momentum theory. A rank-2 tensor decomposes into an isotropic (rank-0) term, $R_{0,0}$ (compare with an s orbital), three rank-1 terms, $R_{1,0}$, $R_{1,\pm1}$ (compare with p orbitals), and five rank-2 terms, $R_{2,0}$, $R_{2,\pm1}$, $R_{2,\pm2}$ (compare with d orbitals). The rank-1 components correspond to the antisymmetric component of the tensor, that is, terms involving $R_{xy} - R_{yx}$, etc. These have no observed effect on NMR spectra and can be safely ignored.

The relationship between the spherical tensor components and the Cartesian representation is simple for a tensor in its principal axis system, that is, with the matrix just containing the diagonal elements:

$$R_{0,0} = R_{\text{iso}} \tag{4.105}$$

$$R_{2,0} = \sqrt{\frac{3}{2}}\zeta \qquad R_{2,\pm1} = 0 \qquad R_{2,\pm2} = \frac{1}{2}\eta\zeta \tag{4.106}$$

where R_{iso} is the isotropic component, and ζ and η are the anisotropy and asymmetry respectively (see inset 2.2). Note how the spherical tensor representation captures the rotational symmetry: the rank-0 isotropic term is independent of orientation, while the anisotropy is contained within the rank-2 terms.

At high field, the only components that survive the secular approximation are the $R_{0,0}^{\text{L}}$ and $R_{2,0}^{\text{L}}$ components of the tensor in the laboratory frame. The isotropic term is, by definition, independent of the axis system, and so the problem of finding the high-field Hamiltonian involves determining $R_{2,0}^{\text{L}}$ after applying a series of transformations to a tensor starting in its principal axis system.

The effect of rotation through Euler angles $\Omega = (\alpha, \beta, \gamma)$ on the rank-2 components is given by

$$R'_{2,m} = \sum_{n=-2}^{2} D_{mn}^{(2)}(\Omega) R_{2,n} \tag{4.107}$$

where $D_{mn}^{(2)}(\Omega)$ are elements of the rank-2 *Wigner rotation matrix*. These are given by

$$D_{mn}^{(2)}(\alpha, \beta, \gamma) = e^{i(m\gamma + n\alpha)} d_{mn}^{(2)}(\beta) \tag{4.108}$$

where the "reduced" matrix elements $d_{mn}^{(2)}(\beta)$ are solely functions of the β angle.[15]

4.6.1 MAGIC-ANGLE SPINNING

As an example, consider a sample under magic-angle spinning. As illustrated in figure 4.5, the various NMR interactions need to be expressed in a *rotor frame* (R), which is aligned along the rotation axis and rotates at the spinning frequency. The transformation from this frame to the final laboratory frame is given by the Euler angles

$$\Omega_{\text{LR}} = (\alpha_0 + \omega_r t, \theta_m, 0) \tag{4.109}$$

---

[15] Table B.2 of Schmidt-Rohr and Spiess in the Further reading contains tables of Wigner functions.

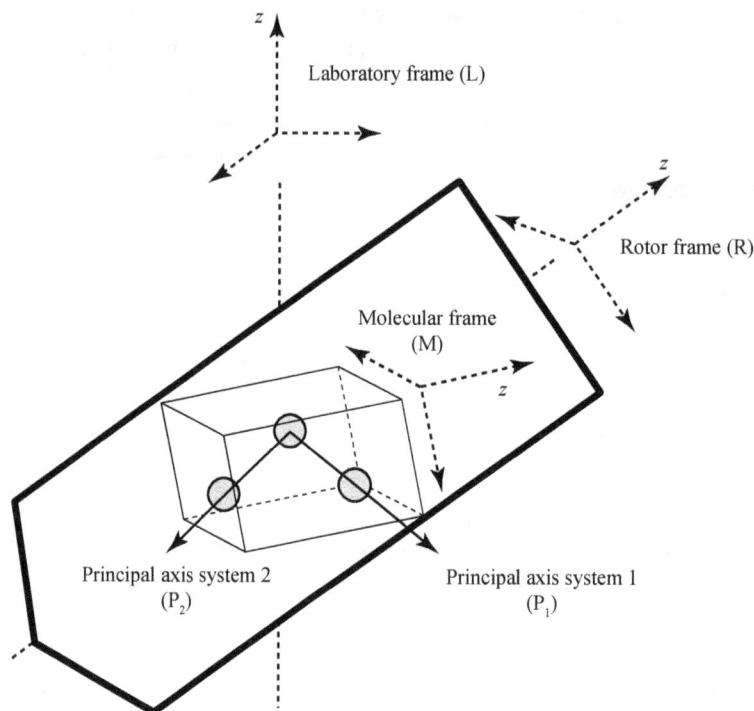

Figure 4.5. Hierarchy of frames of reference for sample spinning problems. Individual principal axis systems (P_1 and P_2 for the two dipolar interactions indicated by solid arrows) are related to a common molecular or crystal reference frame (M). The Euler angle set Ω_{RM} describes the transformation of the molecular frame to a frame fixed with respect to the rotor (R). Finally the rotor frame is related to the laboratory frame (L) via the Euler angle set Ω_{LR} (see equation 4.109).

where the rotor position, a, is time dependent (a_0 and ω_r are the initial rotor angle and spinning frequency respectively). The β angle corresponds to the angle of the rotor with respect to the magnetic field, that is, the magic angle θ_m. The γ angle can be arbitrarily set to zero, as discussed above.

By combining equations 4.108 and 4.109, $R^L_{2,0}$ is given by

$$R^L_{2,0}(t) = \sum_{n=-2}^{2} e^{in(a_0 + \omega_r t)} \, d^{(2)}_{0,n}(\theta_m) R^R_{2,n} \qquad 4.110$$

If there is a single time-dependent interaction (or the different terms of the Hamiltonian commute with each other), then the overall Hamiltonian is said to be "inhomogeneous" and we can consider the Hamiltonians for the individual terms independently; this important distinction between "inhomogeneous" and "homogeneous" Hamiltonians is discussed in inset 4.5. *Average Hamiltonian*

Theory (see section 4.7.1) can then be used to find the average Hamiltonian over a complete rotor period. This will involve the average of $R_{2,0}^{L}(t)$:

$$\left\langle R_{2,0}^{L} \right\rangle = \sum_{n=-2}^{2} d_{0,n}^{(2)}(\theta_m) R_{2,n}^{R} \frac{1}{\tau_r} \int_{0}^{\tau_r} e^{in(a_0+\omega_r t)} dt \qquad 4.111$$

$$= d_{0,0}^{(2)}(\theta_m) R_{2,0}^{R} \qquad (\text{only the } n = 0 \text{ term survives}) \qquad 4.112$$

$d_{0,0}^{(2)}(\beta)$ is the familiar second-order Legendre polynomial, $(3\cos^2\beta - 1)/2$. Hence at the magic angle, $\left\langle R_{2,0}^{L} \right\rangle = 0$, that is, the average Hamiltonian vanishes and there is no evolution of the density matrix.

Unless the spinning rate is much larger than the interaction (in which case the oscillatory terms corresponding to $n = \pm 1$ and $n = \pm 2$ are negligibly small), the $n = \pm 1$ and $n = \pm 2$ in equation 4.110 will lead to some oscillatory evolution of the density matrix. In terms of the NMR signal, this corresponds to the formation of *rotary echoes*, in which the signal evolves during the rotor period, but is "refocused" to create a rotary echo at each complete rotor period (as required by equation 4.112). This series of echoes, with a period equal to the rotor period, corresponds to a set of sharp peaks separated by the spinning frequency in the spectrum, that is, a manifold of spinning sidebands. See figure 3.15 for a practical example of rotary echoes.

4.7 ADDITIONAL ANALYTICAL TOOLS

The approaches outlined above of following the evolution of the density operator using a combination of matrix or symbolic representations of the spin operators are effective for understanding many solid-state NMR problems. There are, however, a large class of experiments where these simple approaches become intractable, and it is necessary either to resort to numerical simulation (which is discussed in appendix D) or to use more sophisticated analytical tools. This section reviews some of these more advanced theoretical tools.

4.7.1 AVERAGE HAMILTONIAN THEORY

Average Hamiltonian Theory (AHT) has been widely used in solid-state NMR to derive exact analytical expressions, but also more qualitatively when the problem is too complex to be reduced to a tractable expression. AHT is useful whenever the Hamiltonian is time-dependent and periodic, that is, the time-dependent component repeats itself after a characteristic "cycle time", τ_c. This may correspond to a repeated RF sequence or sample rotation.

AHT applied to sample rotation has already been discussed in section 4.6. In this simple case, it was sufficient to average over the time dependence to obtain the average Hamiltonian to first order. By contrast, when dealing with strong time-dependent RF, it is generally necessary to transform the Hamiltonian into an *interaction frame* that allows the average Hamiltonian to be readily computed.

Inset 4.5. Homogeneous and inhomogeneous Hamiltonians

As previously discussed in section 2.7, there is an important distinction between "inhomogeneous" interactions, such as the chemical shift anisotropy, which lead to sharp spinning sidebands under magic-angle spinning, and "homogeneous" interactions, notably homonuclear dipolar coupling, that result in centerbands and sidebands with distinct width. The origin of this distinction involves some subtle quantum mechanics. The spin Hamiltonians for shift interactions and heteronuclear couplings at high field only involve z spin operators, for example, $2d_{IS}\hat{I}_z\hat{S}_z$ for the heteronuclear dipolar coupling between spins I and S. Any pair of Hamiltonians involving only z operators must commute, since $[\hat{I}_{jz}, \hat{I}_{kz}] = 0$ for all values of j and k. Hence we can consider their effects on the evolution of the density operator independently. The MAS spectrum associated with an individual interaction consists of sharp centerbands and sidebands, and so the overall spectrum will also consist of sharp features (mathematically, the spectrum is a "convolution" of the spectra associated with the interactions taken individually).

By contrast, if the Hamiltonian involves homonuclear couplings, the flip-flop terms from different spin pairs do not generally commute with each other

$$\left[\hat{I}_{j+}\hat{I}_{k-} + \hat{I}_{j-}\hat{I}_{k+}, \ \hat{I}_{k+}\hat{I}_{l-} + \hat{I}_{l-}\hat{I}_{k+}\right] \neq 0 \quad \text{for} \quad j \neq l \qquad 4.113$$

As a consequence, the evolution under the total Hamiltonian cannot be simply expressed in terms of the individual components. Moreover, the evolution is *not* refocused over a rotor cycle and the centerbands and sidebands have a distinct width.

The presence of a non-commuting term in the Hamiltonian tends to "contaminate" the rest of the Hamiltonian, giving it the same "homogeneous" character. For instance, linewidths in ^{13}C NMR are often limited by the efficiency of the ^{1}H heteronuclear decoupling even though the ^{13}C interactions (at natural isotopic abundance) only involve inhomogeneous terms such as the chemical shift. Unless the ^{13}C nucleus is very efficiently decoupled from the ^{1}H dipolar network, its NMR lines are affected by homogeneous broadening.

The total Hamiltonian is first separated into a term, $\hat{H}_0$, containing the NMR interactions, etc., and a time-dependent component, $\hat{H}_{RF}$, for the RF:

$$\hat{H}(t) = \hat{H}_0 + \hat{H}_{RF}(t) \qquad 4.114$$

The expression for propagators for the interval from $t = 0$ to t is then:

$$U(t) = \hat{T}\exp\left\{-i\int_0^t \hat{H}_0 + \hat{H}_{RF}(t')dt'\right\} \qquad 4.115$$

where $\hat{T}$ is the *Dyson time-ordering operator*—essentially a reminder that we need to evaluate equation 4.115 in small, time-ordered steps to account for the time dependence of $\hat{H}$. If we transform $\hat{H}_0$ into a frame in which time dependence due to the RF has been removed, we can factor the overall propagator into two components:

$$U(t) = U_{RF}(t)\tilde{U}_0(t) \qquad \text{4.116}$$

where
$$U_{RF}(t) = \hat{T}\exp\left\{-i\int_0^t \hat{H}_{RF}(t')\,dt'\right\} \qquad \text{4.117}$$

is the overall evolution under the RF and

$$\tilde{U}_0(t) = \hat{T}\exp\left\{-i\int_0^t \tilde{H}_0(t')\,dt'\right\} \qquad \text{4.118}$$

is the evolution under the system Hamiltonian in the interaction frame (denoted by ~), where

$$\tilde{H}_0(t) = U_{RF}^\dagger(t)\hat{H}_0 U_{RF}(t) \qquad \text{4.119}$$

This may seem obscure, but the rotating frame of reference provides a good example of an interaction frame; $U_{RF}(t)$ in this case is $\exp(i\omega_{NMR}t\hat{I}_z)$, representing rotation about the z axis at the Larmor frequency, ω_{NMR}, and $\tilde{H}_0$ is now the familiar rotating frame Hamiltonian.

Fortunately some simplifications can now be made! First we restrict attention to sequences that are "cyclic", that is, sequences for which there is no net rotation of the axis system over the course of the cycle. This means that $U_{RF}(\tau_c)$ is the identity matrix, and the propagator over a complete cycle is simply $U(\tau_c) = \tilde{U}_0(\tau_c)$. Hence we can describe the effect of the RF pulse sequence over a complete cycle in terms of an "average Hamiltonian," $\bar{H}$, which creates this propagator, $\tilde{U}_0(\tau_c) = \exp(-i\bar{H}\tau_c)$. This still leaves the problem of expressing $\bar{H}$ in terms of $\tilde{H}_0(t)$.

To first order, however, $\bar{H}$ is given by the average of $\tilde{H}_0$ over the cycle period:

$$\bar{H} = \bar{H}^{(0)} + \bar{H}^{(1)} + \cdots \qquad \text{4.120}$$

$$= \frac{1}{\tau_c}\int_0^{\tau_c} \tilde{H}_0(t)\,dt' + \cdots \qquad \text{4.121}$$

The additional terms in this series expansion (termed the *Magnus expansion*) are increasingly complex. However, the first (and successive odd-order) correction terms, $\bar{H}^{(1)}$, etc., evaluate to zero if the interaction-frame Hamiltonian has time symmetry $\tilde{H}_0(t) = \tilde{H}_0(-t)$. Moreover, the correction terms should be small if the interaction frame is appropriate.

The WHH-4 sequence for homonuclear decoupling provides a good illustration of AHT in action (see section 5.5.1 for discussion of homonuclear decoupling sequences). As shown in figure 4.6, the sequence consists of a series of 90° pulses separated by gaps which can be used to detect the NMR signal. The interaction frame during the different periods of the sequence is obtained by applying these 90° rotations to an initial starting set of axes. Note how the frame returns to its starting point at the end of the cycle, satisfying the "cyclicity" condition.

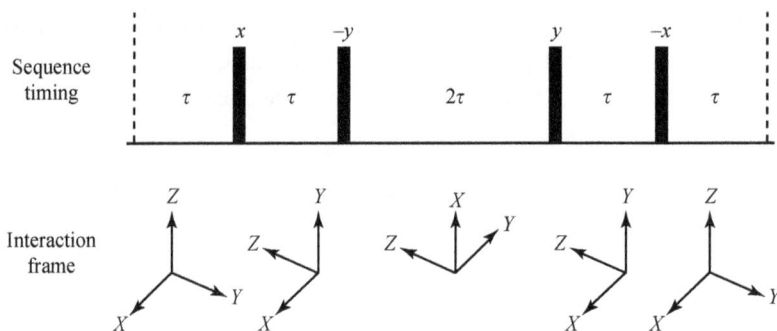

Figure 4.6. The WHH-4 (WAHUHA) pulse sequence for homonuclear decoupling: (top) the pulse sequence and (bottom) the axes of the interaction frame during the different sections of the sequence. Note that x, y, z refer to the laboratory frame, while X, Y, Z refer to the interaction frame. Figure adapted from *Principles of magnetic resonance in one and two dimensions* (see Further reading).

Using $\tilde{H}_{ZZ}$ to denote the interaction-frame Hamiltonian for the homonuclear dipolar coupling between a spin pair at the start of the cycle,

$$\tilde{H}_{ZZ} = d(2\hat{I}_{iz}\hat{I}_{jz} - \hat{I}_{ix}\hat{I}_{jx} - \hat{I}_{iy}\hat{I}_{jy})$$ 4.122

we can write down the Hamiltonians for the successive steps of the cycle. The transformation into the interaction frame for the period between the first two pulses corresponds to relabeling the axes using $y \rightarrow z, z \rightarrow -y$, and so the interaction-frame dipolar Hamiltonian during this period is

$$\tilde{H}_{YY} = d(2\hat{I}_{iy}\hat{I}_{jy} - \hat{I}_{ix}\hat{I}_{jx} - \hat{I}_{iz}\hat{I}_{jz})$$ 4.123

Similarly the Hamiltonian in the middle segment is

$$\tilde{H}_{XX} = d(2\hat{I}_{ix}\hat{I}_{jx} - \hat{I}_{iy}\hat{I}_{jy} - \hat{I}_{iz}\hat{I}_{jz})$$ 4.124

The average Hamiltonian over the entire period is (to first order):

$$\bar{H}^{(0)} = \frac{1}{\tau_c}\int_0^{\tau_c}\tilde{H}_0(t)\,dt$$ 4.125

$$= \frac{1}{6\tau}\left(\tau\tilde{H}_{ZZ} + \tau\tilde{H}_{YY} + 2\tau\tilde{H}_{XX} + \tau\tilde{H}_{YY} + \tau\tilde{H}_{ZZ}\right)$$ 4.126

$$= \frac{1}{3}\left(\tilde{H}_{XX} + \tilde{H}_{YY} + \tilde{H}_{ZZ}\right)$$ 4.127

$$= 0$$ 4.128

that is, the dipolar interaction is suppressed to first order. It is important to note that the average Hamiltonian is only relevant to the density matrix sampling at integer multiples of the cycle period, τ_c. For example, the average Hamiltonian of equation 4.128 implies that the density matrix appears not to evolve when sampled after each complete WHH-4 cycle, even though the system will evolve under the influence of both RF and dipolar couplings in between sampling points.

Since the expression for $\bar{H}^{(0)}$ in equation 4.121 is linear in $\tilde{H}_0$, we can compute the first-order average Hamiltonian for different interactions independently, that is, $\bar{H}_{A+B}^{(0)} = \bar{H}_A^{(0)} + \bar{H}_B^{(0)}$. Hence equation 4.128 will be true for an arbitrary collection of homonuclear coupled spins. This does not apply to higher-order contributions to the average Hamiltonian, and calculating these higher-order "cross-terms" between different components of the Hamiltonian is one of the major challenges to applying average Hamiltonian approaches beyond first order.

4.7.2   AN OVERVIEW OF FLOQUET THEORY

AHT often provides an intuitive summary of complex time-dependent problems. It suffers, however, from a number of restrictions. For example, the evolution is only sampled at complete periods of the time dependence. Just as sampling the NMR signal at too low a rate means that different frequency components may not be distinguished (*aliasing*), so different components of the density operator evolution may be aliased. In the case of magic-angle spinning, for example, the sampling only occurs once per rotor period and so spinning sidebands and centerbands are not distinguished. Hence information about the spinning sideband intensities, or differences in lineshape between centerbands and sidebands, cannot be obtained.

Floquet theory is a widely used approach that generalizes AHT. It is restricted to considering problems where the Hamiltonian can be expressed as a Fourier series:

$$\hat{H}(t) = \sum_m \hat{H}_m e^{2\pi i m t / T} \qquad\qquad 4.129$$

where T is the period (e.g., the rotation period in the case of MAS, or the sequence period in the case of periodic RF) and m is the index of the Fourier series. However, since most functions can be expanded as a convergent Fourier series, this is not a major restriction. Moreover, more than one time dependence (e.g., magic-angle spinning plus time-dependent RF) can be accommodated by expanding with respect to more than one Fourier index.

The density operator, NMR signal, etc., can then all be expressed in terms of Fourier series; for example, the NMR signal for MAS problems becomes

$$s(t) = \sum_m \sum_j A_{m,j} e^{2\pi i (v_j + m v_r t)} \qquad\qquad 4.130$$

where v_j is the centerband frequency for transition j and $A_{m,j}$ is the intensity of sideband m for transition j.

Without going into the mathematical details, Floquet theory provides the machinery to calculate the $A_{m,j}$ coefficients, starting from the Fourier series for the Hamiltonian (plus the initial density matrix). Note how Floquet theory imposes no restriction on the sampling of the NMR signal (equation 4.130 is valid for all t) and naturally distinguishes centerbands and sidebands.

4.7.3 INTRODUCTION TO IRREDUCIBLE SPHERICAL TENSOR OPERATORS

As part of deriving the interaction frame for the WHH-4 sequence above, the effect of the RF pulses was described in terms of "rotations" of the spin operator components of the Hamiltonian. While the 90° rotations of WHH-4 can be readily expressed in terms of Cartesian axes, working with arbitrary rotations in this framework becomes cumbersome. For this reason most work involving average Hamiltonians for RF pulse sequences describes the spin Hamiltonian in terms of a spherical tensor basis, which is analogous to the spherical tensor representation of rotations in physical space. This representation will be only briefly introduced here—see the more advanced texts in Further reading for more information.

In terms of irreducible spherical tensors, the homonuclear dipolar coupling Hamiltonian of equation 4.25 is

$$\hat{H}_D = A_{2,0}\,\hat{T}_{2,0} \qquad\qquad 4.131$$

where $A_{2,0} = \sqrt{6}d$ is the spatial component and

$$\hat{T}_{2,0} = \tfrac{1}{\sqrt{6}}\left[2\hat{I}_z\hat{S}_z - \tfrac{1}{2}(\hat{I}_+\hat{S}_- + \hat{I}_-\hat{S}_+)\right] \qquad\qquad 4.132$$

is the spin operator term.[16]

The principles described in section 4.6 for rotations of spherical tensors in physical space apply equally well to rotations of the spin operators by RF pulses. For example, a π pulse corresponds to a rotation using the Euler angles $(0, \pi, 0)$. From equations 4.107 and 4.108, the rotation of the spin operator $\hat{T}_{2,0}$ by a π pulse will generate

$$\hat{T}' = \sum_{n=-2}^{2} d^{(2)}_{0n}(\pi)\hat{T}_{2,n} \qquad\qquad 4.133$$

The only non-zero $d^{(2)}_{0n}(\pi)$ is $d^{(2)}_{0,0}(\pi) = (3\cos^2\pi - 1)/2 = 1$; hence the spin operator after rotation by π is $\hat{T}' = \hat{T}_{2,0}$. In other words, the homonuclear dipolar Hamiltonian is unchanged by a 180° rotation. The relative ease of expressing rotations in terms of irreducible spherical tensors is a major incentive for their use when developing new RF irradiation schemes.

---

[16] Equation 4.131 is equivalent to equation 4.25, but care is needed to avoid mixing the two representations as the normalization factors (here $1/\sqrt{6}$) are different. Here we have used A to denote a spatial tensor associated with a spherical tensor representation of the spin operators to contrast with R used in section 4.6, where the spatial tensors were associated with a Cartesian representation of the spin operators.

4.7.3.1 Application to Decoupling and Recoupling Sequences

One of the most common applications of AHT is for sequences designed either to suppress homo-nuclear dipolar couplings or to "recouple" the homonuclear couplings that are suppressed by magic-angle spinning.

We can always express the Hamiltonian of the spin system as a sum of terms that are either rank 1 (linear) or rank 2 (bilinear) in spin operators that are relevant to the RF being applied:

$$\hat{H} = \sum_{\lambda} A_{1,0}^{\lambda} \hat{T}_{1,0} + \sum_{\lambda'} A_{2,0}^{\lambda'} \hat{T}_{2,0} \qquad 4.134$$

where $\hat{T}_{1,0} = \hat{I}_z$, and λ and λ' sum over the rank-1 and rank-2 terms respectively. Heteronuclear coupling terms are included in the rank-1 sum, since they contain only operators of the form $\hat{I}_z$ with respect to the spins being manipulated by the RF.

The average Hamiltonian associated with an RF pulse sequence will also be a sum of rank-1 and rank-2 terms. The original spin operators[17] are replaced by new average terms:

$$\hat{T}_{1,0} \rightarrow a_1 \overline{T_1} \qquad \hat{T}_{2,0} \rightarrow a_2 \overline{T_2} \qquad 4.135$$

where $\overline{T_i}$ is a linear combination of the $\hat{T}_{i,n}$ operators, and a_i is a scaling factor. Note how the spin operators contained in the average Hamiltonian are not generally the same as the original operators, $\hat{T}_{1,0}$ and $\hat{T}_{2,0}$.

For *recoupling*, the goal is to design RF sequences that maximize the scaling factor on the homonuclear couplings, that is, a_2. Particularly in quantitative applications it is also desirable to minimize the scaling factor on the rank-1 terms, a_1, in order to create a Hamiltonian that depends only on the homonuclear couplings.

By contrast, homonuclear *decoupling* sequences should *minimize* a_2. In most applications, it is also important to retain the evolution under rank-1 terms (i.e., the chemical shift), and so it is desirable to maximize such rank-1 terms. The spin operator terms of the rank-1 average Hamiltonians will involve

$$a_1 \overline{T_1} = a_1 (a_x \hat{I}_x + a_y \hat{I}_y + a_z \hat{I}_z) \qquad 4.136$$

where $\overline{T_1}$ has been expressed in terms of the conventional Cartesian spin operators, with $a_x^2 + a_y^2 + a_z^2 = 1$. This decomposition effectively defines a new axis system. If the $\hat{I}_z$ operator in this frame is defined by

$$\hat{I}_z' = a_x \hat{I}_x + a_y \hat{I}_y + a_z \hat{I}_z \qquad 4.137$$

---

[17] The focus here is on the spin operators. Obviously the time dependence of the spatial terms must also be included when determining average Hamiltonians if MAS is involved.

then the I-spin magnetization will precess about this z' axis. As discussed further in section 5.5, z' does not usually coincide with the normal rotating frame z axis, resulting in *tilted axis precession*. The scaling factor, α_1, scales the rate of precession about this axis, and a small scaling factor will result in poor spectral resolution. Hence the importance of maximizing α_1 while minimizing α_2.

APPENDICES

Two of the appendices of the book develop themes that are relevant to this chapter, but which are not required in later chapters. Appendix C introduces the topic of *Liouville space* and how this is applied to problems involving relaxation and site exchange. These are rather specialist topics and understanding the theory behind them is not necessary in most applications of solid-state NMR. Appendix D gives an introduction to the numerical simulation of solid-state NMR spectra and experiments. Simulation is a useful way of exploring the ideas developed in this chapter and is often necessary in the development of new techniques and analysis of complex problems in solid-state NMR. The software used for such simulations is, however, subject to change and so the practical description of setting up NMR simulations is confined to a distinct appendix.

FURTHER READING

INTRODUCTORY TEXTS: GENERAL NMR

"*Understanding NMR spectroscopy*", J. Keeler, John Wiley & Sons Ltd. (2005), ISBN 978 0 470 01786 9.
"*Spin dynamics*", M.H. Levitt, John Wiley & Sons Ltd. (2008), ISBN 978 0 470 51117 6.

MORE ADVANCED TEXTS: GENERAL NMR

"*Principles of nuclear magnetic resonance in one and two dimensions*", R.R. Ernst, G. Bodenhausen & A. Wokaun, Oxford University Press (1990), ISBN 0 19 855647 0.
"*Tensors and Rotations in NMR*", L. J. Mueller, Concepts in Magnetic Resonance Part A, **35A** (2011) 221.

MORE ADVANCED TEXTS: SOLID-STATE NMR

"*Introduction to solid-state NMR spectroscopy*", M.J. Duer, Blackwell Publishing Ltd. (2004), ISBN 1 40510 914 9.
"*Multidimensional solid-state NMR and polymers*", K. Schmidt-Rohr & H.W. Spiess, Academic Press (1994), ISBN 978 0 126 26630 6.

CHAPTER 5

Going Further with Spin-$\frac{1}{2}$ Solid-State NMR

5.1 INTRODUCTION

The basic principles of solid-state NMR for dilute spin-$\frac{1}{2}$ nuclei, such as ^{13}C and ^{15}N, were set out in chapter 3. This chapter first considers how the basic techniques of Hartmann–Hahn cross-polarization (CP) and high-power proton decoupling need to be modified under the conditions of high magnetic fields and fast magic-angle spinning (MAS) rates which are now available and which are increasingly important for the study of complex systems, such as microcrystalline proteins.

Although straightforward one-dimensional NMR is usually the most efficient route to solving chemical problems via solid-state NMR, there are inevitably occasions where more "sophisticated" techniques, such as 2D NMR, provide vital information that cannot be obtained from simple 1D spectra. This chapter discusses more complex experiments involving multidimensional NMR and/or homonuclear decoupling techniques (used to improve ^{1}H resolution). The focus here is on experiments that provide answers to essentially qualitative questions, such as "Is site A close to site B?" The more demanding quantitative experiments, for example "What is the distance between site A and B?", are discussed separately in chapter 8.

5.1.1 SPIN-$\frac{1}{2}$ NMR AT HIGH MAGNETIC FIELDS

To the solution-state NMR spectroscopist, a specific section on spin-$\frac{1}{2}$ NMR at high magnetic fields may seem unnecessary, since the effects of increased magnetic fields on spectral resolution and sensitivity are relatively straightforward. However, the question of how linewidths and resolution (and in consequence sensitivity) vary with increasing magnetic field is actually quite subtle in solid-state NMR. Linewidths in spin-$\frac{1}{2}$ NMR are discussed in some detail in section 5.2, but some of the key issues are illustrated in figure 5.1, which shows the ^{13}C spectrum of a solid form of the steroid drug finasteride. The linewidths vary considerably across the spectrum, with some peaks being so sharp that they exhibit truncation artifacts (see section 3.3.2), while the methylene (CH_2) signals in particular are much broader. These differences can be explained by the local strength of the $^{13}C,^1H$ dipolar couplings. Decoupling ^1H from ^{13}C is relatively easy when these local fields are weak, for example for quaternary sites and carbons affected by dynamics, such as methyl groups. By contrast, methylene resonances are particularly difficult to decouple effectively due to the strong homonuclear coupling between the geminal protons, leading to noticeable broadening. Hence a trade-off needs to be made between extending the acquisition (to limit truncation of the narrow lines) and increasing the strength of the decoupling RF field (to narrow the resonances).

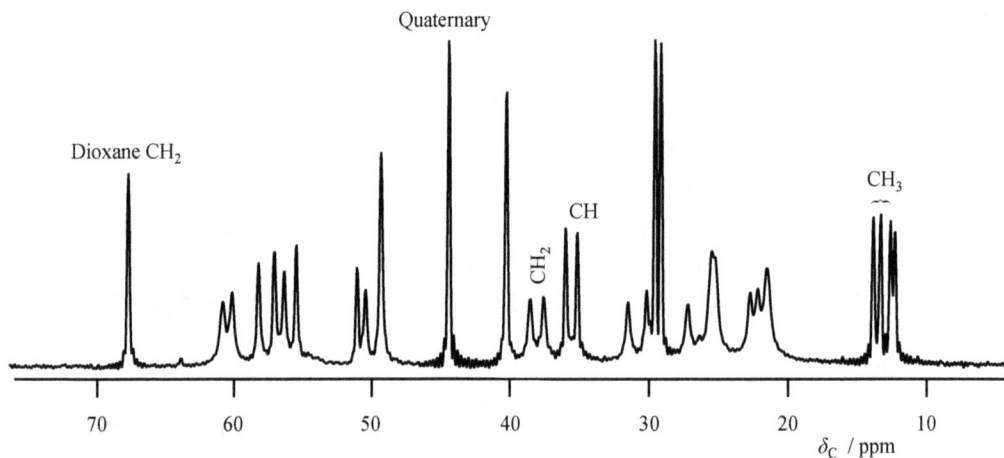

Figure 5.1. Low-frequency section of the ^{13}C spectrum of a solvate form of finasteride with 1,4-dioxane illustrating the interaction between decoupling and widths of selected lines. Two-pulse phase-modulated decoupling (see section 5.1.2) was used during the 30 ms acquisition time (RF nutation rate: 57 kHz). The methyl, quaternary, and dioxane signals are slightly truncated, leading to "sinc-wiggle" artifacts (see section 3.3.2), while the CH_2 resonances are markedly broader than the CH resonances. The resonance of the dioxane is sharp due to the high mobility of the dioxane molecules. (MAS rate: 8.5 kHz. ^{13}C NMR frequency: 125.7 MHz.)

Provided the efficiency of decoupling can be maintained, working at higher magnetic fields is expected to improve both resolution and sensitivity for dilute spins such as ^{13}C. However, obtaining the expected improvements is not as straightforward as it might seem. The scaling of chemical shift anisotropies (CSAs; in frequency units) with the NMR frequency means that the spectra at higher magnetic fields (say over 9.4 T) can become cluttered with excessive spinning sidebands and centerband intensities reduced.[1] Hence MAS rates need to be increased. This is not itself a problem, since the improvements in sensitivity from working at higher magnetic fields allow smaller, faster-spinning rotors to be used. However, as MAS rates exceed ~8 kHz, the implicit assumption that the spinning rate is much smaller than the short-range dipolar couplings (homonuclear and heteronuclear) begins to fail, and key components of the CPMAS experiment, Hartmann–Hahn CP, and continuous-wave (CW) decoupling, become inefficient. The following sections discuss how more sophisticated decoupling and CP schemes allow the full potential of higher magnetic fields to be exploited.

5.1.2 ADVANCED HETERONUCLEAR DECOUPLING

If the effects of MAS can be ignored, ^{1}H decoupling is straightforward in solids. In hand-waving terms, the strong network of dipolar couplings between the spins means that it is sufficient to apply a continuous pulse of RF irradiation to the middle of the ^{1}H spectrum in order to obtain a decoupled spectrum of the dilute spins. The width of the ^{1}H static spectrum (typically ~50 kHz for rigid organic solids), which is due to these same dipolar couplings, means that relatively high RF nutation rates are required in order to obtain useful resolution. However, provided sufficient RF power is used, simple CW irradiation is generally very effective. This contrasts with the situation in solution-state NMR, where the interactions to be decoupled (i.e., J_{CH} couplings) are much weaker. This allows much lower power RF irradiation to be used, but means that the sequences must be effective over the entire width of the ^{1}H spectrum despite the relatively low RF nutation rates. Hence solution-state NMR involves sophisticated phase and/or amplitude modulations to achieve so-called broadband decoupling.

As the MAS rate increases, however, the ^{1}H spectrum begins to break up into spinning sidebands (see section 2.7). Under these conditions, simple CW decoupling starts to lose its efficiency. Since linewidths in dilute-spin NMR are often limited by decoupling efficiency (see section 5.2), this means that lines begin to broaden significantly as the spinning rate increases. The difficulty of maintaining decoupling efficiency at faster spinning rates means, somewhat counterintuitively, that increasing MAS rates tend to *degrade* rather than improve resolution for dilute spins!

A full understanding of ^{1}H decoupling in solids remains a formidable challenge due to the number of interactions involved; homo- and hetero-nuclear dipolar couplings, CSAs and isotropic shifts all influence decoupling efficiency. Fortunately a number of solutions have been found that largely offset the unwelcome effect of MAS on CW decoupling. The simplest of these is the TPPM (*two-pulse phase modulation*) scheme, which involves a continuous train of ^{1}H pulses of constant duration but whose phase switches between two values (see figure 5.4). Empirically it is found that the optimal tip angle of the pulses is generally close to ~170° while the phase modulation, ϕ, is typically of the order of 10–30°. The exact optimal conditions depend on experimental parameters such

---

[1] Spinning sideband suppression techniques, as discussed in section 3.3.6, are only a partial solution as they cancel out the sidebands without increasing centerband intensity.

as MAS and RF nutation rates, but are transferable between similar samples, that is, the decoupling can be optimized on a setup sample and then applied to other samples *under the same conditions*.

These effects are illustrated in figure 5.2. Increasing the strength of the decoupling, by increasing the ^{1}H nutation rate, systematically decreases the ^{13}C linewidths and improves spectral resolution. Replacing CW decoupling with TPPM also leads to significant resolution improvements, and the combination of TPPM decoupling with the higher RF power leads to a much improved spectrum (figure 5.2(d)). As noted above, methylene (CH_2) resonances, such as the one marked with an arrow, are typically very sensitive to the efficiency of the decoupling; the distinctly triangular lineshape of the CH_2 resonances in the spectra of figure 5.2(a–c) is a typical sign of inadequate decoupling.

Other decoupling schemes that have proved popular for solids are SPINAL64 (a more complex modulation, but one that often outperforms TPPM and is simple to optimize) and XiX decoupling

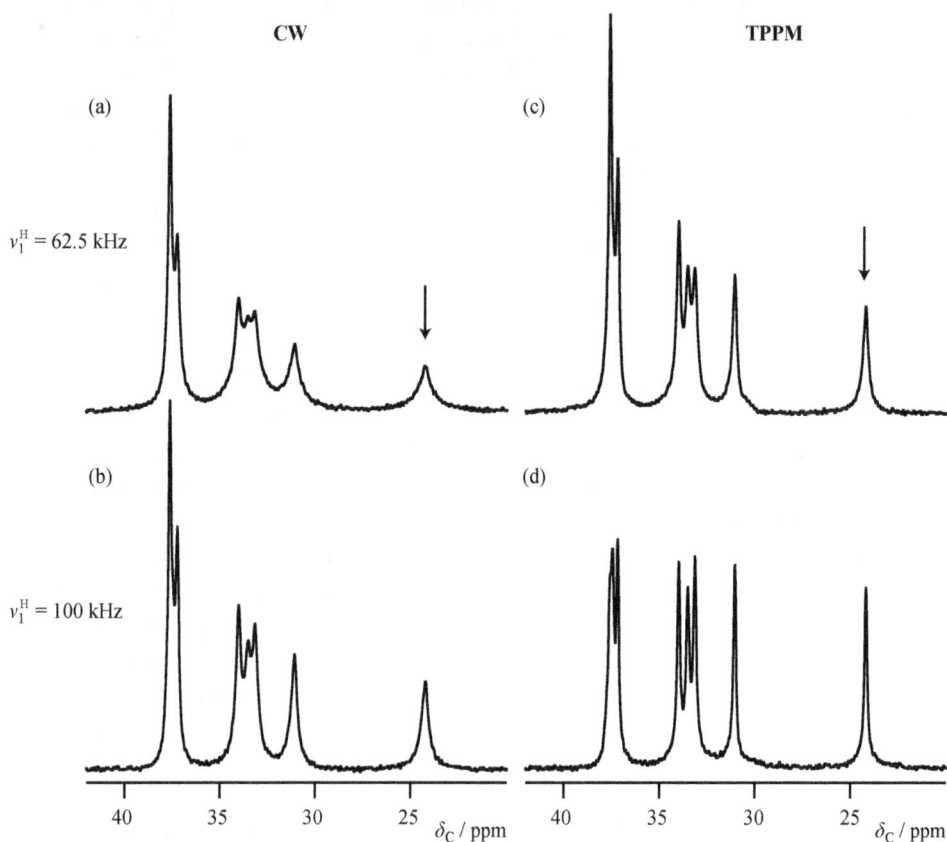

Figure 5.2. Illustration of the effects of ^{1}H decoupling on the ^{13}C NMR spectrum of powdered cortisone, using two different ^{1}H RF nutation rates and continuous-wave (CW) versus two-pulse phase modulation (TPPM) decoupling. (MAS rate: 5 kHz. ^{13}C NMR frequency: 75.40 MHz.)

(which again has a single optimization parameter and often outperforms other decoupling schemes at spin rates above ~30 kHz). However, the development of better heteronuclear decoupling schemes for solids is an area of active research and so "best practice", especially at fast spin rates, can be expected to evolve in the medium term. See Further reading for more information on these sequences and heteronuclear decoupling in general.

5.1.3 ADVANCED CROSS POLARIZATION

Sideband Matching

Cross polarization (CP) using Hartmann–Hahn matching involves matching the nutation rates of the abundant (H) and dilute (X) spins, $v_1^H = v_1^X$. Hartmann–Hahn CP was originally developed for static (non-spinning) samples, but the same matching condition also works for $^1H/^{13}C$ CP under MAS conditions due to the presence of the ^1H homonuclear couplings; spinning at modest MAS rates has negligible impact on the strong homonuclear coupling network, and optimal polarization transfer still occurs at the simple Hartmann–Hahn matching condition.

At faster spin rates, when the ^1H spectrum is itself breaking up into sidebands, or in isolated H,X spin systems, the MAS must be taken into account explicitly, and the matching condition becomes:

$$v_1^H = v_1^X \pm nv_r \qquad\qquad 5.1$$

where v_r is the MAS rate and n is 1 or 2. In other words, the matching condition is no longer at the "centerband," $n = 0$ (when the ^1H nutation rate is matched to that of X), but shifted to "sidebands" either side of the normal matching condition.

This is illustrated in figure 5.3, which shows the *matching profile* for CP from ^1H to ^{13}C at two different spinning rates. When the sample is spinning at 3 kHz, the profile shows a maximum at the normal Hartmann–Hahn matching condition (around 60 kHz). By contrast, the matching profile at 15 kHz has broken up into sidebands, and the efficiency of the CP at the "centerband" matching condition is very poor. As a result, the RF amplitudes need to be modified so that the match is made on one of these sidebands.

The most immediate impact of having to match on a sideband is that the matching is dependent on spinning rate, that is, it is necessary to optimize the CP conditions at the target spin rate. More perniciously, the matching condition is narrower and more delicate than at slow spin rates. Even the "width" of the matching conditions shown in figure 5.3 gives a misleading impression of the degree of tolerance required when matching. Much of this variation is the result of inhomogeneity in the $\boldsymbol{B}_1$ fields of the RF, which results in the matching condition (particularly for a sideband match) being different in different parts of the sample. The narrowness of the matching condition is largely responsible for the observed decrease in maximum CP efficiency; the largest CP signal at

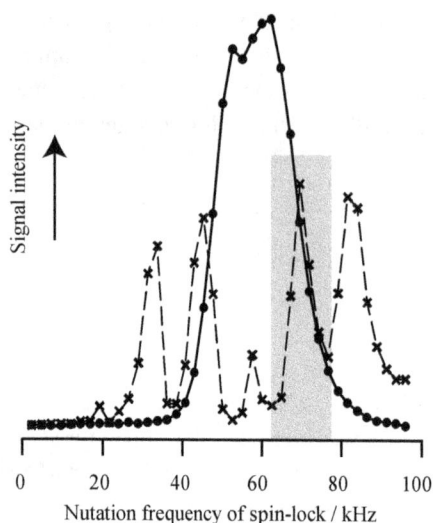

Figure 5.3. A cross-polarization matching profile for hexamethylbenzene at spinning speeds of 3 (solid line) and 15 kHz (dashed line). The profile shows the intensity of the CP signal as a function of the ^{1}H spin-lock nutation rate while the ^{13}C spin-lock field is kept constant. The shaded region indicates a typical range for *ramped CP* (see text).

the faster spin rate is only about half that obtained at the slower rate. The "delicate" nature of the matching condition, and its variation across the sample, makes it increasingly difficult to maintain CP efficiency at increasing MAS rates.[2]

Ramped Cross Polarization

The simplest solution to these problems is to use *ramped CP*. In contrast to simple Hartmann–Hahn matching in which constant amplitude RF is applied to both channels, ramped CP varies the RF nutation rate on one of the channels, sweeping the RF amplitude "through" the matching condition. By arranging the ramp to sweep through a complete sideband in the matching profile (as indicated for the $n = +1$ sideband by the shaded region in figure 5.3), the matching conditions for different parts of the sample will all be achieved at some point in the ramp. Note that the dynamics of ramped CP are subtly different from those of conventional CP due to the time dependence of the RF amplitudes, so the overall duration of the CP period (the contact time) needs to be optimized along with the magnitude of the ramp.[3] In practice, however, ramped CP is considerably more robust with respect to variations in experimental parameters in comparison with conventional matching at fast spin rates. This makes the use of ramped CP particularly valuable in the context of multidimensional NMR involving CP since the reproducibility of CP efficiency is essential in such experiments.

---

[2] Faster MAS also decreases ^{1}H spin diffusion rates (see section 2.7), slowing the rate at which magnetization can be transferred from more distant ^{1}H spins to the X spin. This further diminishes overall CP efficiency.

[3] At first sight, the time dependence of the matching condition in ramped CP might be expected to reduce the effectiveness of polarization transfer at an individual matching condition. In fact, sweeping through a match can (in principle) result in a more complete (*adiabatic*) transfer of polarization from one spin to the other.

Figure 5.4 shows the pulse sequence for a CPMAS experiment suitable for high magnetic field and relatively fast MAS. In comparison with the basic experiment illustrated in figure 3.21, this pulse sequence uses a ramp of the ^{1}H nutation frequency rather than a simple Hartmann–Hahn match, while TPPM is used rather than CW decoupling during the acquisition.

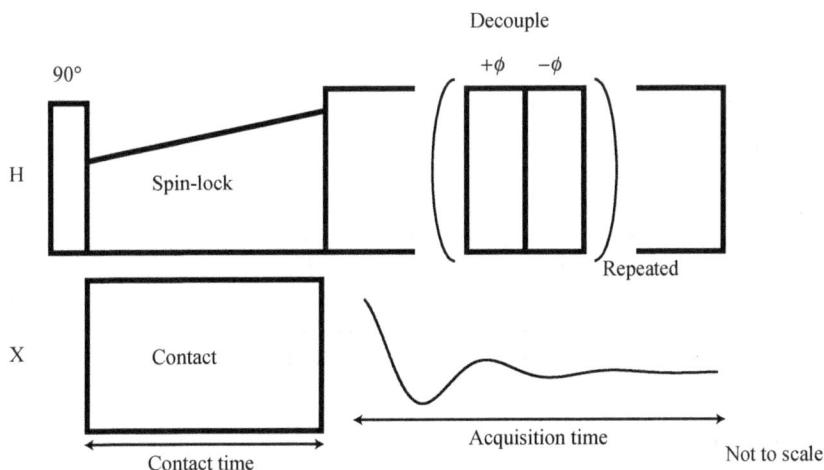

Figure 5.4. Schematic illustration of a typical cross-polarization magic-angle spinning (CPMAS) experiment at high magnetic field and MAS rate, which uses ramped CP and two-pulse phase modulation (TPPM) for ^{1}H decoupling. As shown, TPPM involves alternating the RF phase, with a phase modulation of $\phi \sim 10$–$30°$, with the duration of each pulse being typically a little less than that of a 180° pulse.

5.2 LINEWIDTHS IN SOLID-STATE NMR

The width of resonances in solution-state NMR is not a topic of particular interest since they tend, in the absence of chemical exchange effects, to be more or less independent of sample and magnetic field.

The situation is considerably more complex in solids. For quadrupolar nuclei under MAS, where the resolution is limited by second-order quadrupolar broadenings (see chapter 6), increased magnetic fields offer a clear advantage, since such broadenings (expressed in frequency units) scale inversely with magnetic field, and so the overall resolution increases as the *square* of B_0. However, the resolution for spin-$\frac{1}{2}$ nuclei increases at most linearly with magnetic field, or may not improve at all. To understand why this is the case, the factors that determine spin-$\frac{1}{2}$ linewidths in solids need to be examined in some detail.

Figure 5.5 illustrates how the overall observed lineshape can be decomposed into two fundamental types of contribution: *homogeneous contributions*, which determine the underlying intrinsic linewidth, and *inhomogeneous contributions*. For instance, in solution-state NMR, the homogeneous linewidth would be determined by T_2 (spin–spin) relaxation, while inhomogeneous broadenings

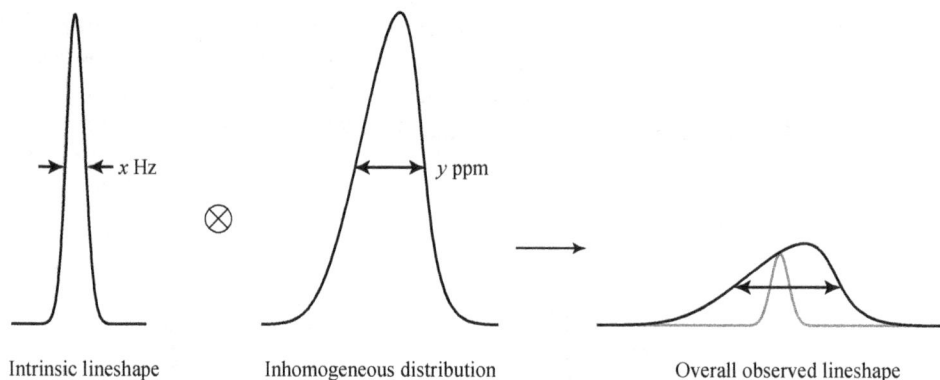

Intrinsic lineshape   Inhomogeneous distribution   Overall observed lineshape

Figure 5.5. Schematic illustration of how the underlying "homogeneous" (intrinsic) lineshape and an inhomogeneous distribution of NMR frequencies combine to give the overall observed lineshape. The intrinsic linewidth is generally dominated by factors, such as couplings, that are independent of magnetic field (and hence expressed in frequency units), while inhomogeneous broadenings scale with magnetic field and are expressed in fractional (parts per million) units. The inhomogeneous distribution spreads the NMR frequencies over a wider range, reducing the maximum signal intensity, while preserving its integral.

would arise from instrumental factors such as inhomogeneity of the magnetic field. This variation of the magnetic field would lead to a distribution of NMR frequencies and hence an overall broadening of the lineshape. NMR in the solid state is subject to additional broadening factors contributing to both the homogeneous and inhomogeneous linewidths.

The homogeneous/intrinsic lineshape is considered first. The limit on the intrinsic linewidth is set by T_2 just as it is in solution-state NMR. However, decay of the NMR signal due to T_2 relaxation is difficult to distinguish from dephasing of the NMR signal due to residual dipolar couplings. For example, ^1H NMR signals do not decay rapidly because of fast spin–spin relaxation but rather from rapid "dipolar dephasing" due to the strong homonuclear couplings. Spin diffusion via the homonuclear couplings rapidly spreads the initial magnetization into other (unobserved) coherences. (This is the basis of the *dipolar dephasing* experiment discussed in section 7.2.4.) Studies in which the rate of dipolar dephasing is modified, for example by changing the MAS rate, suggest that the effects of T_2 relaxation are negligible in comparison with the effects of residual dipolar couplings.[4] Hence the major contribution to the intrinsic linewidth is usually this "residual coherent linewidth" due to dipolar couplings. Heteronuclear decoupling slows down this dephasing process, but this is never 100% effective and dilute-spin linewidths are often limited by the efficiency of the heteronuclear decoupling. The effects of extensive dipolar coupling usually dominate the linewidths

---

[4] It is a common misconception that T_2 relaxation must be fast in solids. Classic NMR relaxation theory assumes that the motional processes occur on a much faster timescale than the NMR frequency and so its results cannot be simply extrapolated to the slow-motion limit; indeed T_2 relaxation must be slow in the absence of motional processes to drive it. At the other end of the scale, the only difference between a plastic crystalline phase and a liquid is the absence of translational motion. Hence it would be expected that $T_1 = T_2$ for such "soft solids" as for the fast-motion limit in solution-state NMR.

of abundant spins themselves, for example, ^{1}H, ^{19}F, and often ^{31}P, even under fast MAS. Techniques for obtaining useful resolution in these cases are discussed separately in section 5.5.

In terms of inhomogeneous contributions, there are again factors that are particular to solid-state NMR. Dissolving a sample means that the NMR responses of all the molecular entities of a given type are identical, and they perfectly sum together to give sharp resonances. For well-crystalline samples (i.e., samples in which a negligible fraction of sites are close to surfaces and/or defects), crystalline symmetry means this should also be true for solid NMR. In many cases, however, solid-state NMR is being used precisely because it can provide information from samples that are not well-crystalline, for example, amorphous solids. In such samples, the local environment of different molecules is thus subtly different, and so a distribution of chemical shifts will be observed for each site rather than a single well-defined shift.

Although poor crystallinity generally reduces the quality of solid-state NMR spectra, it is often the case that careful sample preparation does not eliminate inhomogeneous line broadening. This is typically observed in solids involving π-stacked aromatic rings. The bulk magnetic susceptibility of such samples is strongly anisotropic, since the induced magnetic field associated with the stacked rings (due to the aromatic ring current effect) is strongly dependent on the orientation of the stack with respect to the external magnetic field. This induced magnetic field changes the effective field in the sample, uniformly shifting the NMR frequencies for a given crystallite. (This overall bulk sample effect is distinct from the local changes in magnetic field at the nuclei due to shielding.) The *anisotropy of the bulk magnetic susceptibility* (ABMS) results in shifts that depend on the crystallite orientation. While the isotropic component of the bulk magnetic susceptibility is eliminated by MAS, the ABMS broadening is only partially averaged, and its effect can be significant for certain samples, for example about 1 ppm for hexamethylbenzene (both ^{1}H and ^{13}C resonances).

As illustrated in figure 5.5, the final lineshape arises from the combination (mathematically, a *convolution*) of the intrinsic/homogeneous lineshape with any inhomogeneous distribution of the NMR frequencies. As a result, the resolution of dilute-spin solid-state NMR spectra depends crucially on the nature of the dominant source of linewidth. For samples such as amorphous solids, where the dominant line broadening is inhomogeneous in nature, the lineshape simply scales in width with NMR frequency, and so increased magnetic fields offer no advantage in resolution and only modest improvements in sensitivity. By contrast, linewidths for crystalline samples are often limited by homogeneous residual dipolar coupling effects. For example, steroids generally form well-defined microcrystalline samples with negligible ABMS broadenings. As shown previously in figure 5.1, their ^{13}C NMR spectra typically show extremely narrow signals for non-protonated carbons, while the signals from the CH$_2$ region are significantly broader and are very sensitive to the ^{1}H decoupling efficiency. This is characteristic of linewidths limited by the residual influence of dipolar coupling.

5.3 EXPLOITING INDIRECT (J) COUPLINGS IN SOLIDS

Much of the power and flexibility of modern solution-state NMR flows from the complementarity of information provided by chemical shifts and through-bond (indirect) couplings; the chemical shift is key to resolving and identifying sites with different chemical functionality, while indirect couplings (often loosely termed J couplings) provide direct information on the chemical connectivity between the different sites.

The situation is rather different in solids. The dipole–dipole interactions (particularly those involving ^{1}H) are much larger, and techniques such as MAS and high-power proton decoupling are necessary to suppress the dipolar interactions in order to achieve useful chemical shift resolution. Even after this, the linewidths observed in solids are often too broad to allow other indirect couplings to be directly observed, with the exception of spectra involving heavy atoms where J couplings are naturally large; for example, coupling constants between two directly bonded ^{195}Pt nuclei may be ~9 kHz.

Fortunately, it is often possible to exploit J couplings in the solid state even when they are not resolved in the spectrum. As is illustrated in figure 5.6, a spin-echo experiment (see section 4.4.2) refocuses the evolution under inhomogeneous factors and is often sufficient to reveal evolution

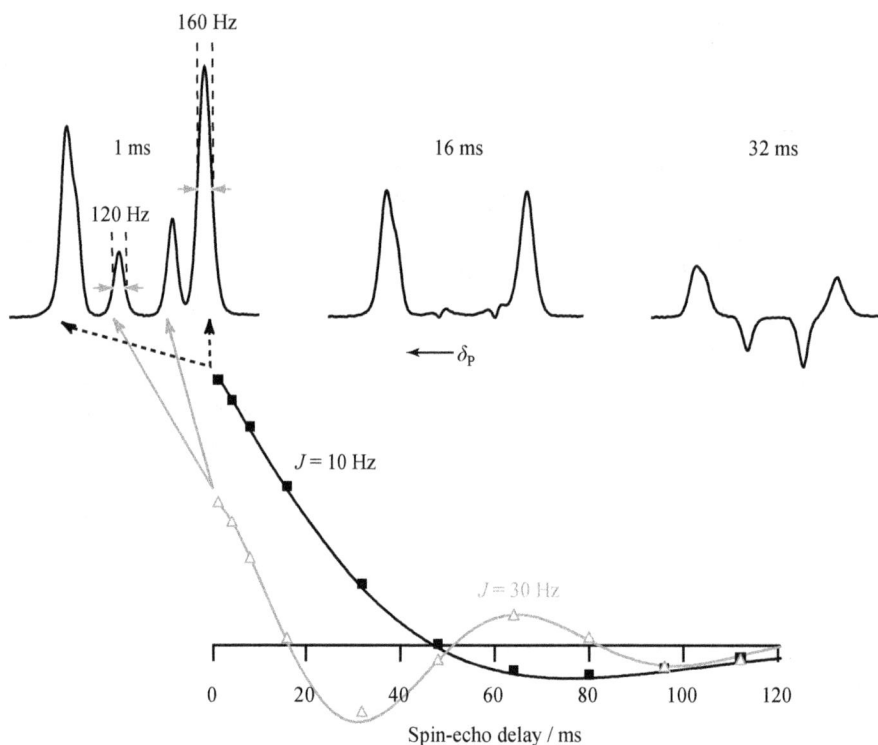

Figure 5.6. ^{31}P NMR spectra of a molybdenum pyrophosphate after spin-echo periods of different durations. The peak intensities are modulated by the $^2J_{PP}$ couplings (which are not refocused by the spin echo), and the solid lines are fits to equation 5.2. These couplings, and hence the corresponding P–O–P angle in the pyrophosphate (P_2O_7) units, are very different for the inner and outer pairs of peaks, providing direct structural information. The spectral linewidths must be dominated by inhomogeneous effects, since couplings as small as 10 Hz are measured despite linewidths being over an order of magnitude larger. (Figure adapted from results published in Lister *et al.*, *Inorg. Chem.* **49** (2010) 2290.)

under J couplings, either homonuclear or heteronuclear.[5] This evolution appears as an oscillation superimposed on the decay due to relaxation and other homogeneous line-broadening factors; that is, the signal for a given spin coupled to one other will fit to:

$$S(2\tau) = A\cos(2\pi J\tau)\exp(-2\tau/T_2') \qquad\qquad 5.2$$

where 2τ is the total spin-echo period, J is the coupling constant, and T_2' is the phenomenological time constant for the decay (the prime distinguishing it from true T_2 spin–spin relaxation). T_2' sets the ultimate limit on the size of couplings that can be measured; if the coupling is significantly smaller than $1/\pi T_2'$, then the NMR signal will have decayed before evolution under the coupling can be observed or exploited. Particularly in non-protonated samples, where dipole–dipole couplings are relatively modest, T_2' values can be relatively long (tens or even hundreds of milliseconds), allowing couplings as small as a few hertz to be observed in some cases.

By contrast, observation of J couplings involving ^{1}H is usually very challenging for solids due to the rapid decay of the ^{1}H coherences under the influence of the homonuclear dipolar couplings. For example, naively omitting heteronuclear decoupling in order to observe a "coupled" ^{13}C spectrum does not reveal the J_{CH} couplings but just gives a very poor quality spectrum. However, by applying strong ^{1}H *homonuclear* decoupling, the dipolar dephasing can be suppressed to the point where the effects of $^1J_{CH}$ couplings (which are typically in the range 125–250 Hz) can be observed. This is illustrated in figure 5.7, which compares the ^{13}C spectra of adamantane under conditions of ^{1}H heteronuclear and homonuclear decoupling. When the ^{1}H homonuclear couplings are suppressed, the CH_2 and CH resonances split into a triplet and doublet respectively

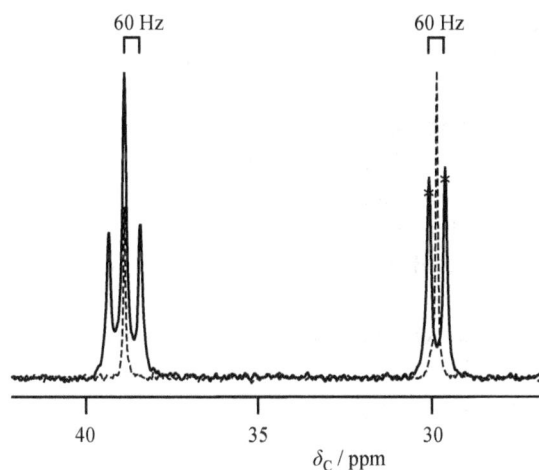

Figure 5.7. ^{13}C cross-polarization magic-angle spinning (CPMAS) spectrum of powdered adamantane acquired with (dashed line) CW ^{1}H heteronuclear decoupling and (solid line) BLEW-12 ^{1}H homonuclear decoupling.

---

[5] The heteronuclear case requires simultaneous π pulses on *both* nuclei.

under the effect of the $^1J_{CH}$ couplings. Note that the apparent couplings (about 60 Hz) are scaled by the homonuclear pulse sequence; this scaling is discussed further in sections 4.7.3 and 5.5.1. As explained in inset 7.8, adamantane is a somewhat special case since molecular motion strongly reduces the magnitude of the dipolar couplings, which means they can be suppressed relatively effectively. However, with efficient decoupling strategies, $^1J_{CH}$ couplings can be resolved in more typical organic solids, allowing an important class of solution-state NMR experiments to be applied to these samples. The resulting correlation experiments are discussed in more detail below.

$^3J_{HH}$ couplings, which are vital to many classic solution-state NMR experiments such as COSY and TOCSY, are at least an order of magnitude weaker than $^1J_{CH}$. There is little immediate prospect of such small couplings being resolved in the context of typical rigid organic solids. However, ^2J couplings are often accessible: $^{2h}J_{NN}$ couplings measured across hydrogen bonds have been used to probe intermolecular interactions, and a number of researchers have investigated the dependence of $^2J_{XX}$ couplings on bond angle in X–O–X units (as in figure 5.6).

5.4 SPECTRAL CORRELATION EXPERIMENTS

5.4.1 BASIC PRINCIPLES OF TWO-DIMENSIONAL NMR

A key class of "advanced" solid-state NMR experiments involves *spectral correlation*, that is, two (or more) dimensional experiments that correlate one-dimensional spectra of the same or of different nuclei. The basic principles of 2D correlation spectroscopy are identical in solid- and solution-state NMR so just the key principles are presented here as a reminder; see Further reading for more detailed discussions.

Figure 5.8 schematically illustrates the basic principles of 2D NMR. An additional (variable) evolution time t_1 is inserted after the initial "excitation," during which the magnetization evolves. An optional mixing time may then, say, allow exchange of magnetization between sites, etc. The NMR signal is then detected during t_2. The evolution during t_1 is thus observed indirectly; crucially, the amplitude[6] of the NMR signal at the start of t_2 is "modulated" by the evolution in t_1 (plus mixing time). The two-dimensional data set is created by repeating the experiment but regularly incrementing t_1 in order to sample the evolution in this *indirect dimension*. Double Fourier transformation of this data set (with respect to both t_1 and t_2) results in a two-dimensional spectrum.

The nature of indirect detection means that careful thought is required when setting up two-dimensional experiments. When the NMR signal is detected directly, there is little penalty for sampling more data points than strictly required, either by using very short dwell times (leading to a larger spectral width than necessary) or acquiring far out into the free-induction decay (FID) where the NMR signal has disappeared. When sampling indirect dimensions, however, the duration of the experiment increases directly with the number of sampling points, and so the dwell time

---

[6] Generally, the *amplitude* of the signal is modulated at the start of t_2, rather than its *phase*. "Phase modulation" usually results in "twisted" lineshapes in 2D spectra, which cannot be phased into all-positive peaks. These subtle, but important, issues in 2D NMR are discussed in more detail in the Further reading.

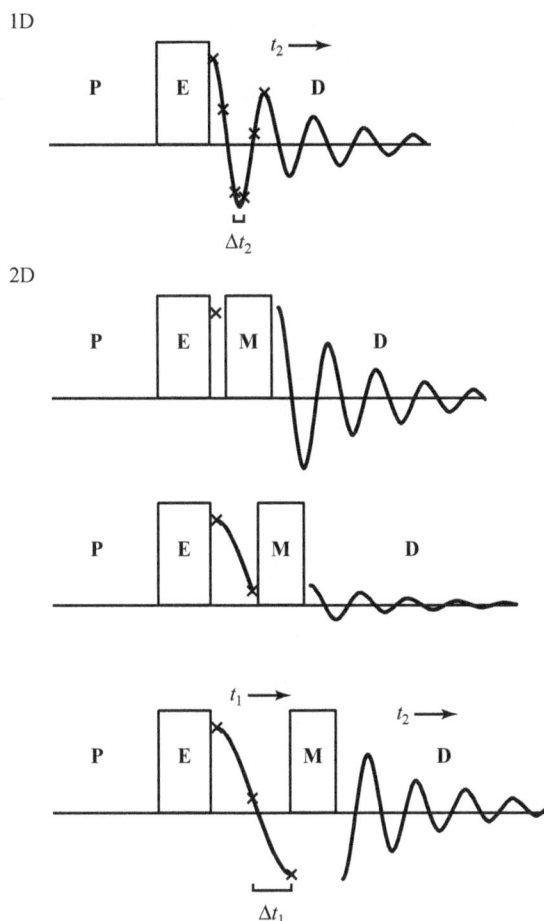

Figure 5.8. Schematic illustration of 1D and 2D NMR pulse sequences. In 1D NMR, the NMR signal is **D**etected directly after an **E**xcitation block (such as a 90° pulse or cross polarization). Detection involves direct sampling of the NMR signal at intervals of Δt_2 (the dwell time). (**P** refers to **P**reparation—e.g., a period for the magnetization to recover toward thermal equilibrium.) In 2D NMR, an additional time period, t_1, is added after excitation, during which the magnetization evolves. The signal is then detected during t_2 after an optional "**M**ixing block" of RF pulses. The t_1 time is regularly incremented to create a 2D data set with each row corresponding to a different t_1 value.

and maximum t_1 need to be chosen carefully to keep total experiment times reasonable. A second difference between direct and indirect dimensions is the treatment of frequencies outside the set spectral width. In the direct dimension, filters largely suppress frequencies (both signal and noise) outside the spectral width,[7] although some frequencies outside the set width may "fold" around

---

[7] In some cases, it can be useful to "open up" the filters and deliberately allow extensive folding in the direct dimension, for example to superimpose all the peaks of a spinning sideband manifold.

(see figure 3.13). By contrast, there is no means to suppress frequencies outside the spectral width set by the sampling rate in indirect dimensions; all the frequencies outside the spectral width will be folded back in. It is then often necessary to set indirect dimension spectral widths to be a multiple of the spinning rate so that centerbands and sidebands are folded on top of each other rather than cluttering up the spectrum.

A final point of contrast between directly and indirectly detected dimensions is the handling of "sign discrimination". As discussed in sections 3.4.1 and 4.4.1, both the x and y components of the precessing magnetization are typically measured when acquiring an NMR signal, allowing frequencies with different signs to be discriminated. There are different schemes of achieving such quadrature detection in indirect dimensions, but a common approach is to acquire *two* FIDs per t_1 increment, which are analogous to the x and y components of the directly detected signal. Details of these schemes, together with other practical details of implementing 2D NMR and processing 2D spectra can be found in the Further reading.

The power of 2D NMR lies in the flexibility of the basic scheme; changing the different "blocks" of the pulse sequence allows a wide variety of spectra to be created. This chapter focuses on using 2D NMR to establish correlations between one-dimensional spectra, for example peaks in the 2D spectrum indicate resonances that are linked by indirect or dipolar couplings. The key step in such correlation experiments is the mixing period (**M**), during which magnetization that has evolved on one spin during t_1 is transferred to another spin prior to detection of the NMR frequencies in t_2. This transfer is usually via J couplings in solution-state NMR experiments, although the dipolar couplings are used indirectly to provide proximity information in NOESY-type experiments (the dipolar coupling being largely responsible for the nuclear Overhauser effect). The strengths of dipolar couplings in the solid state make them a natural first choice for correlation experiments, but J couplings can also be exploited in favorable cases (as previously discussed). Often both dipolar-based and J-based experiments are necessary when correlating complex spectra or when the underlying structure is unknown.

5.4.2 *TRANSFER OF MAGNETIZATION VIA DIPOLAR COUPLINGS*

Dipolar couplings are in general much larger than corresponding indirect (J) couplings, and so it ought to be relatively straightforward to exploit dipolar coupling to transfer magnetization between the spins before the magnetization decays away. Indeed, for ^{1}H spins in typical organic solids, it is sufficient to use a short mixing period of the order of milliseconds; z magnetization is efficiently exchanged between the spins via the dipolar couplings (i.e., spin diffusion). The rate of spin diffusion may be reduced by MAS, but it is still effective even at the highest achievable spinning rates.

Transfer of magnetization via the dipolar couplings becomes more difficult when the coupling network is weaker, as these couplings are then efficiently suppressed by the MAS required for good spectral resolution. For example, the rate of direct ^{13}C, ^{13}C spin diffusion under MAS conditions is generally too slow to be useful even in fully ^{13}C labeled compounds.[8] In addition, the rate of spin

---

[8] Carbon-carbon proximity information must be obtained via efficient ^{1}H spin diffusion. Such experiments, such as *proton-driven spin diffusion*, are only viable for labeled compounds and are not considered further.

diffusion is likely to depend strongly on the sites involved—the greater the frequency difference between the sites (relative to their linewidths), the slower the rate of spin diffusion. This complicates the interpretation of correlation peaks, since their intensities are not simply related to spatial proximity.

In order to get around this problem, RF pulse sequences can be used to *recouple* the dipolar interactions of interest, while using MAS to suppress the unwanted interactions. In the absence of RF, evolution under individual dipolar couplings is refocused over a rotor period, forming rotary echoes at multiples of the rotor period (see figure 3.15 and section 4.6.1). Unless the coupling is sufficiently large for spinning sidebands to be observed, it is not then possible to observe or exploit the coupling. However, addition of RF pulses, synchronized with the rotation, disrupts the averaging by MAS. Fully disrupting the MAS averaging to reintroduce CSAs as well as dipolar couplings is obviously counterproductive, and so the RF pulse sequences are designed to selectively recouple the interactions of interest.

Many such sequences have been devised, for recoupling both homonuclear and heteronuclear interactions. For example, *radiofrequency-driven recoupling* (RFDR) involves applying a train of rotation-synchronized π pulses to recouple (see figure 5.12 for an example). More complex schemes such as POST-C7 are designed to create a well-defined "average Hamiltonian" over the rotation period that depends only on the dipolar couplings. (Note that such sequences often require fixed relationships between the spin rate, v_r, and the relevant RF nutation rate, v_1, for example, the condition $v_1 = 7v_r$ must be met for C7-based sequences.) In practice, the particular sequence used is not critical for the qualitative applications described here.

For most applications, it is desirable for these sequences to be "broadband," that is, the efficiency of the dipolar recoupling should be approximately constant across the spectrum. This ensures that the intensity of any cross peak in a correlation experiment provides a reasonable measure of the strength of the coupling between the pair of spins involved. The magnetization transfer will be fastest between spins with the strongest couplings, but as the mixing period increases, magnetization will equilibrate between all the coupled spins. Such nonselective transfer is usually appropriate for correlation experiments. Site-selective exchange of magnetization is usually more relevant when the goal is measurement of specific internuclear distances. These applications are discussed in section 8.3.

5.4.3 HETERONUCLEAR CORRELATION

Heteronuclear correlation via the dipolar interaction is particularly straightforward since the necessary magnetization transfer can generally be achieved using CP. Figure 5.9 shows such a heteronuclear correlation, or HETCOR, experiment applied to ^{1}H and ^{29}Si. The spectrum shows correlation peaks between the ^{1}H signal at ~17 ppm and the silicate framework, allowing this resonance to be assigned to SiOH groups. No correlations are observed with the signal at ~5 ppm, which is assigned to water.

Obtaining useful correlations from HETCOR and related experiments is more difficult if abundant, strongly coupled spins, such as ^{1}H, are involved. Firstly, homonuclear decoupling (see section 5.5)

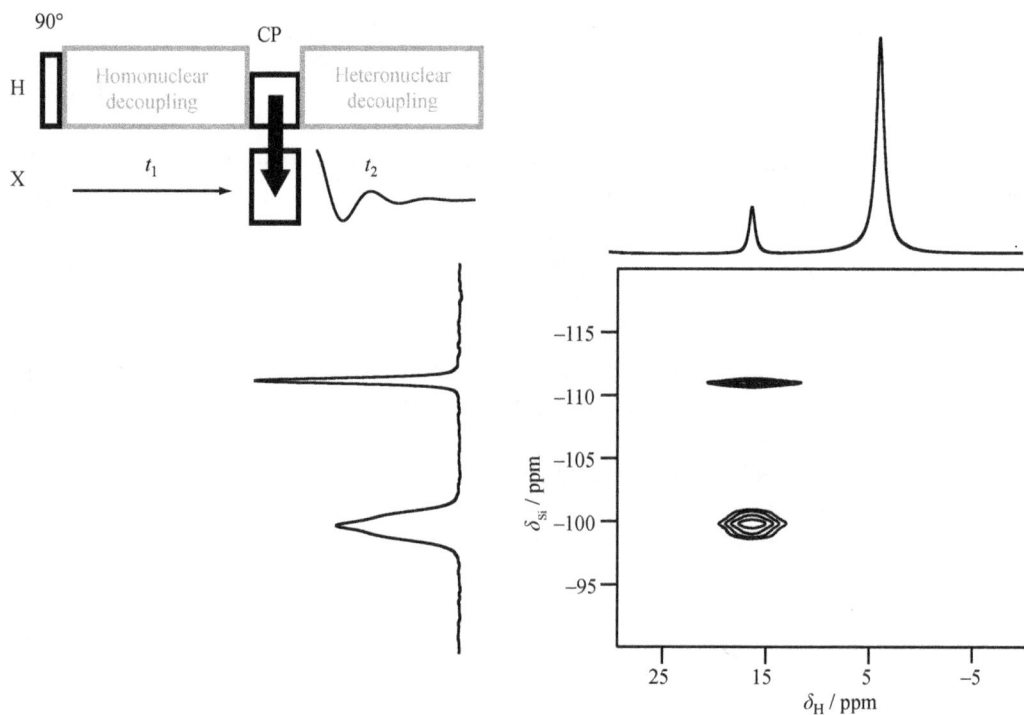

Figure 5.9. (Left) Schematic of an H/X HETCOR experiment: ^{1}H magnetization evolves at the ^{1}H NMR frequencies during t_1, before being transferred (usually via CP to the X spins) where it is detected. (Right) An ^{1}H/^{29}Si HETCOR spectrum of octosilicate acquired in about 14 h. In this case, the ^{1}H spins are sufficiently weakly coupled to each other that ^{1}H homonuclear decoupling during t_1 was unnecessary. The traces at the top and left-hand side are separate 1-D spectra, not projections.

is usually necessary in order to narrow the abundant spin linewidths.[9] More subtly, any spin diffusion occurring during the magnetization transfer step (CP) will compromise the correlation information in the final spectrum. For example, if dilute spin X has a significant (heteronuclear) coupling to abundant spin A but not to B, the HETCOR would ideally only show a cross peak between A and X. If, however, magnetization is exchanged between A and B due to spin diffusion during the mixing time, then some magnetization starting on B will finish on X at the end of the mixing time, leading to a potentially misleading (but sometimes useful) additional *relay* peak between B and X. Keeping the mixing time short limits the build-up of such relay peaks, as well as cross peaks arising from longer-range (weaker) dipolar couplings, but at the expense of overall sensitivity. Indeed, as illustrated in figure 5.10, it is useful to record HETCOR experiments at both short and longer mixing times to distinguish between the correlations that build up quickly (and so correspond to truly short distances)

---

[9] The choice of probe and MAS rotor for heteronuclear experiments is generally determined by the least sensitive nucleus. As this will be the dilute spin (for unlabeled samples), the rotor diameter will tend to be relatively large, limiting the maximum MAS rate that can be used.

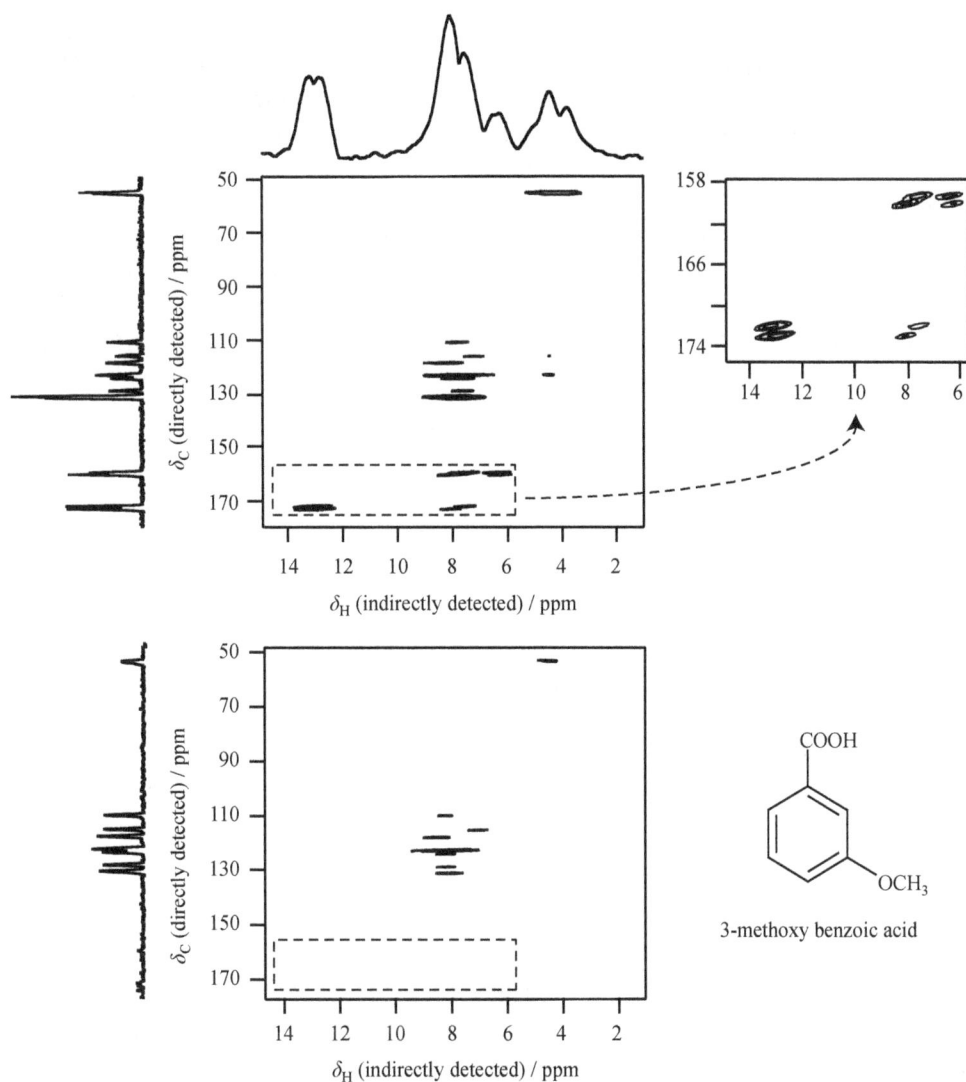

Figure 5.10. ^{1}H/^{13}C HETCOR experiment on 3-methoxy benzoic acid using cross-polarization contact times of (bottom) 100 µs and (top) 400 µs. At short mixing (contact) times, correlations are only observed between protonated carbon sites and their associated H resonances. At longer times, additional correlations are observed, for example, between the carboxylic acid resonance (at ~173 ppm, split by crystallographic inequivalence) and the acid protons (~13 ppm). Experiment time: ~14 h for each spectrum.

and those that build up more slowly. Another technique for reducing the impact of spin diffusion on such correlation experiments is to use a modified CP sequence that suppresses the homonuclear couplings while maintaining polarization transfer via the *heteronuclear* couplings. The simplest of these techniques, Lee-Goldburg (LG) CP, is discussed in the context of homonuclear decoupling in section 5.5.1.

One important application of correlation spectra is to assist the assignment of one-dimensional NMR spectra, by connecting known peaks in one spectrum to unknown peaks in the other. However, for complex organic systems, the ambiguities introduced by ^{1}H spin diffusion often limit the usefulness of the simple HETCOR experiment. Experiments based on indirect couplings may provide more clear-cut answers in these cases. As discussed in section 5.3, it is possible to resolve $^1J_{CH}$ couplings in rigid organic solids by applying homonuclear decoupling to the ^{1}H spins. This allows techniques familiar from the solution state to be adapted to the solid state, such as the heteronuclear multiple quantum coherence experiment illustrated in figure 5.11. Because the correlation information is provided by through-bond couplings and is not compromised by spin diffusion, such 2D experiments provide unambiguous correlations between ^{1}H and dilute-spin spectra. Note how the 1D ^{1}H spectrum of this system would be strongly overlapped and uninterpretable. The 2D correlation is effective in "pulling apart" these overlapped resonances, allowing the chemical shift of the hydrogens attached to each carbon to be determined. On a practical note, however, these experiments involve extended

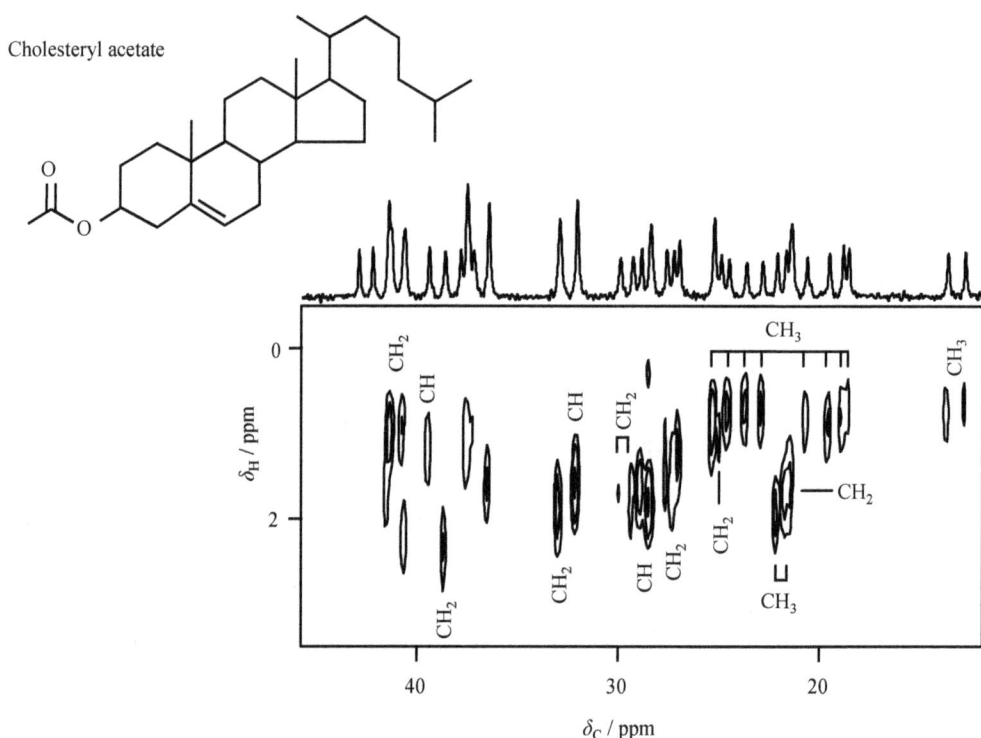

Figure 5.11. Solid-state NMR version of the heteronuclear multiple quantum coherence experiment, which correlates the ^{13}C and ^{1}H NMR spectra via J couplings, as applied to cholesteryl acetate. This alkyl section of the 2D spectrum shows how each of the carbon sites in the ^{13}C spectrum can be connected with the ^{1}H resonance of the attached hydrogens. (Figure adapted from results published in Lesage et al., *J. Am. Chem. Soc.* **120** (1998) 13194.)

periods of continuous homonuclear and heteronuclear decoupling and are technically demanding; careful optimization of the homonuclear decoupling is required to ensure that the $^1J_{CH}$ couplings can be resolved while at the same time avoiding overstressing the probe.

5.4.4 HOMONUCLEAR CORRELATION

Homonuclear correlation experiments are particularly valuable for abundant nuclei such as ^{19}F and ^{31}P, as illustrated in figure 5.12. In contrast to ^{1}H, where specialized homonuclear decoupling is often necessary, simple MAS is usually sufficient for these nuclei to resolve multiple sites, although homonuclear recoupling sequences (such as RFDR or C7) are needed to ensure "broadband" recoupling of the dipolar interactions.

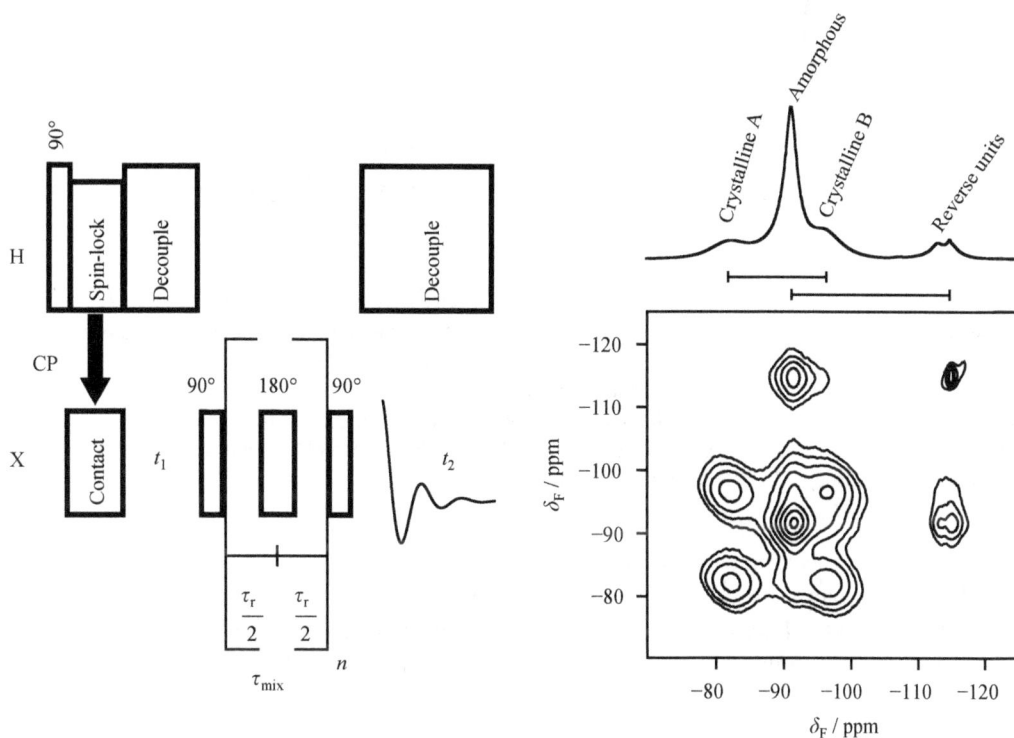

Figure 5.12. (Left) Schematic of a homonuclear correlation experiment using radiofrequency-driven recoupling (RFDR) (dipolar recoupling). X magnetization (generated from CP) evolves during t_1 before the magnetization is transferred to the z direction. Application of 180° pulses in the middle of each rotor period "recouples" the homonuclear dipolar interactions, allowing magnetization to be exchanged during τ_{mix}. The magnetization is then returned to the xy plane for detection. (Right) ^{19}F correlation spectrum of the semicrystalline polymer polyvinylidene fluoride. The 2D plot unambiguously shows that the "defect" (reverse unit) sites are associated with the amorphous regions of the polymer (experiment time: ~18 h).

As for the heteronuclear case discussed earlier, correlations derived from J couplings are a useful complement to dipolar-based homonuclear experiments. Figure 5.13 shows the pulse sequence for the refocused INADEQUATE experiment which is widely used to acquire J-based homonuclear correlation spectra in solids.[10] This experiment is directly analogous to the INADEQUATE experiment in solution-state NMR,[11] but it is more appropriately termed a double quantum/single quantum (DQ/SQ) correlation when dealing with abundant spins such as ^{31}P; the Incredible Natural Abundance part of the INADEQUATE acronym is hardly appropriate for 100% abundant nuclei!

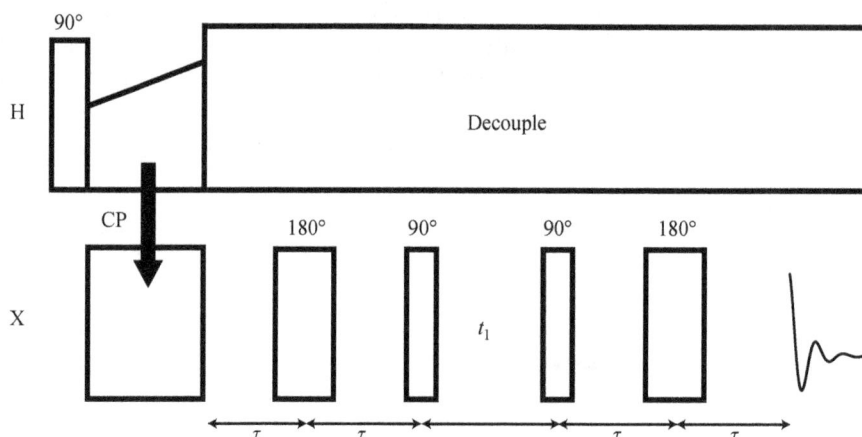

Figure 5.13. The (refocused) INADEQUATE pulse sequence as used for solids. During the time 2τ following CP, the magnetization evolves under the influence of (J) couplings to create double-quantum coherences between coupled spins; the 180° pulse refocuses evolution due to chemical shifts, and synchronizing the τ period to multiples of the rotor period avoids modulation of the signal by MAS. The double-quantum coherences evolve during the indirect acquisition time t_1 before being "refocused" to simple single-quantum coherences prior to signal acquisition. Phase cycling is essential to ensure that the detected signal has passed through double-quantum coherences.

Figure 5.14(a) shows a DQ/SQ correlation spectrum which exploits the $^2J_{PP}$ couplings within pyrophosphate groups, $P_2O_7^{2-}$. Although this coupling is small (typically 10–30 Hz) and is not resolved in the 1D NMR spectrum, sufficient double-quantum coherence can be created to result in a high-quality 2D spectrum. The appearance of the spectra in figure 5.14 is markedly different from

---

[10] "Refocused" refers to the refocusing period after t_1 which converts the anti-phase doublets normally observed in solution-state INADEQUATE experiments into in-phase doublets. The lower resolution in typical solid-state NMR spectra makes the acquisition of anti-phase doublets highly undesirable, since the overlap of adjacent signals with opposite signs results in extensive signal cancellation.

[11] The original solution-state ^{13}C INADEQUATE experiment is rarely used due to its extremely low sensitivity (the signal arises from the tiny fraction of C–C bonds in which both nuclei are ^{13}C). Although an even more demanding experiment in the solid state, it may be the only means to completely assign a ^{13}C spectrum. The situation is considerably eased if samples can be globally enriched in ^{13}C.

(a)

(b)

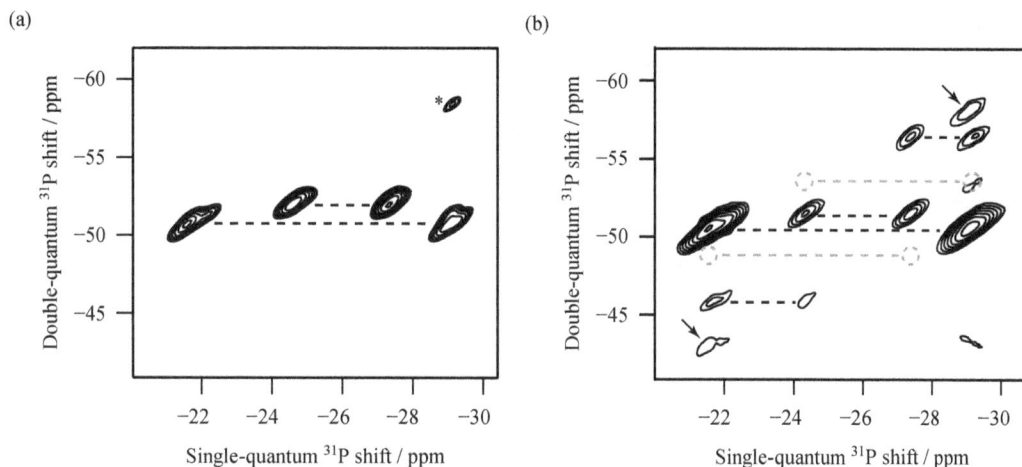

Figure 5.14. ^{31}P DQ/SQ homonuclear correlation experiments on a pyrophosphate material. (a) The correlations in the J-based experiment (refocused INADEQUATE) show which pairs of resonances are part of the same P_2O_7 unit. The diagonal peak (*) is the result of a "rotational resonance" and is not the result of J coupling (see text). (b) The dipolar-based experiment (using POST-C7 recoupling) shows the correlations within the pyrophosphate units, and also weaker longer range couplings which reflect interactions between nearby P_2O_7 units. Dashed circles mark *absent* correlations and arrows mark auto-correlation peaks (see text). (Figure adapted from results published in Lister *et al.*, *Inorg. Chem.* **49** (2010) 2290.)

the previously shown single-quantum/single-quantum correlation experiments since the indirect dimension plots *double-quantum* frequencies. Such coherences involve pairs of spins and evolve at a frequency that is the sum of the individual NMR (i.e., single-quantum) frequencies. A "cross peak" in a double-quantum/single-quantum correlation experiment consists of a pair of horizontal peaks linking two single-quantum frequencies (direct dimension) with the double-quantum frequency (indirect dimension), which is the sum of the two SQ frequencies. The presence of this correlation implies that a double-quantum coherence was created involving the linked spins. Figure 5.14 compares the information content from J- and dipolar-based spectra on the same sample: the J-based spectrum, figure 5.14(a), identifies ^{31}P resonances related by bonding, that is, those part of the same pyrophosphate unit, while the dipolar-based experiment, figure 5.14(b), gives information on spatial proximity between P sites. There is the risk with dipolar-based correlation experiments in particular, that so many correlations are observed that little concrete information can be derived. In these cases, it is often useful to look for *absent* correlations, indicated by dashed circles in figure 5.14(b), which identify sites that must be relatively far apart.

Interpretation of Cross Peaks in Correlation Spectra

In most cases, the presence of a cross peak (or pair of cross peaks in double-quantum/single experiments) can be interpreted straightforwardly in terms of close spatial proximity of the spins involved (dipolar-based experiments) or in terms of chemical bonding (experiments based on indirect (J) couplings). There are, however, some subtle issues that must be borne in mind when analyzing correlation spectra.

Figure 5.15 illustrates schematically the different types of DQ/SQ and SQ/SQ correlation experiments. The information content of the double-quantum and single-quantum variants is essentially the same (and the same recoupling sequences can be used for both). The DQ/SQ spectra are, in general, less crowded since peaks along the diagonal are not usually present (but see the discussion of autocorrelation peaks below), whereas diagonal peaks are always present in simple SQ/SQ correlation spectra.[12] This makes it easier to observe cross peaks between sites of similar NMR frequency.

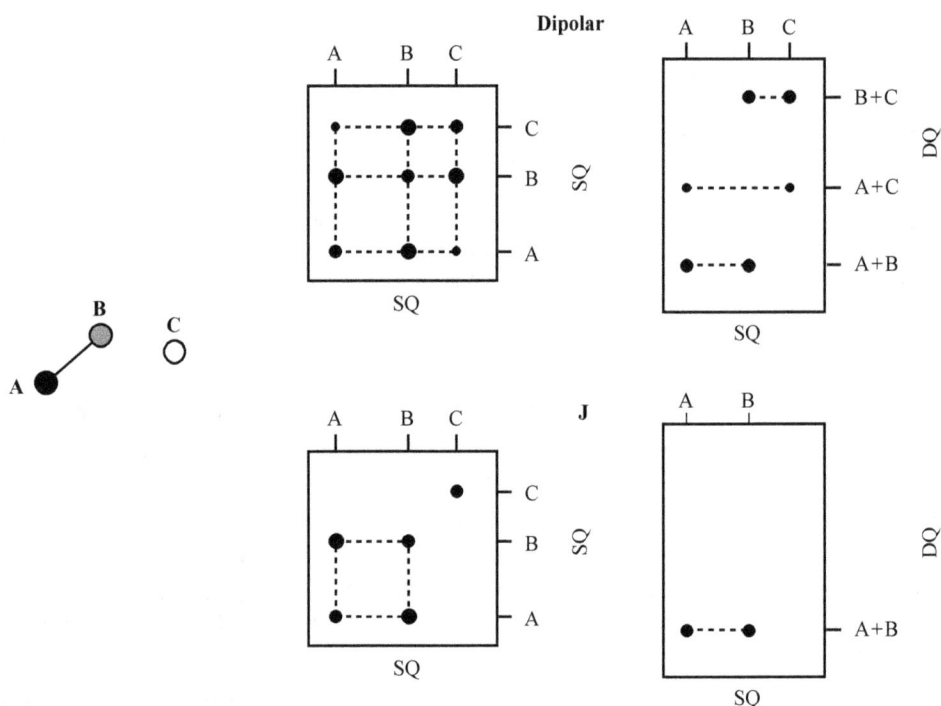

Figure 5.15. Schematic illustration of different correlation experiments for a system of three inequivalent spins, A–C, in which only sites A and B are linked by indirect (J) coupling. The relative size of the spots indicates the strength of the correlation; the stronger the coupling, the more intense the cross peak at a given mixing time.

---

[12] With appropriate phase cycling, it is possible to modify SQ/SQ correlation experiments so that only pathways involving double-quantum coherences are selected. The information content of such double-quantum-filtered experiments is then equivalent to the DQ/SQ experiment, with the only significant difference lying in the form of the spectra.

In contrast to J-based experiments, dipolar-based experiments will tend to show cross peaks between all sites in the same phase as the mixing time increases, as a result of spin diffusion (see section 2.7). While the cross peaks between close spins (A/B and B/C) will build up more rapidly, cross peaks are also observed between A and C, corresponding to a combination of slower magnetization transfer via the weak A,C coupling together with relayed transfer of magnetization between A and C via site B. Since it is also impossible to distinguish between transfer via inter- and intra-molecular couplings, care is required when interpreting correlation peaks in dipolar-based experiments. As illustrated in figure 5.10, it is advisable to obtain correlation spectra at both short and longer mixing times in order to identify the longer-range correlations. Spin diffusion can be used to good effect to distinguish physically distinct components in heterogeneous samples, as previously shown in figure 5.12.

Interpreting J-based experiments is generally more straightforward, since (subject to an important caveat given below) the presence of a cross peak unambiguously identifies sites with a non-negligible indirect coupling, which generally corresponds to pairs connected by a bonding pathway. Care is required, however, in interpreting *autocorrelation peaks*, which involve pairs of sites with the same (isotropic) chemical shift. If the sites are fully equivalent, then no double-quantum coherence will be created via the J coupling (just as coupling between equivalent spins has no visible effect on the conventional NMR spectrum). This is in contrast to the dipolar coupling-based experiment, in which double quantum coherence can be created between both equivalent and inequivalent sites[13], for example, the arrowed peaks in figure 14(b). Note that the equivalence of spins is a somewhat subtle issue in solid-state NMR; this is discussed in inset 5.1.

Inset 5.1. Equivalence in solid-state NMR

In the solid state, two sites are only truly equivalent if they are related by a center of inversion or simple translation; in this case, all the tensors for the NMR interactions are always identical. If, however, there is a symmetry relationship between the sites which is less than this, then only the isotropic components of the NMR interactions are strictly identical. The sites will have the same isotropic NMR frequency, but will be inequivalent at a general crystallite orientation as a result, for example, of CSA. Coupling between such sites will result in a so-called $n = 0$ rotational resonance under sample spinning, leading to unusual NMR lineshapes and unexpected autocorrelation peaks in 2D spectra (see figure 8.7 for an example of a related type of rotational resonance). Such rotational resonance effects can be suppressed by "spinning out" the anisotropic interactions, but they can, with care, be used to elucidate quite subtle questions of crystal symmetry.

---

[13] The reason for the difference lies in the different form of the Hamiltonians for the rank-0 (isotropic) and rank-2 tensor components of NMR couplings (see section 4.2.3). Figure 2.4 shows the spectrum of a pair of equivalent spins coupled by dipole–dipole coupling (pure rank-2); the equivalent spectrum for a pair of identical spins coupled by an isotropic J (indirect dipole–dipole) interaction (pure rank-0) would be a simple singlet!

5.5 HOMONUCLEAR DECOUPLING

The suppression of the strong dipolar couplings between protons in the solid state has been a repeated theme in many of the experiments described above. This section explores this important topic in greater detail.

As discussed in section 2.7 and inset 4.5, the Hamiltonian for a collection of spins coupled by homonuclear dipolar couplings behaves "homogeneously" under MAS; in contrast to other anisotropic interactions such as the CSA, MAS does not break up the static lineshape into a set of sharp spinning sidebands, but rather the sidebands (and centerband) have a finite width that decreases only modestly with increasing spin rate. (To a good first approximation, the linewidths decrease inversely with increasing spin rate.)

The problem of limited resolution is particularly acute in ^{1}H NMR. This is firstly due to the strength of the homonuclear interactions; ^{1}H has a high magnetogyric ratio and individual dipolar couplings are large (e.g., over 20 kHz for the two protons in a rigid CH_2 group), resulting in a large overall homogeneous linewidth (typically ~50 kHz for a rigid solid). Secondly, the ^{1}H chemical shift range is limited to about 15 ppm, which corresponds to a range of frequencies of only 7.5 kHz at an ^{1}H Larmor frequency of 500 MHz. As a result, even the fastest MAS is typically only able to resolve a handful of distinct resonances. The situation is much easier for other abundant spins such as ^{19}F and ^{31}P since overall coupling strengths are lower and the range of NMR frequencies spanned is much larger, particularly at higher magnetic fields. It is usually much easier when working at current magnetic field strengths[14] to use fast MAS rather than homonuclear decoupling for these nuclei. The remainder of this section focuses exclusively on homonuclear decoupling applied to ^{1}H.

Achieving useful resolution in a ^{1}H spectrum requires the effects of the homonuclear couplings to be suppressed by more than two orders of magnitude. This is a demanding requirement, and even small experimental imperfections can become noticeable when such high levels of performance are needed. Moreover, homonuclear decoupling is subject to a variety of experimental factors that are often hard to control or characterize (e.g., variation of B_1 across the sample, finite rise and fall time of pulses). The higher the resolution expected from the homonuclear decoupling, the more critical the experimental set-up.

A significant complication associated with all the widely used decoupling sequences (see below) is that the ^{1}H magnetization precesses about a tilted axis during the homonuclear decoupling, rather than precessing about the z axis as is the case in the absence of RF irradiation (figure 5.16). Moreover, the exact position of the precession axis is not always well defined, as a result of experimental imperfections. Sampling the raw x and y magnetization leads to *quadrature errors* when the signal is Fourier transformed, that is, artifact peaks appear at frequency $-f$ for each genuine signal at frequency f. The usual solution is to perform the experiment off-resonance; the transmitter frequency is placed to one end of the ^{1}H spectrum making it unnecessary to acquire a full quadrature signal. This may degrade

---

[14] However, increased magnetic fields also mean that the contribution of CSAs and isotropic chemical shift differences to the overall Hamiltonian can become substantial for nuclei such as ^{19}F. As a result, decoupling sequences derived for ^{1}H NMR often perform poorly when applied to other nuclei.

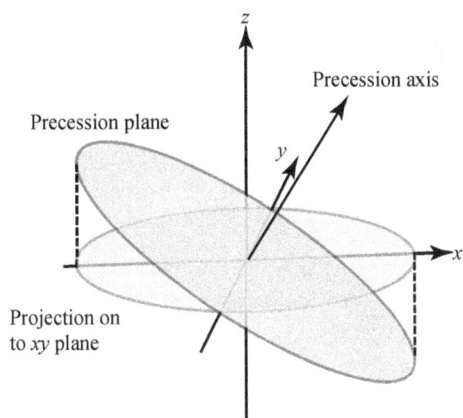

Figure 5.16. The effective Hamiltonian under RF homonuclear decoupling in general corresponds to precession in a plane tilted away from the z axis. This tilting means that the amplitude of the motion has a different projection on the x and y axes, which leads to quadrature artifacts if the magnetization is sampled directly.

the effectiveness of decoupling for peaks furthest from the transmitter, but the limited frequency range of the ^{1}H spectrum means that off-resonance detection is usually preferred.

Homonuclear decoupling does not remove other anisotropic components of the Hamiltonian. Hence if homonuclear decoupling is being applied to obtain resolved ^{1}H spectra, it is usually combined with MAS to suppress the other anisotropic interactions, particularly the CSA. This combination of RF homonuclear decoupling and MAS is known as CRAMPS—*combined rotation and multiple-pulse spectroscopy*.

Figure 5.17 illustrates the resolution that can be achieved on ^{1}H using either fast MAS alone to suppress the homonuclear couplings or using MAS at a moderate rate plus RF decoupling.

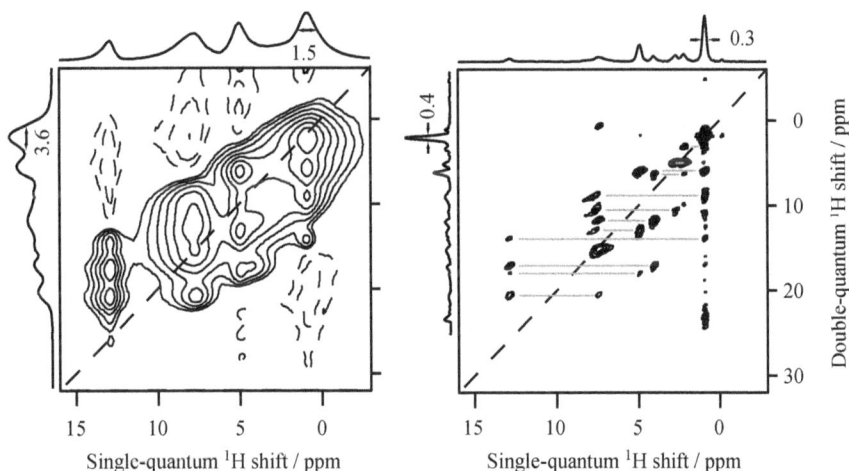

Figure 5.17. DQ/SQ ^{1}H correlation experiments on a small dipeptide using (left) 30 kHz magic-angle spinning (MAS) and (right) 12.5 kHz MAS + homonuclear decoupling in both dimensions. The improvements in resolution are substantial (about a factor of 5). (Figure adapted from results published in Brown *et al., J. Am. Chem. Soc.* **126** (2004) 13230.)

The simple MAS experiment provides some useful correlation information, especially in the less crowded high-frequency regions of the spectrum, while the experiment using RF decoupling provides correlation information for all the ^1H resonances.

5.5.1 OVERVIEW OF HOMONUCLEAR DECOUPLING SEQUENCES

Much of the early development of NMR involved abundant spins in solid samples, and so the problem of reducing linewidths in dipolar-coupled systems was addressed at an early stage. One of the first sequences to emerge was the WHH-4 (or WAHUHA) sequence previously discussed in section 4.7.1. As shown in figure 4.6, the repeating element of WHH-4 involves four 90° pulses separated by delays, with the double-length delay providing a convenient *window* for acquiring the NMR signal. Such *windowed* sequences can be used in the direct dimension of an NMR experiment. In subsequent developments (e.g., MREV-8, BR-24, and others), this basic pattern was extended by "supercycling" to achieve higher quality decoupling and make the sequences more robust with respect to experimental imperfections.

To work efficiently, windowed sequences require both the pulses and the delay periods to be as short as possible without overloading the probe or causing excessive breakthrough of the tail of a pulse with the signal acquisition (see section 3.3.5). Hence, acquiring the NMR signal during homonuclear decoupling requires careful balancing of the experimental parameters. These considerations are largely redundant if homonuclear decoupling is being applied in indirect dimensions, that is, it is being applied during a t_1 period of experiments such as $^1H/^{13}C$ HETCOR. This allows the use of *windowless* sequences that use continuous RF irradiation,[15] for example, the classic BLEW-12 sequence as well as more recently developed sequences such as PMLG (see below) and DUMBO (see Further reading).

Lee-Goldburg (LG) Decoupling

One particularly important windowless sequence is *LG decoupling*. This involves using off-resonance RF irradiation, where the RF offset has been chosen such that the effective nutation axis is at the magic angle with respect to the static magnetic field (see section 3.2.1 for discussion of off-resonance irradiation). That is, if the nutation rate due to on-resonance RF is v_1, the RF transmitter is set $\Delta v = v_1/\sqrt{2}$ off-resonance. The nutation axis is then at

$$\tan^{-1}(v_1/\Delta v) = \tan^{-1}(\sqrt{2}) = \theta_m (54.7°)$$ 5.3

---

[15] Windowless sequences (which use continuous RF) will tend to decouple more efficiently than windowed sequences consisting of isolated pulses. Indeed, recent work that applies homonuclear decoupling in directly acquired dimensions has generally inserted acquisition windows into new windowless sequences (such as PMLG, DUMBO) rather than using classic windowed sequences.

with respect to the z axis. Magnetization precesses[16] about this tilted axis and the homonuclear couplings are suppressed (to first order).

Simple LG decoupling is not particularly competitive for achieving high-resolution spectra since it only suppresses the dipolar line broadening to lowest order. In order to suppress "higher order" terms and hence achieve more effective decoupling, *frequency-switched Lee-Goldburg* (FSLG) decoupling (figure 5.18) alternates the sign of the frequency offset and switches the RF phase. This significantly improves decoupling performance. Note that FSLG is frequently implemented using continuous linear ramps of the RF phase, avoiding the need for frequency switching, in which case it is usually termed *phase-modulated Lee-Goldburg* (PMLG) decoupling.

Although there are close analogies to physical MAS of the sample, it is important to note the differences. In MAS, it is the *spatial* component of the Hamiltonian (the interaction tensor) that is being modulated by the spinning. Hence isotropic components (such as the isotropic chemical shift) are unaffected by the averaging process. In LG decoupling, however, it is, in effect, the *spin* components of the Hamiltonian that are being modulated. "Bilinear" interactions such as the homonuclear dipolar couplings are averaged (to first order). But "linear" terms, such as the chemical shift Hamiltonian or heteronuclear couplings, are also partially averaged—they are scaled by a factor of $\cos\theta_m = 0.577$ for LG decoupling. Hence if a spin would precess at frequency Ω in the absence of RF, it will precess at a frequency $\Omega\cos\theta_m$ under LG decoupling.

As discussed in more detail in section 4.7.3, this scaling of the precession frequencies is an unavoidable consequence of using RF irradiation to suppress a component of the spin Hamiltonian, and it is important to correct for this scaling before extracting chemical shift information from CRAMPS spectra. Although the ideal scaling factor for a given sequence will be known, observed scaling factors are quite sensitive to experimental factors, and it is good practice to measure the actual scaling factor on a known set-up sample.

Figure 5.18. Schematic illustration of the Lee-Goldburg (LG) and frequency-switched Lee-Goldburg (FSLG) decoupling sequences. LG decoupling involves continuous radiation of fixed phase (here x), but with the transmitter off-resonance such that the precession axis is at the magic angle, θ_m, with respect to z. The frequency-switched variant involves simultaneously switching the phase by 180° and swapping the sign of the resonance offset. The switching period usually corresponds to a full 360° rotation about the tilted precession axis.

---

[16] *Precession* rather than *nutation* is used here as the spin evolution over complete cycles of the Lee-Goldburg irradiation rather than the rapid nutation during the LG cycle is being considered. See the discussion of WHH-4 decoupling in section 4.7.1 for more details.

LG irradiation is frequently used as a simple way to reduce ^{1}H spin diffusion during the CP mixing time of HETCOR-type experiments (see section 5.4.3); off-resonance irradiation is used to spin-lock the ^{1}H magnetization along an axis inclined at the magic angle.[17]

Interaction Between Homonuclear Decoupling and MAS

A final important practical consideration is the interaction between homonuclear decoupling and MAS. The classic decoupling sequences were developed for static or slowly spinning samples and their performance strongly degrades when the MAS rotation period becomes comparable to the repeat period of the decoupling sequence. This is a particular problem at higher magnetic fields since faster spinning rates are necessary to spin out interactions such as the CSA, making it difficult to avoid destructive interaction between the homonuclear decoupling and MAS. Hence sequences with short repeat periods such as PMLG/FSLG have distinct advantages over many of the classic decoupling approaches.

5.6 USING CORRELATION EXPERIMENTS FOR SPECTRAL ASSIGNMENT

Assignment of the peaks in an NMR spectrum to chemical sites is straightforward if the resonances are clearly distinguished and can be identified using well-established correlations between isotropic chemical shifts and structural groups, for example using correlation charts derived from solution-state studies. The differences in chemical shifts between the solution and the solid states are usually small, and so assignments of the solution-state NMR spectrum can be useful guides when assigning solid-state spectra. However, ambiguities often arise, either because of spectral complexity or from substantial differences between solution- and solid-state chemical shifts, for example due to intermolecular hydrogen bonding. As discussed in section 8.8.2, first principles calculations can be used to predict chemical shifts in crystalline solids starting from the atomic coordinates in the crystallographic unit cell. Since these account for intermolecular interactions, they are a useful complement to solution-state assignments, where the crystal structure is known. In complex cases, however, assignment of closely spaced peaks may still not be possible, and further experiments may be required.

Useful assignment information for dilute spins in organic solids can often be provided by relatively straightforward one-dimensional NMR methods that probe the local dipolar environment. For example, *dipolar dephasing* experiments, discussed in section 7.2.4, selectively observe sites with weak dipolar interactions to ^{1}H, allowing CH and CH_2 sites to be distinguished from quaternary carbons and methyl groups (where molecular motion strongly reduces the net dipolar couplings).

---

[17] Note that the effective nutation rate under off-resonance irradiation is given by $\sqrt{v_1^2 + (\Delta v)^2}$, which corresponds to an increase by a factor of $\sqrt{3/2}$ at the LG condition. The matching condition needs to be adjusted accordingly.

In more difficult cases, however, two-dimensional correlation experiments may be necessary. For example, ^{13}C/^{1}H HETCOR experiments are useful for assigning ^{13}C resonances if the ^{1}H spectrum can be at least partially assigned. However, homonuclear correlation experiments involving well-resolved spectra provide the most powerful tools for assignment. Thus a ^{13}C INADEQUATE experiment can solve a common problem of assignment in cases where there is more than one molecule in the crystallographic asymmetric unit. In such cases, each chemical type of carbon gives rise to multiple signals, and it is very difficult to determine which individual signal is associated with which individual molecule in the crystallographic asymmetric unit. Thus for two doublets, A and B say, there is no simple way of determining whether the low-frequency A signal is for the same molecule as the low-frequency B signal or not. However, the INADEQUATE spectrum allows such linkages to be made.

Figure 5.19 shows part of the ^{13}C INADEQUATE spectrum of the α-form of testosterone, which has two molecules in the crystallographic asymmetric unit. The horizontal and vertical lines

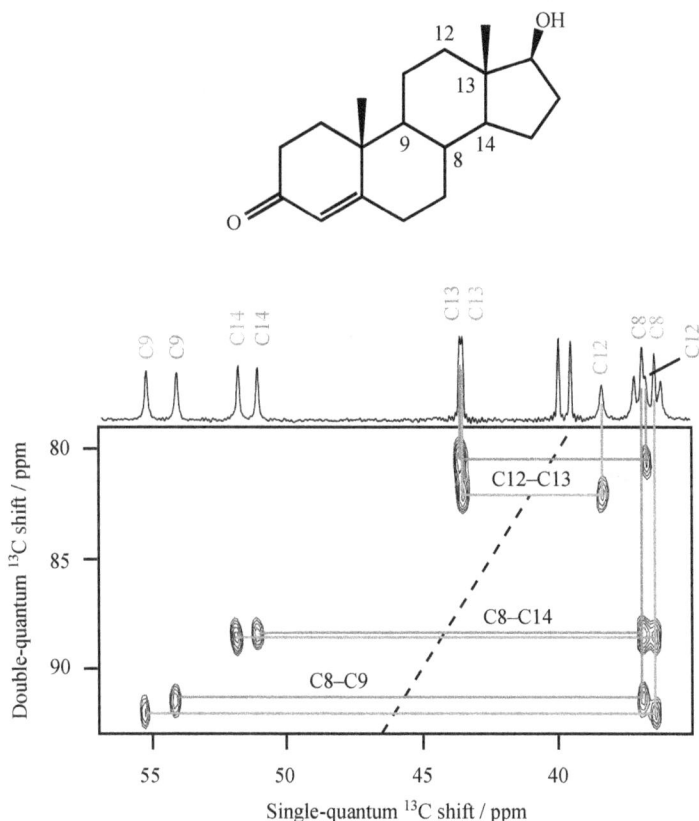

Figure 5.19. INADEQUATE ^{13}C spectrum for α testosterone (see text). This experiment, performed at natural isotopic abundance, required about 3 days of continuous spectrometer time on a 500 MHz instrument! (Figure reproduced, with permission, from Harris *et al.*, *Phys. Chem. Chem. Phys.* **8** (2006) 137).

link signals for bonded carbons, showing that the high-frequency C9 line is in the same independent molecule as the low-frequency C8 line. Of course, connectivity obtained by this method for an organic molecule is interrupted when heteroatoms intervene (a problem that does not occur for testosterone). Similarly tracing out the connectivity is complicated if the pairs are not clearly resolved.

The major weakness of INADEQUATE-style experiments for dilute spins is that it requires observation of signals from adjacent pairs of NMR-active nuclei. For ^{13}C at natural isotopic abundance, this involves only ~0.01% of the molecules for each atom pair. Thus this experiment is highly demanding both of spectrometer time (often requiring several days!) and of hardware—high-power proton decoupling is being applied for relatively long periods. By contrast, such DQ/SQ correlation experiments are straightforward for abundant spins such as ^{31}P (see figure 5.14).

5.7   FURTHER APPLICATIONS

5.7.1   LABELED SYSTEMS

This chapter has concentrated on experimental protocols that are practical for "normal" samples with natural isotopic abundance. However, for some systems, such as large biomolecules, working at natural abundance is not viable due to the complexity of the spectra and the small quantities of samples that can be produced affordably. Hence NMR studies of large molecular systems, such as biomolecules, are usually only feasible after isotopic enrichment. Nonselective or partially selective labeling strategies are used to enhance the concentration of spin-active nuclei, typically ^{13}C and ^{15}N, and, increasingly, deuteration schemes also are being used to dilute the concentration of ^{1}H spins. The latter reduces the strength of the proton–proton coupling network, greatly improving resolution (particularly on ^{1}H itself, but also for the dilute spins), and reduces the problems caused by spin diffusion. Indeed, with a high-quality microcrystalline sample, the ^{13}C resolution is often limited by $^{1}J_{CC}$ couplings, which can no longer be ignored in enriched samples; pulse sequences may need to be adapted to include "homonuclear decoupling" of ^{13}C or to selectively observe individual components of J multiplets.

Isotopic enrichment of other dilute spins has a major influence on the viability of various solid-state NMR experiments. In particular, methods involving double-quantum coherences between spin pairs become much more accessible, as illustrated by the INADEQUATE experiment discussed above. Especially if ^{1}H resolution has been significantly improved via partial deuteration, the suites of techniques and protocols used for the assignment and analysis of proteins in the solution state can then be adapted for the solid state. More information on this important area can be found in the Further reading.

5.7.2   QUANTITATIVE APPLICATIONS

This chapter has focused on applications of 2D NMR to provide resolved spectra that can be used in an essentially qualitative manner to answer chemical problems, for example, tracing connectivity in a framework material or a molecular solid. Once these questions have been answered, however, it is often possible to use NMR to address quantitative problems of structure and dynamics in solid materials.

Chapter 7 discusses how molecular motion affects NMR spectra and introduces a range of NMR experiments that can be used to quantify parameters such as exchange rates and activation barriers. Finally, chapter 8 builds directly on the techniques introduced here in order to provide quantitative information on NMR parameters and structure. In particular, dipolar recoupling, often combined with selective labeling to mark individual sites or pairs of sites, can be used to provide accurate measurements of dipolar couplings and hence internuclear distances.

FURTHER READING

2D AND MULTI-DIMENSIONAL NMR APPLIED TO THE SOLUTION STATE

"*Understanding NMR spectroscopy*", J. Keeler, Wiley (2005), ISBN 978 0 470 01786 9.
"*High-resolution NMR techniques in organic chemistry*", T.D.W. Claridge, Elsevier (2009), ISBN 978 0 080 54818 0.

HETERONUCLEAR DECOUPLING

"Heteronuclear decoupling in the NMR of solids", P. Hodgkinson, *Prog. NMR Spectry.*, **46** (2005) 197. DOI: 10.1016/j.pnmrs.2005.04.002.

HOMONUCLEAR DECOUPLING AND ^{1}H NMR

"High-resolution ^{1}H NMR spectroscopy of solids", P. Hodgkinson, *Ann. Rep. NMR Spectry.*, **72** (2011) 185–223.
"Probing proton–proton proximities in the solid state", S.P. Brown, *Prog. NMR Spectry.*, **50** (2007) 199. DOI: 10.1016/j.pnmrs.2006.10.002.

BIOMOLECULAR APPLICATIONS

"*NMR spectroscopy of biological solids*", Ed. A. Ramamoorthy, CRC Press (2005), ISBN 978 1 574 44496 4.

CHAPTER 6

Quadrupolar Nuclei

6.1 INTRODUCTION

This chapter is intended to give the reader an understanding of the unique characteristics of the quadrupolar interaction, which come into play when it is significant in magnitude compared with the Zeeman interaction, giving *second-order* effects. The reader is also introduced to the special techniques that are frequently necessary to obtain good-quality, well-resolved spectra of quadrupolar nuclei.

As briefly mentioned in chapter 1, quadrupolar nuclides are distinguished by the fact that they have spin quantum numbers greater than $\frac{1}{2}$. Their NMR properties are substantially different from those of spin-$\frac{1}{2}$ nuclides. The width of the spectrum is usually dominated by quadrupolar coupling, the magnitude of which varies very widely. As explained in section 2.6, this effect is caused by the existence of an electric quadrupole moment, eQ, which is characteristic of the nuclide in question. The value of Q ranges from -0.0808 fm^2 for ^{6}Li to 85.6 fm^2 for ^{189}Os and is over 200 fm^2 for isotopes of Hf, Ta, and Re. Values for all stable nuclides (except those of the lanthanides and actinides) are to be found in appendix B. Some of the more important ones are given in table 6.1. Note that quadrupole moments can be either positive (prolate) or negative (oblate),[1] though this is usually of little consequence for NMR. Now the broader a resonance, the lower the sensitivity in terms of S/N, so that it becomes more difficult to resolve signals from the same nuclide in different chemical situations. Since Q strongly affects spectral widths, nuclides with small Q are the easiest to study, other things being equal. The ^{2}H, ^{6}Li, and ^{133}Cs nuclides have particularly small quadrupole moments and their NMR spectra are usually relatively simple. The ratio Ξ/Q^2 (with Q in fm^2 and Ξ in %—see table 6.1) strongly influences the occurrence of second-order effects (see section 6.4) and so it gives a measure of the viability of obtaining good resolution (other factors being comparable).

---

[1] The existence of a quadrupole moment corresponds to a nonspherical (but axially symmetrical) distribution of positive charge within the nucleus. If the charge is greater along the spin axis than perpendicular to it, the distribution is prolate. If the reverse is true, the quadrupole moment is oblate.

Table 6.1. Quadrupole moments of some nuclides

Nuclide	Spin	Q/fm^2	Ratio[a]	Nuclide	Spin	Q/fm^2	Ratio[a]
^{2}H	1	0.2860	187.7	^{35}Cl	$\frac{3}{2}$	−8.165	0.147
^{6}Li	1	−0.0808	2254.1	^{37}Cl	$\frac{3}{2}$	−6.435	0.197
^{7}Li	$\frac{3}{2}$	−4.01	2.417	^{39}K	$\frac{3}{2}$	5.85	0.136
^{10}B	3	8.459	0.150	^{51}V	$\frac{7}{2}$	−5.2	0.973
^{11}B	$\frac{3}{2}$	4.059	1.947	^{59}Co	$\frac{7}{2}$	42.0	0.014
^{14}N	1	2.044	1.730	^{63}Cu	$\frac{3}{2}$	−22.0	0.055
^{17}O	$\frac{5}{2}$	−2.558	2.072	^{81}Br	$\frac{3}{2}$	26.2	0.039
^{23}Na	$\frac{3}{2}$	10.4	0.245	^{93}Nb	$\frac{9}{2}$	−32.0	0.024
^{27}Al	$\frac{5}{2}$	14.66	0.121	^{127}I	$\frac{5}{2}$	−71.0	0.004
^{33}S	$\frac{3}{2}$	−6.78	0.167	^{133}Cs	$\frac{7}{2}$	−0.343	111.5

[a] Magnitude of the ratio of $\varXi$ (the resonance frequency of the reference compound for the nuclide in question as a % of that of the protons in TMS) to Q^2/fm^4.

From this point of view, it can be seen that ^{2}H, ^{6}Li, and ^{133}Cs are expected to give particularly good spectra. It is also clear that ^{37}Cl may be preferred to ^{35}Cl (though the relative isotopic abundances indicate that the latter would give the higher intensity).

Factors other than Q can be equally or more important for spectra of quadrupolar nuclei. In particular, the electric field gradient (EFG) at the nucleus—which is a tensor—plays a strong role. In fact, a quadrupolar spectrum depends on the relevant *quadrupolar coupling constant*, χ (see equation 2.25, where it is defined in frequency units).[2] This represents the anisotropy of the quadrupolar tensor (see section 2.6) and depends on both eQ and eq_{ZZ}, where the latter is the largest (in magnitude) principal component of the EFG.

---

[2] Alternatively, the symbol C_Q can be used for the quadrupole coupling constant, though χ is recommended by IUPAC.

The value of the EFG depends (unlike the quadrupole moment) on chemistry. Simple ions, such as are frequently found for compounds of the alkali metals and halogens, have, in principle, zero EFGs at the nucleus (though in practice, the environment of an ion in a crystal may reduce the symmetry substantially) and therefore give narrow resonances. This fact, combined with the low value of Q for ^{133}Cs, leads to this nucleus being regarded as having pseudo-spin-$\frac{1}{2}$ status! Atoms at sites of cubic, octahedral, or tetrahedral symmetry will also have very low values of χ. For instance, ^{33}S in sulfate ions gives relatively sharp resonances; the narrowest appears to be for ammonium aluminum sulfate dodecahydrate, for which the linewidth is 18 Hz.[3] In other chemical circumstances the EFG can be large, leading to broad lines. Finally, as shown by equation 2.27, there is a strong influence of the spin quantum number on the first-order quadrupolar energy.

As discussed in section 2.6, nuclides with spin quantum numbers, I, which are odd multiples of $\frac{1}{2}$, have relatively sharp *central* transitions (i.e., $\frac{1}{2} \leftrightarrow -\frac{1}{2}$), whereas the other (*satellite*) transitions are spread over a range of frequencies for polycrystalline samples. Each type of transition may be said to give a *subspectrum*. On the other hand, nuclides with integral spin quantum numbers have no central transitions and their spectra are inevitably spread over a range of frequencies. Fortunately, of such nuclei only ^{14}N raises any real problems for the NMR spectroscopist; its quadrupole coupling constants are frequently of the order of 3 MHz and are therefore too large to treat simply.

In general, spectra of quadrupolar nuclei will be influenced by shielding anisotropy and asymmetry as well as the quadrupolar effects themselves. However, in this chapter the interplay between the two tensors will be ignored and the quadrupolar effects treated in isolation. Chapter 8 will discuss some more complicated cases.

6.2 CHARACTERISTICS OF FIRST-ORDER QUADRUPOLAR SPECTRA

Magic-angle spinning (MAS) is commonly used to record spectra of quadrupolar nuclei. MAS can, in principle, average first-order quadrupolar interactions to zero, but in most cases accessible spin rates are insufficient for the result to be a single line for each chemically distinct site. In consequence, it is normal for a substantial number of spinning sidebands to appear. Figure 6.1(a) shows a complete (central and satellite transitions) MAS spectrum of a spin-$\frac{3}{2}$ nucleus with zero asymmetry. The quadrupole coupling constant is only modest in value (150 kHz) so that first-order theory (section 6.3) suffices to explain the spectrum. The strong central transition is the most prominent feature (as for the static case shown in figure 2.9). To record the whole of such spectra, the maximum available spin rate should normally be used (unless the quadrupole coupling constant is small).

Figure 6.2 shows an experimental $I = 1$ (^{2}H) spectrum (for a nearly axially symmetrical situation, as is usual for C–D bonds). The slight lack of symmetry in the ^{2}H spectrum illustrated here probably arises for technical reasons (e.g., digitization limitations), although there may be some effect from shielding anisotropy.

---

[3] T.A. Wagler, W.A. Daunch, M. Panzner, W.J. Youngs & P.L. Rinaldi, *J. Magn. Reson.* **170** (2004) 336.

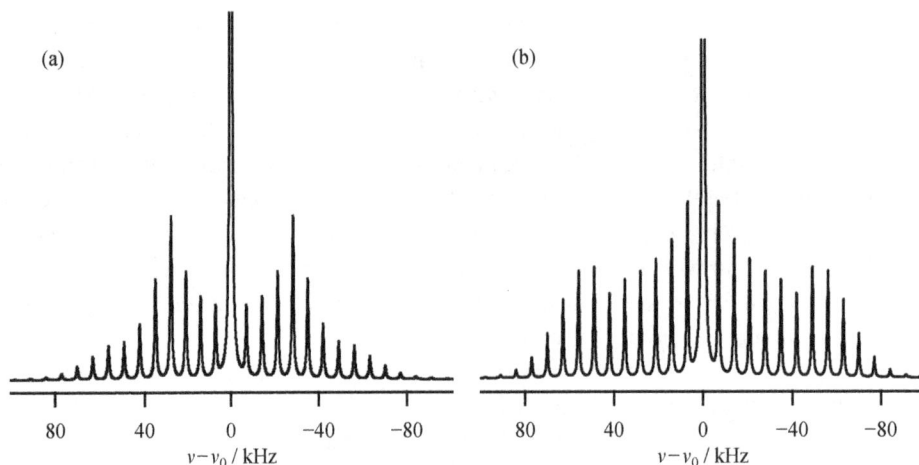

Figure 6.1. Simulated magic-angle spinning (MAS) spectra for a first-order spin-$\frac{3}{2}$ nucleus with $\chi = 150$ kHz: (a) $\eta = 0$, and (b) $\eta = 1$. Shielding anisotropy is ignored. The intensity of the central transition is truncated.

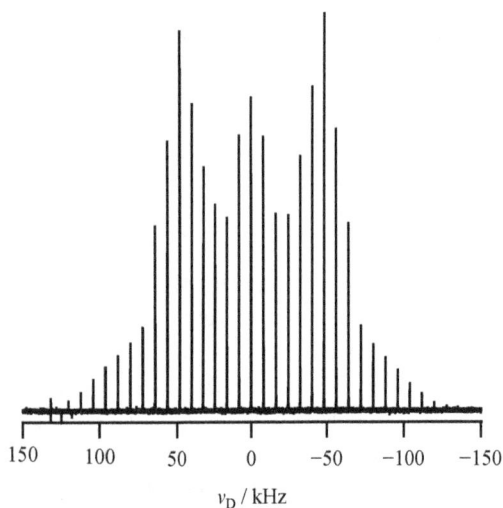

Figure 6.2. Deuterium magic-angle spinning (MAS) spectrum at 7.05 T and a spin rate of 8 kHz for 2,3-d_2-fumaric acid. The quadrupolar parameters are $\chi \cong 160$ kHz and $\eta \cong 0$.

The strong central feature of spectra for nuclides with I an odd multiple of $\frac{1}{2}$ can be readily used to determine the isotropic chemical shift for both static and MAS cases (but see later for complications in cases with large quadrupole coupling constants). For nuclides with integral spin, such a measurement is more difficult and it may be necessary to take the full bandshape into account. However, in both cases, it is feasible to fit the total bandshape to give the quadrupole coupling constant; suitable computer programs are available for simulating and iteratively fitting both static and MAS spectra, in the latter case using the intensities in the spinning sideband manifolds. The results, in turn, give information on the electronic environment of the nucleus and thus on the bonding situation. Quantum theory is often used to compute quadrupole coupling constants from crystal structure information, thus providing a sensitive test for both theory and experiment.

Thus far, it has been tacitly assumed that axially symmetric sites are being discussed. However, in general, the quadrupole coupling tensor is asymmetric. Asymmetries affect the appearance of static spectra (and hence of MAS spectra also). Figure 6.1(b) shows an example for a spin-$\frac{3}{2}$ nucleus with $\eta = 1$, for comparison with the $\eta = 0$ case of figure 6.1(a). The computer programs mentioned above take account of asymmetries and thus these parameters can be derived by fitting static spectra or spinning sideband manifolds.

The accurate setting of the magic angle is particularly important for quadrupolar nuclei because of the frequently very wide range of the resonance. A very small offset from the magic angle is immediately manifested in line broadening for the spinning sidebands of the satellite transitions. In fact, the usual way of setting the magic angle is to optimize the linewidths of spinning sidebands for the ^{79}Br spectrum of KBr. The optimization is best done by observing the duration of the free-induction decay (FID), as shown in figure 6.3.

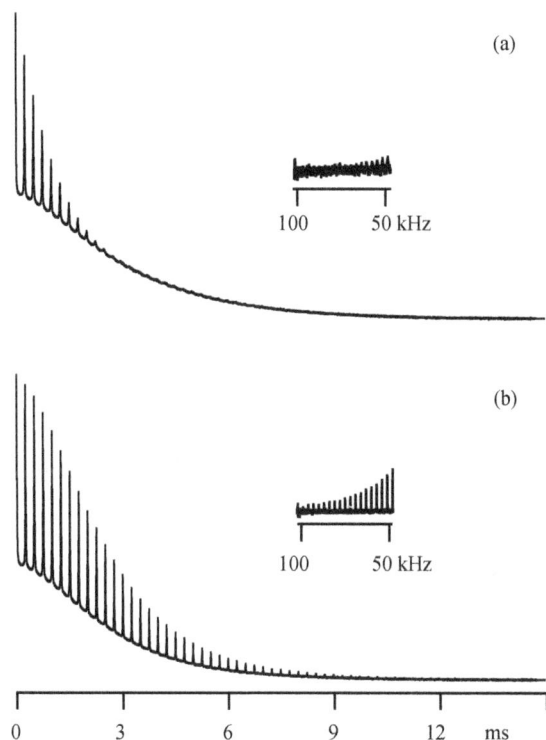

Figure 6.3. Bromine-79 free-induction decays (FIDs) of KBr, spinning at 3 kHz (a) off (by <1°) and (b) on the magic angle at a magnetic field of 9.4 T. The insets show portions of the corresponding spectra away from the centerband.

The total intensity in a given subspectrum (central or satellite transition) is governed by the raising and lowering operators (I_+ and I_-, respectively; see section 4.2.1) and is, therefore, for the transition $m_I \rightarrow m_I - 1$, proportional to:

$$I(I+1) - m_I(m_I - 1) \qquad\qquad 6.1$$

See inset 6.1 for subspectra intensity ratios[4] as a function of I.

Inset 6.1. Intensity ratios for subspectra of different spins	
I	Ratio
$\frac{3}{2}$	$3:4:3$
$\frac{5}{2}$	$5:8:9:8:5$
$\frac{7}{2}$	$7:12:15:16:15:12:7$
$\frac{9}{2}$	$9:16:21:24:25:24:21:16:9$

6.3 FIRST-ORDER ENERGY LEVELS & SPECTRA

Quadrupolar coupling is unique among NMR interactions in that quadrupole coupling constants are frequently several megahertz (and may be up to several hundred megahertz) in magnitude, which is significant in relation to the Zeeman energy. In many cases (especially at high magnetic fields B_0), the Zeeman term is still dominant, so quadrupolar coupling can be seen as mixing the spin states established by the Zeeman interaction. For relatively low quadrupole coupling constants, perturbation theory is applicable. For high-spin nuclei, such theory is adequate even for moderately large quadrupole coupling constants. Perturbation theory allows the contribution of the quadrupolar energy to be expressed as a sum of terms of decreasing significance. When quadrupole coupling constants are very high, a full theoretical treatment will be necessary, but in most cases of interest, second-order perturbation theory is sufficient. Therefore, only the first two quadrupolar Hamiltonian terms will be retained in this book:

$$\hat{H}_Q = \hat{H}_Q^{(1)} + \hat{H}_Q^{(2)} \qquad\qquad 6.2$$

---

[4] These ratios are correct only for nonselective pulses (see subsection 6.7.4).

In many cases (with low quadrupole coupling constants), a first-order perturbation treatment of quadrupolar effects suffices, so that equation 2.27 for the quadrupolar energy is applicable. Such situations have been fully discussed in section 2.6 and readers should digest the information therein before tackling the rest of the present chapter. The most important result is that the energy levels $m_I = \pm\frac{1}{2}$ are affected equally by first-order perturbations and therefore the central transition remains as a single line (independent of the EFG orientation in the magnetic field) at the isotropic chemical shift. However, the first-order approach is often inadequate, and the levels $m_I = \pm\frac{1}{2}$ are shifted to different extents (see figure 6.4) so that the resonance frequency of the central transition becomes influenced by quadrupolar effects (and depends on the orientation of the quadrupolar tensor with respect to B_0). Thus, when the quadrupolar energy is above about 5% of the Zeeman energy, extending the perturbation treatment to second order is required.

Figure 6.4. Energy-level diagram for a spin-$\frac{3}{2}$ nucleus in a magnetic field B_0, showing the effects of zero (left-hand side), first-order (center), and second-order (right-hand side) quadrupolar perturbations, the last leading (in general) to a decrease in the resonance frequency for the central transition.

As mentioned in section 4.2.3, the full Hamiltonian for a coupling interaction between spins I_1 and I_2 is $\hat{H} = \hat{I}_1 \cdot R \cdot \hat{I}_2$, where R is the tensor for the coupling in question. In the case of the quadrupolar effect, only one spin is involved and the Hamiltonian becomes:

$$\hat{H}_Q = \frac{e^2 Q}{2I(2I-1)} \hat{I} \cdot q \cdot \hat{I} \qquad\qquad 6.3$$

As discussed in section 2.6, only two parameters are in principle needed to characterize the magnitude of q in any given situation, namely its anisotropy and asymmetry (η). The chosen parameters for quadrupolar interactions are the quadrupole coupling constant, χ (which is proportional to the anisotropy of q) and the quadrupolar asymmetry (identical to the asymmetry of q). These are defined in the principal axis system (PAS) in equations 2.25 and 2.26.

The anisotropy χ (alternative symbol C_Q) is normally expressed in frequency units but η is dimensionless ($0 \leq \eta \leq 1$). It may be shown that, in terms of these parameters, equation 6.3 becomes (in the quadrupolar PAS):

$$h^{-1}\hat{H}_Q = \frac{\chi}{4I(2I-1)}\left[3\hat{I}_z^2 - \hat{I}^2 + \tfrac{1}{2}\eta\left(\hat{I}_+^2 + \hat{I}_-^2\right)\right] \qquad 6.4$$

where $\hat{I}_+^2$ and $\hat{I}_-^2$ are the raising and lowering operators, respectively (see section 4.2.1).

Before this equation is developed here, the reasons why the results are often mathematically complicated should be understood. The first is that all Hamiltonian terms must be given in the same axis system. However, equation 6.4 is expressed in the principal axis frame of the EFG tensor, whereas the Zeeman energy is determined by B_0, which lies in the laboratory frame of reference. Therefore, a conversion of the spin operators from one frame to the other is necessary, which involves rotations, generally over two angles (see section 4.6 for details). In fact, when MAS is employed, the rotor axis must also be considered and an additional rotation of axes is necessary. Converting Hamiltonian terms by rotations is relatively simple when the tensors are axial, since then only one angle is involved (e.g., that between the Z EFG axis and B_0).[5] Thus, when asymmetry is neglected, equation 6.4 becomes, in the laboratory frame:

$$h^{-1}\hat{H}_Q = \frac{\chi}{4I(2I-1)}\left[\begin{array}{l} \tfrac{1}{2}\left(3\cos^2\theta - 1\right)\left(3\hat{I}_z^2 - \hat{I}^2\right) \\ +\tfrac{3}{2}\sin\theta\cos\theta\left\{\hat{I}_z\left(\hat{I}_+ + \hat{I}_-\right) + \left(\hat{I}_+ + \hat{I}_-\right)\hat{I}_z\right\} + \tfrac{3}{4}\sin^2\theta\left(\hat{I}_+^2 + \hat{I}_-^2\right) \end{array}\right] \qquad 6.5$$

This is the origin of the familiar $3\cos^2\theta - 1$ factor, which is averaged to zero by MAS (though proper consideration involves transformation to the axes of the rotation). However, the angular dependence in the second and third terms within the square brackets cannot be averaged by MAS and is the source of difficulties for high-resolution spectra of quadrupolar nuclei, as discussed in the next section. Significant additional complications result if the quadrupole tensor is asymmetric, since three angles are then needed to describe rotations of axes.

The second reason for complications lies in the nature of the spin operators. Terms such as $3\hat{I}_z^2 - \hat{I}^2$ are *secular* (i.e., affect the energies of states but do not mix states significantly); application of such operators gives *first-order* energy changes. Terms such as $\hat{I}_+^2$ and $\hat{I}_-^2$ (see section 4.2.1), on the other hand, mix states (in the sense that the wavefunctions are significantly changed), requiring substantial mathematical manipulation. This gives rise to *second-order* contributions to the energies.

In what follows in this section and later sections, the text will deal with various relatively simple situations, starting with the effects of asymmetry on first-order spectra.

---

[5] Since the rotor and laboratory frames involve unique directions, only the unique angle between them is involved in changing operators between these frames.

QUADRUPOLAR NUCLEI • 149

When asymmetry is included, the value of q_{zz} in the laboratory frame is found to be:

$$q_{zz} = \tfrac{1}{2}(3\cos^2\theta - 1 + \eta\cos2\phi\sin^2\theta)q_{ZZ} \qquad 6.6$$

(Capital letters for the axes indicate the principal components, as usual.) If only the secular terms are retained (i.e., at relatively low quadrupole coupling), the truncated quadrupolar Hamiltonian becomes:

$$h^{-1}\hat{H}_Q^{(1)} = \frac{\chi}{8I(2I-1)}\left[3\hat{I}_z^2 - \hat{I}^2\right]\left(3\cos^2\theta - 1 + \eta\cos2\phi\sin^2\theta\right) \qquad 6.7$$

Application of the Hamiltonian 6.7 gives the first-order quadrupolar correction to the energy, including the effect of asymmetry, which has been discussed in chapter 2 (see equation 2.27). Equation 2.28 expressed the transition frequencies when the asymmetry is zero, while figure 2.9 showed an experimental spectrum for a static sample of a zero-asymmetry case. Equation 6.8 gives the transition frequencies when there is non-zero asymmetry of the EFG

$$\nu_Q^{(1)} = \frac{3(2m_I - 1)}{8I(2I-1)}\left(1 - 3\cos^2\theta - \eta\cos2\phi\sin^2\theta\right)\chi \qquad 6.8$$

Figure 6.5 shows schematically the effect of asymmetry on the static powder pattern for a spin-$\frac{3}{2}$ nucleus. The two subspectra for the transitions $\frac{3}{2} \to \frac{1}{2}$ and $-\frac{1}{2} \to -\frac{3}{2}$ are indicated. In all cases the spectrum is symmetrical (in the absence of shielding anisotropy). Thus the subspectra for

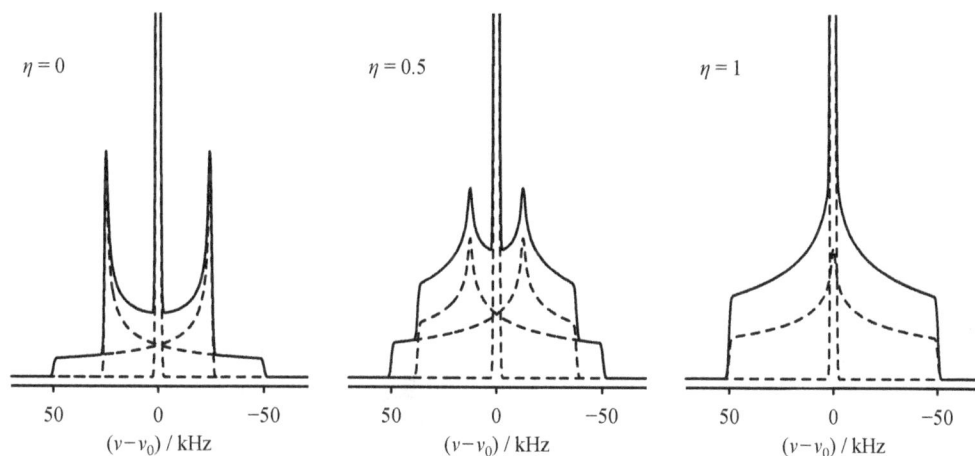

Figure 6.5. Simulated full spectra for a static sample with $I = \frac{3}{2}$ and varying asymmetry, showing the subspectra (dashed lines). The satellite transitions are identical when $\eta = 1$. The quadrupole coupling constant used is 100 kHz. The central transition has been cut off below its full intensity.

the satellite transitions are mirror images for all values of the asymmetry parameter; they completely overlap when $\eta = 1$.

Magic-angle spinning increases the S/N but results in the appearance of spinning sideband manifolds, which may span moderately large ranges for the subspectra (except for the central transition). Simulated spectra of this type were displayed in figure 6.1.

While equation 6.7 applies equally to nuclei with integral spin quantum numbers and to those with quantum numbers an odd multiple of $\frac{1}{2}$, this chapter will mainly address spectra for the latter. However, section 6.8 will deal briefly with nuclides of integral spin quantum number.

The first-order quadrupolar effect is independent of the applied magnetic field, $\boldsymbol{B}_0$. Therefore, as the applied field increases, the ratio of quadrupolar to Zeeman energy decreases and spectra become less complicated by the second-order effects to be described below. Generally speaking, this results in better resolution (enhanced because chemical shifts increase in frequency units). There are also gains in signal intensity with increase in $\boldsymbol{B}_0$.

6.4   SECOND-ORDER ZERO-ASYMMETRY CASES

6.4.1   TRANSITION FREQUENCIES

When quadrupole coupling constants are significant with respect to the Zeeman term, it becomes necessary to include perturbation to second order, which causes a number of complications to appear in the spectra. This is because terms involving the raising and lowering operators cause mixing of states, as mentioned above, which means that the eigenstates are no longer pure Zeeman states. This leads to second-order energy contributions and thence to shifts in the spectra. The evaluation of the relevant equations can be tedious unless advanced methods (spherical harmonics) are used (see chapter 4), and the results are complicated. Here only an equation for the second-order quadrupolar energy, $E_Q^{(2)}$, of a static sample in an axially symmetrical case will be given:

$$h^{-1}E_Q^{(2)} = \frac{m_I}{2\nu_0}\left(\frac{\chi}{2I(2I-1)}\right)^2\left[\begin{array}{l}\frac{3}{5}\left\{I(I+1)-3m_I^2\right\}+\frac{3}{14}\left\{8I(I+1)-12m_I^2-3\right\}\\ P_2(\cos\theta)-\frac{9}{70}\left\{18I(I+1)-34m_I^2-5\right\}P_4(\cos\theta)\end{array}\right] \qquad 6.9$$

where θ is the angle between the EFG symmetry axis and $\boldsymbol{B}_0$, while ν_0 is the Larmor frequency, and m_I is the spin component quantum number of the energy level in question. In equation 6.9, the three terms in the square brackets are referred to as zeroth-, second-, and fourth-rank contributions. They involve the zeroth-, second-, and fourth-degree *Legendre polynomials*, respectively. The zeroth-degree polynomial is unity and the other two are:

$$P_2(\cos\theta) = \frac{1}{2}\left(3\cos^2\theta - 1\right) \qquad 6.10$$

$$P_4(\cos\theta) = \frac{1}{8}\left(35\cos^4\theta - 30\cos^2\theta + 3\right) \qquad 6.11$$

These two factors arise from the requirement to rotate the spin operators between different axis systems (in this case those of B_0 and the EFG). The second-degree polynomial will be recognized as the one involved in MAS, as discussed in chapter 2.

Since equation 6.9 depends on m_I as well as m_I^2, the energies of the levels $m_I = \pm\frac{1}{2}$ are changed differentially. In fact the proportionality of E_Q to m_I shows that they are affected to equal but opposite extents. Therefore, the splitting between these levels (the central transition) is altered (in contrast to the zero effect of the first-order quadrupolar term)—see figure 6.4. Moreover, the magnitude of this effect depends on the orientation of the EFG as expressed by θ. Therefore, the central transition will be a powder pattern when polycrystalline or amorphous materials are studied.

6.4.2 CENTRAL-TRANSITION SPECTRA: STATIC SAMPLES

The simplest form for the resonance frequency of the central transition ($m_I = \frac{1}{2} \rightarrow m_I = -\frac{1}{2}$) for an axially symmetric case, which can be derived from equation 6.9, is:

$$v = v_0 - \frac{9\chi^2}{16v_0} f(I) \left(1 - \cos^2\theta\right)\left(9\cos^2\theta - 1\right) \qquad 6.12$$

where the spin-dependent factor, $f(I)$, is given by:

$$f(I) = \frac{I(I+1) - \frac{3}{4}}{\left[2I(2I-1)\right]^2} = \frac{2I+3}{(4I)^2(2I-1)} \qquad 6.13$$

Equation 6.12 makes a number of spectral features of the second-order quadrupolar effect on the central transition clear:

- There is a general shift away from v_0 for all orientations except for $\theta = 0°$, $180°$, and $\cos^{-1}\left(\pm\frac{1}{3}\right)$ ($\theta = 70.53°$ and $109.47°$).
- Since this shift depends on the orientation of the EFG in B_0, a microcrystalline sample will give a powder pattern, in contrast to the first-order perturbation (which left the central transition as a single line).
- The center of gravity of the powder pattern will not be at v_0.
- The effect is inversely proportional to v_0 and will therefore become smaller if B_0 is increased.
- The magnitude of the effect decreases with I, other things being equal (see inset 6.2).
- Since the second-order term as a whole is not proportional to $\left(3\cos^2\theta - 1\right)$, MAS cannot eliminate the effect.

Inset 6.2. Dependence of the second-order effect on the nuclear spin quantum number for the central transition

The value of the spin-dependent factor, $f(I)$, is as follows:

I	$\frac{3}{2}$	$\frac{5}{2}$	$\frac{7}{2}$	$\frac{9}{2}$
$f(I)$	$\frac{1}{12}$	$\frac{1}{50}$	$\frac{5}{588}$	$\frac{1}{216}$
	(0.0833)	(0.020)	(0.0085)	(0.0046)

This factor means that, for comparable quadrupolar coupling constants, second-order effects get significantly smaller as I increases, which is perhaps counterintuitive. Thus, second-order broadening of the central transition for ^{87}Sr ($I = \frac{9}{2}$) is only 0.055 that of ^{87}Rb ($I = \frac{3}{2}$) in spite of the fact that the former has the larger quadrupole moment (33.5 vs. 13.35 fm^2).

The upper spectrum of figure 6.6 shows an example of second-order quadrupolar effects: the central ^{27}Al transition for a static sample of microcrystalline solid aluminum acetylacetonate at 7.0 T. The effects of the first three of the points bulleted above can be readily seen in the top spectrum of that figure. The EFG for this compound is nearly axially symmetric ($\eta = 0.16$), with a modest quadrupole coupling constant ($\chi = 3.0$ MHz).

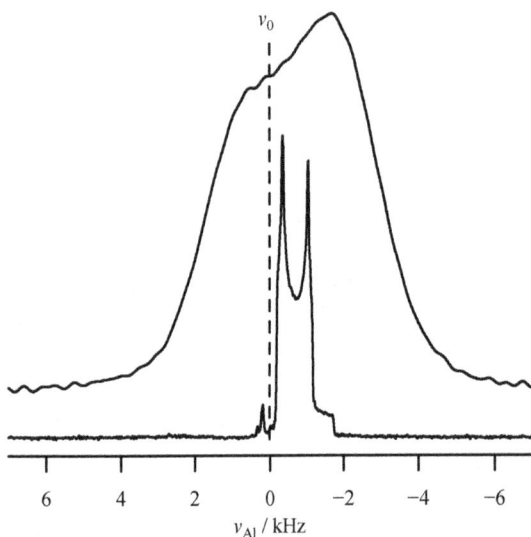

Figure 6.6. Central-transition ^{27}Al spectra of polycrystalline aluminum acetylacetonate. Top: static sample. Bottom: Sample rotated at the magic angle at 9.26 kHz. The position of the isotropic chemical shift is indicated as v_0. The small peak just to high frequency of v_0 arises from the centerband of the inner satellite transition ($\pm\frac{1}{2} \leftrightarrow \pm\frac{3}{2}$).

While equation 6.12 is useful for descriptive purposes, greater insight can be obtained by recasting it into a form obtained by use of Legendre Polynomials (see equation 6.9):

$$v_0^{(2)} = -\frac{\chi^2}{v_0} f(I)\left[\tfrac{3}{10} + \tfrac{6}{7}P_2(\cos\theta) - \tfrac{81}{70}P_4(\cos\theta)\right] \qquad 6.14$$

The center of gravity of the powder pattern for the central transition is determined by an *isotropic second-order shift*, given (for zero asymmetry) by the first term of equation 6.14:

$$-\frac{3\chi^2}{10v_0}f(I) \qquad 6.15$$

Thus, as mentioned earlier, the center of gravity is shifted to low frequency, though (for a static sample) the powder pattern extends both sides of the isotropic shielding position.

6.4.3   CENTRAL-TRANSITION SPECTRA: RAPID SAMPLE SPINNING

When a sample is spun about an angle β to $\boldsymbol{B}_0$, there is some (though not complete) averaging of the second-order contribution to the resonance frequency. Equation 6.14 shows that the average values of $P_2(\cos\theta)$ and $P_4(\cos\theta)$ are required. This involves another change of axes and therefore brings in new Legendre polynomials. The necessary average of equation 6.14 can be shown to be:

$$v_0^{(2)} = -\frac{\chi^2}{v_0} f(I)\left[\tfrac{3}{10} + \tfrac{6}{7}P_2(\cos\beta)P_2(\cos\Theta) - \tfrac{81}{70}P_4(\cos\beta)P_4(\cos\Theta)\right] \qquad 6.16$$

where the angle β is that between the rotor axis and $\boldsymbol{B}_0$, while Θ is the angle between the Z axis of the EFG and the rotor axis (which can take any value for a powder sample). The factors $P_2(\cos\beta)$ and $P_4(\cos\beta)$ act to scale the relevant terms for the powder bandshape. At the magic angle, $\beta = \cos^{-1}\left(\tfrac{1}{\sqrt{3}}\right)$, the middle term is eliminated, while the final term is scaled by $-\tfrac{7}{18} = -0.389$.

Thus the second-order shift under conditions of MAS (including the isotropic part) becomes:

$$v_0^{(2)} = -\frac{3\chi^2}{10v_0} f(I)\left[1 + \tfrac{3}{2}P_4(\cos\Theta)\right] = -\frac{3\chi^2}{32v_0} f(I)\left(5 - 18\cos^2\Theta + 21\cos^4\Theta\right) \qquad 6.17$$

This term describes a powder pattern, with low-frequency shifts for all values of Θ (unlike the static case). The center of gravity (equation 6.15) is invariant to the spinning. The lower spectrum in figure 6.6 shows the case of aluminum acetylacetonate under MAS conditions. Clearly, the second-order effects depend on the ratio χ^2/ν_0 and thus inversely on the magnitude of the applied magnetic field $\boldsymbol{B}_0$. Figure 6.7 shows such a variation of the central transition for a zero-asymmetry case. The bottom spectrum contains sharp spinning sidebands from the outer transitions, which become broadened as χ^2/ν_0 increases. At higher χ^2/ν_0 ratios, spinning sidebands of the central transition appear.

Direct measurements from spectra such as that shown in figure 6.6 can yield information about both the isotropic shielding and the quadrupolar coupling constant, the latter being strongly related to the electronic environment of the nucleus in question. Inset 6.3 indicates how the relevant parameters may be derived from the MAS powder pattern of the central transition for an axially symmetric case. Optimum choice of operating magnetic field depends on the aim of the experiment. Resolution is best for the highest $\boldsymbol{B}_0$ (and this makes determination of isotropic chemical shifts relatively easy). However, when quadrupole coupling constants are small, it may be advantageous to obtain the quadrupolar parameters from the central-transition bandshape at relatively low applied magnetic field so as to increase the magnitude of the second-order broadening. Alternatively, such parameters may be derived from analysis of the spinning sidebands from the satellite transitions at low χ^2/ν_0.

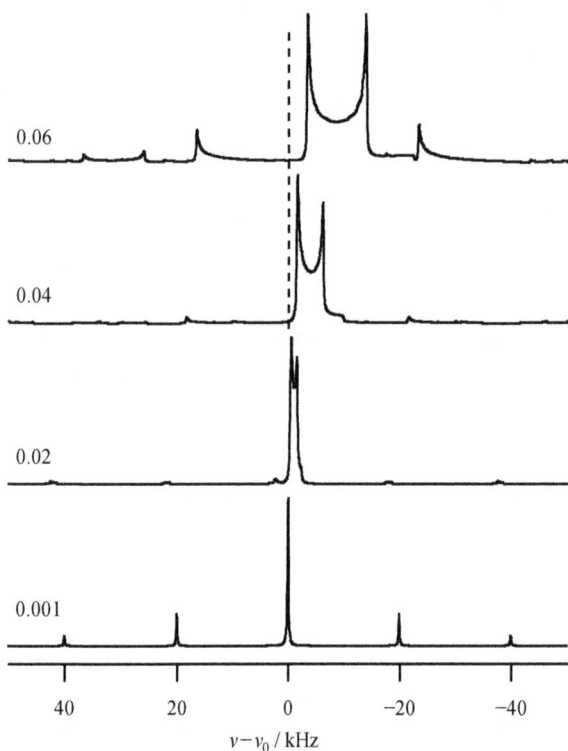

Figure 6.7. Variation of the central-transition region of a spin-$\frac{3}{2}$ spectrum (simulated) with the ratio of the quadrupolar coupling constant to the Larmor frequency, χ^2/ν_0. The dotted line indicates the isotropic chemical shift.

Inset 6.3. Analysis of the central transition for a second-order MAS powder pattern

Figure 6.8 shows a schematic powder pattern for the central transition of a quadrupolar nucleus under MAS conditions, with the positions of the three turning points indicated. Expressions for these turning points may be obtained by differentiating equation 6.17. They are found to be at $\Theta = 0°, 90°,$ and $49.1°$ (with values of $5 - 18\cos^2\Theta + 21\cos^4\Theta$ equal to 8, 5, and $\frac{8}{7}$, respectively). The positions are given in figure 6.8 in units of K, where:

$$K = \frac{3\chi^2}{4\nu_0} f(I) \qquad\qquad 6.18$$

For a zero-asymmetry case with a well-defined powder pattern, the quadrupole coupling constant can be readily derived from the splitting between the prominent horns of the powder pattern, which is $\frac{27}{56} K$. In the case of aluminum acetylacetonate (see figure 6.6), this splitting is 670 Hz, leading to a value of 2.7 MHz for χ. The powder pattern may be reproduced in full by suitable computer programs and the values of χ and ν_0 obtained more accurately by iterative fitting of the full bandshape. For the aluminum acetylacetonate case, a value of 3.0 MHz for χ was obtained from a fit of the full bandshape. Although assumed to be zero in the approximation above, the fit gave an asymmetry value of $\eta = 0.16$.

Figure 6.8. Schematic magic-angle spinning (MAS) NMR powder pattern for the central transition of a second-order spectrum for a quadrupolar nucleus with an axially symmetric EFG. The angle $\Theta = 22.2°$ gives rise to a resonance at the same frequency as the turning point at 90°.

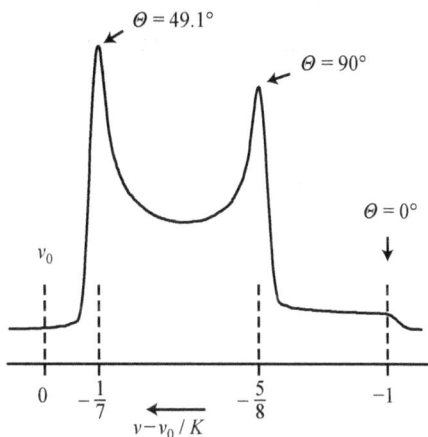

6.5 SPECTRA FOR CASES WITH NON-ZERO ASYMMETRY: CENTRAL TRANSITION

The relevant equations for cases with $\eta \neq 0$ are significantly more complex and will not be quoted here. The reader is referred to other texts, as listed in the Further reading section, for details. However, the general effects of $\eta > 0$ for the MAS case are shown in figure 6.9. For a few purposes, the effect of asymmetry is simply introduced by a factor of $3+\eta^2$. For example, in equation 6.15 for the isotropic second-order effect, the factor 3 must be replaced by $3 + \eta^2$. Since $\eta \leq 1$, the influence of this factor in such cases is rather small.

Of course, real examples do not often look like the ideal cases of figure 6.9. Thus figure 6.10 shows the central-transition portion of the ^{27}Al ($I = \frac{5}{2}$) spectrum for a natural aluminosilicate clay. The relatively featureless spectrum is typical of those from many aluminosilicate materials that contain some disorder. The two intense centerband resonances are from aluminum octahedrally ($\delta_{Al} \cong 1$ ppm) and tetrahedrally ($\delta_{Al} \cong 58$ ppm) coordinated by oxygen. The remaining signals are

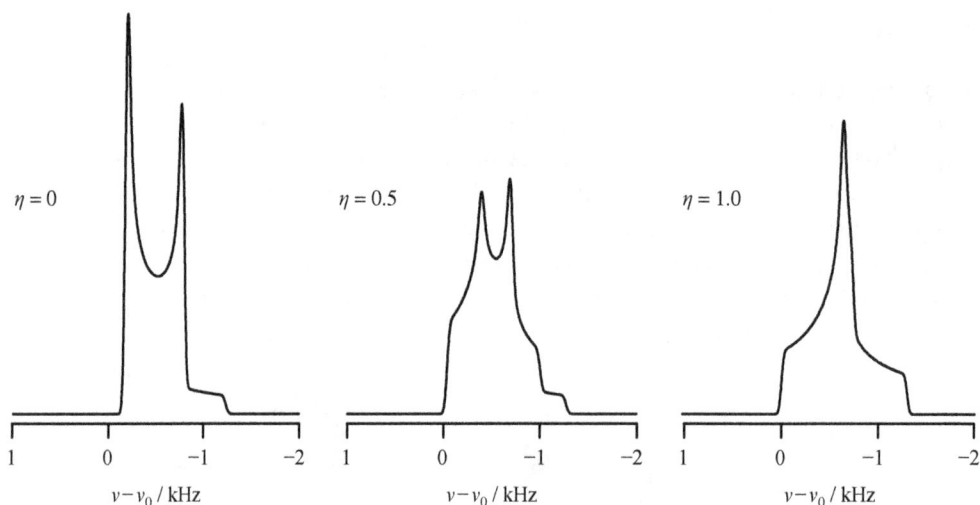

Figure 6.9. Simulated second-order magic-angle spinning (MAS) quadrupolar powder patterns for the centerband of a spin-$\frac{3}{2}$ nuclide as a function of the asymmetry factor for $\chi = 1$ MHz and $\nu_0 = 52.9$ MHz.

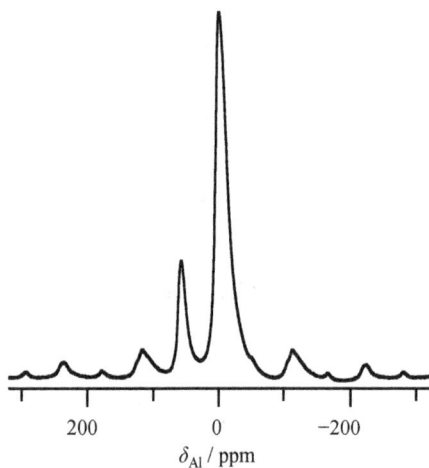

Figure 6.10. Aluminum-27 magic-angle spinning (MAS) spectrum, obtained at 9.4 T with a spin rate of 12 kHz, for an aluminosilicate clay.

spinning sidebands. While the theoretical shapes of the resonances are not apparent, the isotropic chemical shifts (including the second-order quadrupolar contribution) can be derived from the centers of gravity of the signals. If the spectrum is measured as a function of magnetic field, the true isotropic shift (i.e., excluding quadrupolar effects) can be obtained by plotting the center of gravity as a function of the inverse of the Larmor frequency and extrapolating to zero.

6.6   RECORDING ONE-DIMENSIONAL SPECTRA OF QUADRUPOLAR NUCLEI

For most samples, spin–lattice relaxation times of quadrupolar nuclei are short since they are dominated by the quadrupolar mechanism and this is normally efficient (a consequence of the large quadrupolar interactions!) provided there is some molecular-level mobility. Therefore direct-excitation methods (repeated single pulses, each followed by an FID) are normally used (for both central and satellite transitions). Recycle delays following the FIDs do not need to be lengthy (in contrast to the case for spin-$\frac{1}{2}$ nuclei). This implies that the optimum signal-to-noise ratio may be obtained from pulse angles less than 90°, so it may be worth using the Ernst angle (see section 3.3.1). Moreover, there are other reasons for keeping the pulse angle low (see below). Proton decoupling may or may not be required, depending on the chemical system involved.

An example of the potential value of obtaining quadrupole coupling information is provided by the ^{23}Na spectrum of a mixture of sodium chloride and sodium nitrite shown in figure 6.11. It is immediately obvious that the high-frequency signal has a negligible quadrupole coupling constant and so the sodium atoms must be at a site of very high symmetry. This establishes the assignment as that of sodium chloride. The other signal, which must be from sodium nitrite, has the typical appearance arising from an axially symmetric site; simulation confirms this and gives $\chi = 1.09$ MHz. This is consistent with the crystal structure and would have given structural information in the absence of diffraction data.

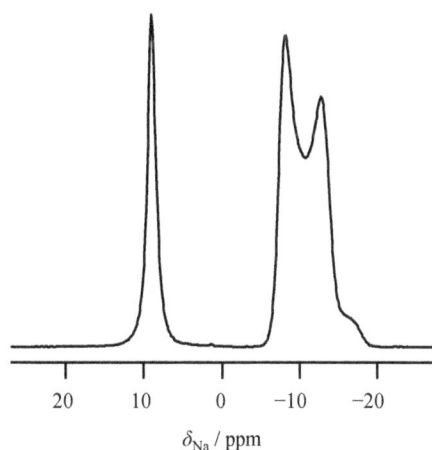

Figure 6.11. Sodium-23 direct-excitation spectrum (centerband region), obtained at 79.34 MHz, of a mixture of sodium chloride and sodium nitrite using a 15° pulse angle and a 1 s recycle delay. The spin rate was 4.5 kHz.

Cross-polarization (CP) is rarely required for the purpose of gaining signal intensity, though it may sometimes be valuable to identify signals of nuclei near to protons (i.e., to obtain selectivity). Moreover, there are some exceptions to the general rule that T_1 is short for quadrupolar nuclei, especially for nuclides with low quadrupole moments. In such cases, CP may be valuable in increasing signal intensity. However, the match condition for CP to a quadrupolar nucleus does not have the simplicity of equation 3.8 but now involves the quadrupolar coupling constant (and therefore depends on both the sample and the nuclear environment). Determining the optimum match condition is best done empirically on the system of interest by varying either of the RF field strengths. The experiment is highly selective.

When quadrupole coupling constants get significantly large (say, >5 MHz), the central transition becomes hundreds of kilohertz wide, even under MAS conditions, and the S/N is reduced. This problem can be at least partially overcome by the use of the quadrupole Carr–Purcell–Meiboom–Gill (QCPMG) method. As the name implies, this technique is analogous to the CPMG method for spin-$\frac{1}{2}$ nuclei described in section 7.2.3. It starts with a two-pulse sequence ($\pi/2$ - τ_1 - π) which causes an echo signal centered at a time τ_1 following the π pulse. Repeated π pulses then give a train of echoes, which are detected. The experiment can be done on static samples or with MAS. In the latter case, it is essential to have rotor synchronization throughout. In the quadrupolar case, the pulse durations are scaled from the nominal $\pi/2$ and π angles by the Rabi factor (see the next section). The normal powder-pattern signal is broken up into a series of discrete spikelets (analogous to the occurrence of the sidebands caused by MAS), separated by the inverse of the π-pulse separations (see figure 6.12). In consequence, the S/N significantly increases. The quadrupolar parameters can be obtained by analysis of the spikelet envelope. For cases with really large quadrupole coupling constants, it is necessary to use co-added frequency-stepped experiments in order to get a true representation of the bandshape.

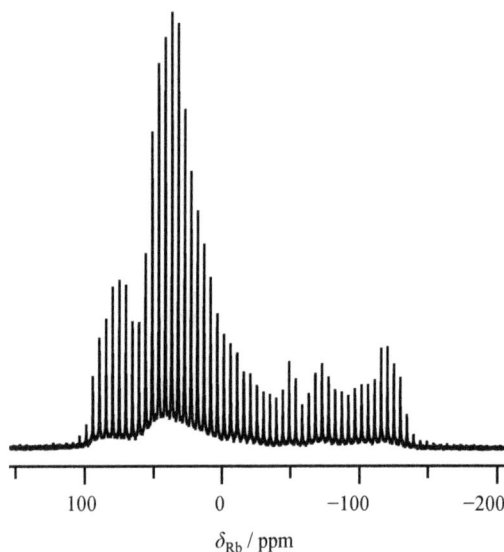

Figure 6.12. Rubidium-87 quadrupole Carr–Purcell–Meiboom–Gill (QCPMG) spectrum of a static sample of Rb_2SO_4, obtained at 9.4 T. There are two rubidium environments with chemical shifts 42 and 15 ppm. The quadrupolar parameters are: $\chi=2.8$ and 5.7 MHz; $\eta=0.95$ and 0.15, respectively. The spikelet separation is 625 Hz.

In practice, spectra will be influenced by shielding anisotropy and asymmetry. Since the quadrupole coupling constant is independent of the applied magnetic field and second-order quadrupole effects decrease as the field increases, whereas chemical shift anisotropy increases (when expressed as a frequency), separately determining the two tensors is facilitated by operating at two or more magnetic fields and computer-simulating the spectra.

6.6.1 NUTATION

However, the existence of large quadrupole coupling constants raises new issues for obtaining spectra. During solid-state experiments on spin-$\frac{1}{2}$ nuclei, the RF pulse energy E_{RF} (generally expressed as the equivalent angular frequency, γB_1) is sufficiently high in relation to the strengths of dipolar and shielding interactions that they can be ignored during the pulse. However, that is not necessarily the case for the quadrupolar interaction. In the situation where the quadrupolar energy is much less than the RF energy, $E_Q \ll E_{RF}$, *nutation* (rotation of the magnetization under the influence of the RF; see section 3.2.1) will occur as in a case with zero quadrupole coupling, that is at an angular frequency γB_1. This is known as a *hard pulse* and results in all the resonances (central and satellite transitions) being observed.

On the other hand, in the extreme case that $E_Q \gg E_{RF}$, nutation for the central-transition magnetization occurs at an angular frequency $\left(I + \frac{1}{2}\right)\gamma B_1$, where $\left(I + \frac{1}{2}\right)$ is known as the *Rabi factor*. In such a case, the RF duration required for a 90° pulse is shorter than that for an equivalent situation for a solution; for an $I = \frac{3}{2}$ case, for example, the Rabi factor is 2. Such a situation is said to involve a *soft pulse* and the effect is selective, which means that only the central transition of the quadrupolar energy-level system will be fully affected. Both soft pulses and hard pulses have uses for quadrupolar spectra. For intermediate values of the pulse power (sometimes also referred to, rather confusingly, as "hard," but better called simply *nonselective*),[6] the nutation situation is complex, and different coherences (see section 4.3) are created. Figure 6.13 shows the development of two such coherences, namely single quantum (central transition, SQ) and triple quantum (TQ) for a spin-$\frac{3}{2}$ system following such a pulse. Because TQ coherence cannot be created directly but only via the creation of SQ coherence, there is an apparent "induction period" prior to the appearance of the former, as can be clearly seen in figure 6.13.

Of course, a particular pulse may be hard in respect of some nuclei but soft in relation to other nuclei of the same isotope in the same sample. Variation of the pulse duration, with two-dimensional Fourier transformation, gives a *nutation spectrum* that can sharply reveal differences of quadrupolar interaction strengths in cases where resonances overlap in one-dimensional spectra. However, such differences produce problems for determining relative intensities between signals in well-resolved one-dimensional spectra. The easiest way to obtain good results semiquantitatively is to employ small pulse angles (≤8°), since deviations between the various cases are relatively small under such conditions. However, this does not optimize the signal intensity for any of the nuclear sites.

---

[6] For a discussion of the conditions for selective excitation, see the article by Freude and Haase in Further reading.

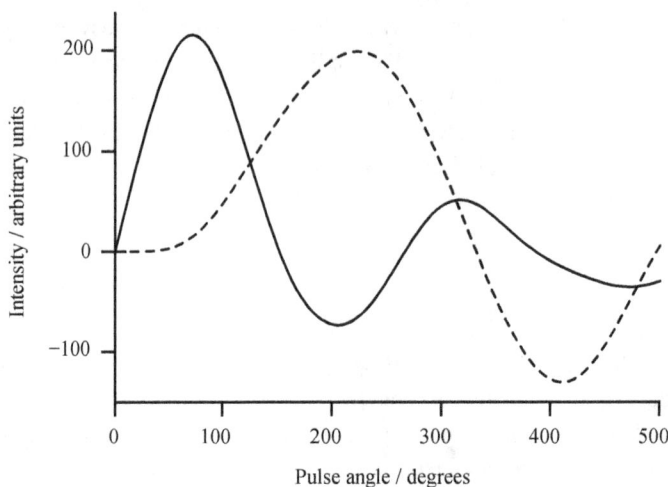

Figure 6.13. Development of single-quantum (central transition) and triple-quantum coherences for an $I = \frac{3}{2}$ spin system by means of a RF pulse with a power-equivalent of 100 kHz for a system with a quadrupole coupling constant of 0.5 MHz. The solid line is for SQ coherence, whereas the dashed line refers to TQ coherence. (Figure supplied courtesy of Dr. Sharon Ashbrook.)

6.7   MANIPULATING THE QUADRUPOLAR EFFECT

Although it can be useful for the derivation of quadrupolar parameters, the broad bandshapes caused by the second-order quadrupolar effects introduce problems of spectral resolution, making analysis complicated if there is more than one relevant site for the nucleus in the crystals in question. In principle, it would be useful to be able to eliminate the second-order effects so as to improve resolution and to obtain chemical shifts more readily. One obvious method is to use high magnetic fields, but these are frequently not accessible and, in many cases, spectrometers operating at sufficiently high fields do not exist. There are several more sophisticated ways round this problem by manipulating the quadrupolar effect, as described in the next few subsections. However, these techniques are largely confined to cases with relatively small quadrupole coupling constants since impossibly high spin rates would be required otherwise.

6.7.1   VARIABLE-ANGLE SPINNING

Spinning at the magic angle does not eliminate the second-order quadrupolar effect, since it is not, as a whole, proportional to $(3\cos^2\theta - 1)$. Some narrowing will be achieved by spinning about any angle, β, to B_0. The effect can be seen by examination of equation 6.16. The isotropic term will, of course remain under variable-angle spinning (VAS). The other two terms will be scaled by the averages of $P_2(\cos\beta)$ and $P_4(\cos\beta)$, respectively. The first of these is only zero for rotation at the magic angle.

The term in $P_4(\cos\beta)$ is zero at two angles, namely 30.56° and 70.12°. Unfortunately, the value of using these angles is much reduced by the fact that shielding anisotropy and dipolar coupling effects are not eliminated, nor are the contributions to the first- and second-order quadrupolar terms involving $P_2(\cos\beta)$—in fact, they are only scaled by 0.6122 and –0.3265, respectively. However, analysis of spectra obtained at different spinning angles may enable shielding anisotropies and asymmetries to be separated from quadrupolar effects. More importantly, such considerations form the basis of the next technique to be discussed.

6.7.2 DOUBLE ROTATION

Spinning at an angle of 54.74° eliminates shielding anisotropy and dipolar interactions, plus the first- and second-order quadrupolar terms that involve $P_2(\cos\beta)$. However, use of the angles 30.56° or 70.12° removes the second-order quadrupolar effects dependent on $P_4(\cos\beta)$ so that simultaneous rotation about 54.74° and 30.56°, say, should remove all problems arising from anisotropies. This will result in a single line for the central transition, which will be at the Larmor frequency plus the isotropic second-order quadrupolar shift. Spinning sidebands will also be seen. This experiment requires a rotor within a rotor, with independent rotations. It is a difficult experiment to implement, but it can be done. Figure 6.14 shows the general arrangement. The outer rotor is fixed at an angle of 54.74° to B_0 while the angle between the two rotor axes is set to 30.56°.

Figure 6.15 compares a ^{17}O double rotation (DOR) experimental spectrum of L-alanine with that of a simple MAS experiment. The former reveals the signals for the two crystallographic sites, which overlap severely in the latter. A major limitation of DOR lies in the low value of the feasible rates of rotation, especially for the outer rotor; spinning sidebands can reduce the effective resolution.

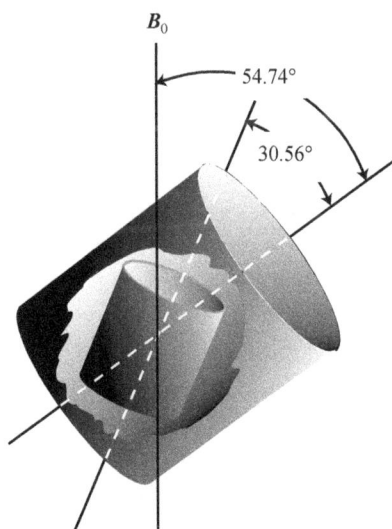

Figure 6.14. The arrangement of rotors for double rotation (DOR).

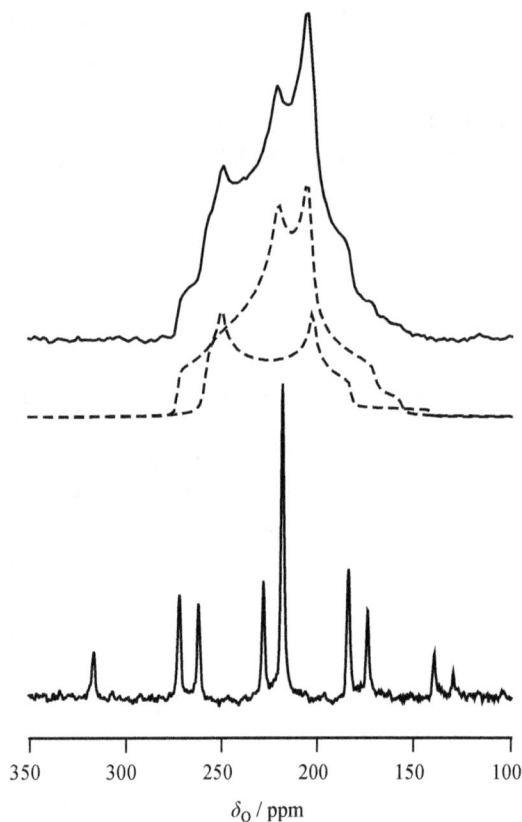

Figure 6.15. Oxygen-17 spectra of L-alanine at 81.3 MHz. Top: magic-angle spinning (MAS) at ~16 kHz; bottom: double rotation (DOR) with outer rotor rate 1.8 kHz and inner rotor rate ~8.5 kHz. Simulations of the spectra of the two individual sites are plotted underlying the upper spectrum. (Figure supplied courtesy of Professor Ray Dupree.)

6.7.3 DYNAMIC-ANGLE SPINNING

This technique provides two-dimensional spectra (see section 5.4.1 for a general discussion of 2D operation), giving isotropic spectra in one dimension and quadrupolar-broadened anisotropic spectra in the other. The experiment consists of spinning at different angles during two times t_1 and kt_1, where the constant k is calculated so as to provide the elimination of the anisotropic effects. A CT-selective 90° pulse is applied prior to each of these times and the magnetization is stored in the z direction (by another 90° pulse) between them while the rotor angle is changed. The FID is recorded during time t_2 following the end of kt_1. As is usual in two-dimensional experiments, t_1 is incremented and the dataset $S(t_1, t_2)$ is doubly Fourier transformed. In the f_1 dimension, the second-order quadrupolar broadening is eliminated (though the second-order isotropic effect remains). There are many pairs of angles that can be used, but none eliminate shielding anisotropy or dipolar interactions, so resolution remains a problem. The best pair of angles is 0° and 63° (which requires $k = 5$), since the latter is not far from the magic angle (so that in t_2 the first-order anisotropic effects are significantly reduced). This method is also technically demanding, since it requires rapid changes of rotor angle, but (in contrast to DOR) there is no difficulty about obtaining high spin rates because only one rotor is involved.

6.7.4 MULTIPLE QUANTUM MAGIC-ANGLE SPINNING

The two-dimensional multiple quantum magic-angle spinning (MQMAS) approach now forms the most popular method of studying quadrupolar nuclei, so it will be discussed in some detail. The operation of some of the pulse sequences involved will be introduced and some of the practical issues will be explored in depth.

As discussed in sections 4.3, 4.4.2, and 6.6.1, coherences between energy levels differing by more than one in m_I can be created by RF pulses, though they are not directly detectable. In practice, while MQ and SQ coherences are qualitatively similar in operation, the former do have one distinction from single-quantum coherences, namely in the way they are affected when the phase of a pulse is changed. *Coherence order, p,* is defined as the difference between the values of m_I for the energy levels linked by the coherence. Pulses change coherence orders. When the phase of a pulse changes by ϕ, say, a given pulse-induced change of coherence order, Δp, acquires a phase shift of $-\Delta p \cdot \phi$. This is consistent with simple ideas of single-quantum $\Delta p = \pm 1$ transitions induced by an RF pulse: In this case, inverting the phase of a $\pi/2$ (x) pulse from $+x$ to $-x$ will change the phase of the resulting magnetization vector from y to $-y$ (i.e., phase shift of $-\pi$).

The creation of coherences other than $p = \pm 1$ is relatively simple for quadrupolar nuclei. All that is necessary is to apply a single nonselective (but not truly "hard") pulse ($E_Q > E_{RF}$). The number and strengths of the various coherences change with the pulse duration (see section 6.6.1). The difference in phase behavior between different coherences allows phase cycling to be used to select a particular coherence order. The coherence selected can then be allowed to freely precess. After a period of time, a second pulse can transfer the selected coherence into observable single quantum coherence.

Such a sequence of two pulses, with a variable time between them (as illustrated in figure 6.16(a)) forms the basis of the two-dimensional MQMAS technique.[7] Phase cycling is used (section 3.4.1) to select those transitions caused by the first pulse that are symmetrical in quantum number m_I, that is, $+m_I \leftrightarrow -m_I$. For instance, with $m_I = \frac{3}{2}$, a triple-quantum transition is involved (3QMAS NMR), which is the only possibility for $I = \frac{3}{2}$. As for the $\frac{1}{2} \leftrightarrow -\frac{1}{2}$ case, all such symmetrical transitions are unaffected by first-order quadrupolar interactions.

Figure 6.16 involves the coherence transfer pathways for the triple-quantum/single-quantum (3QMAS) correlation experiment, which is the one most commonly encountered. The two pathways ($p = 0 \rightarrow +3 \rightarrow -1$ and $p = 0 \rightarrow -3 \rightarrow -1$) can be combined to give a signal that is *amplitude-modulated* (with respect to t_1), which, with a two-dimensional Fourier transformation,

---

[7] Actually, there are innumerable variations in the pulse sequences used for MQMAS, giving different efficiencies and having various advantages and disadvantages. The subject is not for the faint-hearted! A detailed evaluation is to be found in the article by Goldbourt and Madhu (see Further reading).

(a)

+3
0
−1
−3

(b)

+3
0
−1
−3

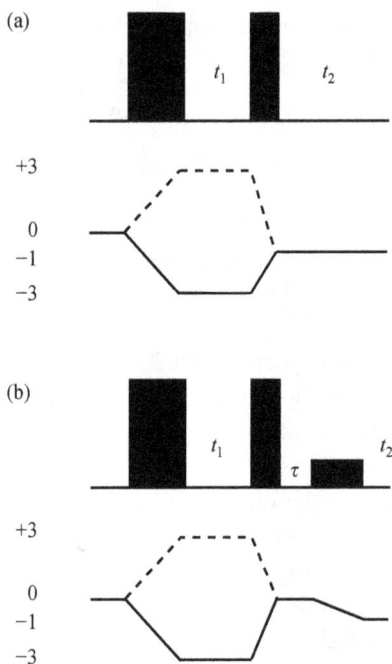

Figure 6.16. The pulse sequence for (a) a two-pulse and (b) a three-pulse, z-filtered MQMAS experiment along with the coherence transfer pathways for triple-quantum experiments. t_1 is the usual evolution time in a two-dimensional experiment and data acquisition starts in t_2 after a dead-time delay following the final pulse. In (b), the third pulse is a low-power selective pulse (on the $\frac{1}{2} \leftrightarrow -\frac{1}{2}$ transition). With a phase-cycle to select the appropriate coherence transfer, the delay τ simply needs to be long enough for the spectrometer to implement the large change in RF power level between the second and third pulses.

gives (under the right conditions; see later in this subsection) a pure absorption lineshape.[8] The phase cycle for this experiment is chosen to achieve this. (See section 3.4.1 for a general discussion of phase cycling.)

The effect of the first pulse depends on the size of the quadrupole coupling, so it is sample and environment dependent. The implications of this, for samples containing more than one environment, will be discussed later. After an evolution time, t_1 (which can be rotor synchronized although this is not essential), the ±3 coherence order is converted to an observable −1 coherence order with the second pulse. As usual for two-dimensional experiments, FIDs are collected over a time t_2 for various values of time t_1 to produce a two-dimensional data set involving two time variables. Double Fourier transformation then gives a two-dimensional plot involving two frequency dimensions, f_2 and f_1, respectively. The *direct dimension* (frequency f_2) may be labeled MAS or 1Q since a projection onto this axis results in the normal one-dimensional spectrum. The *indirect dimension* (frequency f_1) is labeled MQ (or, specifically, 3Q, 5Q, etc., as appropriate). The two coherences ($p=\pm3$) selected by the first (*excitation*) pulse are produced with equal efficiencies. However, the second (*reconversion*) pulse, which is arranged to give observable $p=-1$ coherence, acts unsymmetrically, that is the effectiveness of the $\Delta p=-4$ and $\Delta p=+2$

---

[8] An alternative approach is to select only one pathway and generate a *phase-modulated* signal with respect to t_1. See the review articles in Further reading for more information on this class of experiment.

coherence transfer steps is different. If the duration of this pulse is chosen badly, then there will be an imbalance between the signals generated by each pathway, which results in a poor final lineshape. Conversely, for the $I = \frac{3}{2}$ case only, if the duration of the pulse is chosen so that the two transfer steps produce an equal amount of signal then a pure absorption lineshape will result. For higher spins, a pulse duration can be selected to give a good approximation to a pure-phase spectrum, as shown in figure 6.17(b)[9] for a spin-$\frac{5}{2}$ case.

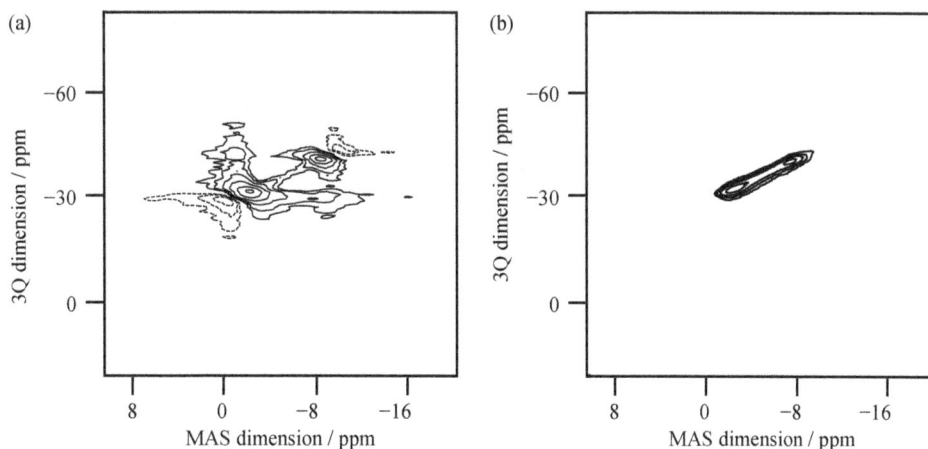

Figure 6.17. Triple-quantum ^{27}Al MQMAS spectra from aluminum acetylacetonate. The spectra were obtained at 9.4 T, with an excitation-pulse duration of 3.6 μs at an RF field equivalent to 110 kHz. The reconversion pulse durations were 0.8 and 2.0 μs for (a) and (b), respectively. (a) Contour plot illustrating the poor lineshape (containing a large dispersive component) obtained when the two coherence transfer pathways are combined unequally. The dashed lines represent negative intensity. (b) The same experiment but with coherence transfer pathways combined approximately equally.

The resonance shown in figure 6.17(b) does not lie parallel to the axis in the directly detected dimension. This occurs because the second-order effects influence the spectra in both dimensions. A process known as *shearing* is used to compensate for this so that the resonance will then lie parallel to the MAS dimension (figure 6.18). This matter was not mentioned when multidimensional spectra were first introduced (in chapter 5) because it is rarely important for spin-$\frac{1}{2}$ spectra. The Fourier transform routines of the spectrometer cope with shearing automatically. (This matter is described in detail in the review articles listed in the Further reading section.) After shearing, one

---

[9] In some articles, the scale in the 3Q dimension is divided by three to make it comparable to the scale of the MAS dimension. However, this practice is not followed in comparable spin-$\frac{1}{2}$ 2D spectra and we believe it is unhelpful.

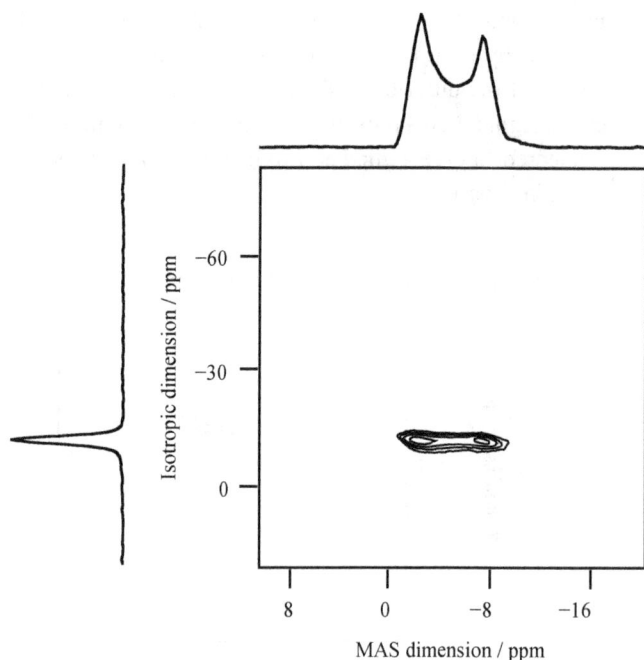

Figure 6.18. The triple-quantum ^{27}Al MQMAS spectrum from aluminum acetylacetonate, as shown in Figure 6.17 (b) but after shearing has been applied.

of the axes is of mixed 1Q and MQ character. It has become common to label this as the "isotropic" axis, though this is potentially confusing (see later in this subsection).

The possible imbalance between the two coherence pathways, mentioned above, gives rise to practical difficulties. In particular, for more complex systems containing species with differing quadrupolar interaction strengths, it may not be possible to produce pure absorption lineshapes for all the resonances in the spectrum. Modifying the experiment, so that it includes a *z-filter* (see Further reading), can solve this problem. The experiment, with its coherence transfer pathway is shown in figure 6.16(b). The second pulse in this sequence is used to create a period with coherence order zero, while the third pulse (a selective $\pi/2$) finally produces observable $p = -1$ coherence. Because the $\Delta p = \pm 3$ coherence transfer steps are now equivalent, the experiment produces pure absorption bandshapes for all environments. The result, with the exception of some loss of signal, is as illustrated in figure 6.18. Note that a shearing transformation is still required.

The two-dimensional character of plots such as figure 6.18 enables good resolution to be obtained. The f_1 "isotropic" dimension gives a high-resolution spectrum free of second-order effects. The position of the resonance along this axis involves the strength of the quadrupole interaction (as well as being dependent on the chemical shift). The f_2 bandshapes (slices in the "MAS" dimension), which retain the second-order broadening (though in practice it may be somewhat distorted), can be computer-fitted to extract the quadrupole coupling parameters and isotropic chemical shifts for each resonance. While, in practical terms, the MQMAS experiment is, perhaps, superfluous on a system with only one environment (except to prove that this is the case when the bandshape is not so clear-cut as it is here), there is potential for improving resolution in more complex systems where there is an overlap of bandshapes in a one-dimensional spectrum. The utility of the MQMAS experiment for deciphering complex spectra consisting of overlapping bandshapes is illustrated in figure 6.19. However, because the effectiveness of the coherence transfer steps in this experiment are potentially different for different environments within a sample, the MQMAS experiment cannot be considered to be generally quantitative.

The practical steps required to successfully carry out an experiment of the type discussed above are described in inset 6.4. Referencing the indirectly detected (isotropic) axis is discussed in inset 6.5.

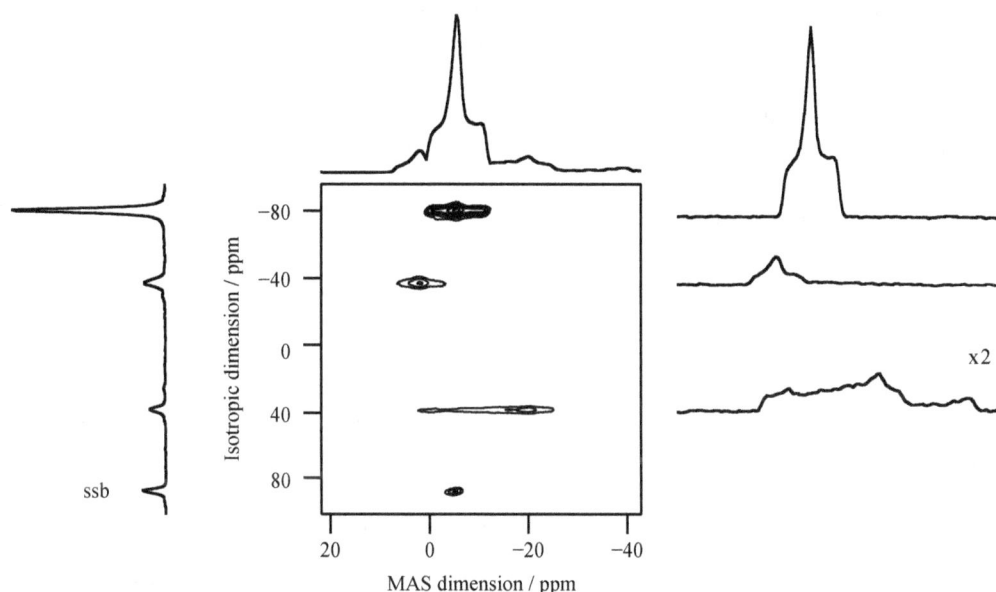

Figure 6.19. Sodium-23 triple-quantum MQMAS spectrum from a mixture of polymorphs of sodium tripolyphosphate ($Na_5P_3O_{10}$). A three-pulse z-filter experiment was used at 9.4 T with pulse durations of 4.8 and 2 μs (at an RF field equivalent to 100 kHz) for the first two pulses. The selective third pulse had a duration of 12 μs (at an RF field equivalent to 10 kHz). The delay between the second and third pulses was 3 μs. The recycle delay was 2 s and 48 (complex) increments in t_1, with 192 repetitions for each, were utilized. The spectral width in the isotropic dimension was 20 kHz (before scaling; see inset 6.5) and the sample spin rate was 10 kHz. Note that spinning sidebands (ssb) can occur in the isotropic dimension just as they can in the MAS one. Cross-sections through each resonance are shown on the right.

Inset 6.4. MQMAS in practice

The practical steps necessary to set up a successful three-pulse, z-filtered 3QMAS experiment illustrate some of the features that are specific to quadrupolar nuclei and multiple-quantum excitation. Here, a step-by-step approach (see figure 6.20) to this set-up is described for a spin-$\frac{3}{2}$ case (i.e., ^{23}Na). The appearance of signal height vs. pulse duration plots will be sample (χ) dependent. The system under study in this illustration contains multiple sodium sites and the largest χ value is ~2 MHz. The magnetic field is 9.4 T.

(1) Using a standard solution (0.1 M NaCl in the case described here) calibrate a 90° ^{23}Na pulse, at the highest RF power available within the limits of the probe. Reference the zero point on the chemical shift axis (see section 3.3.8).

(2) (Optional) On the sample of interest determine the response to a single pulse at the higher RF power (figure 6.20(a)). Note the shape of the nutation curve compared with that for the spin-$\frac{1}{2}$ case shown in figure 3.5. The 90° pulse duration on the solution at this RF field was 2.5 μs. If not already known, determine an appropriate recycle delay.

(3) Carry out a pulse calibration at low RF power and note the observed (or *inherent*) 90° pulse duration (figure 6.20(b)). This selective pulse behaves more like the spin-$\frac{1}{2}$ case.

(4) Using a three-pulse z-filter pulse sequence (figure 6.16(b)), with $t_1 = 0$, set the third pulse duration to be the low-power 90° calibrated in step (3) (8 μs here). A good starting value for the second pulse angle is 60° at the high RF power (relative to the 90° determined on the solution). Vary the duration of the first pulse and note the value that gives maximum signal (figure 6.20(c)). The "signal height" scale can be compared with that in figures 6.20(a) and 6.20(b). In this case the experiment now gives about 10% of the maximum signal height obtained in response to a single pulse. If the signal intensity is low, change the duration of the second pulse and repeat the experiment.

(5) Set the duration for the first pulse to the value determined in step (4), then vary the duration of the second pulse and again note the maximum (figure 6.20(d)).

(6) (Optional) Using the values determined from the previous two steps, the duration of the third pulse can also be optimized (figure 6.20(e)).

(7) The final step is to choose a spectral width for the indirect dimension. This can be rotor synchronized (although it is not essential). If it is set equal to the spin rate, it is possible that it will not be large enough to accommodate the full spectrum (so the potential for folded-back resonances to occur must be considered; see section 3.3.4). Choose the number of increments in t_1 to avoid truncating the signal.

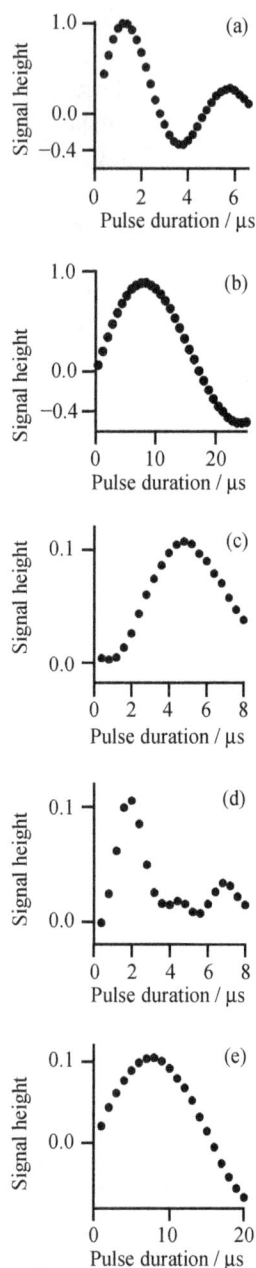

Figure 6.20. The steps required to set up a 3QMAS experiment.

Numerous other experiments purport to improve the experiment. Some are classed as *shifted-echo* experiments (but may not work if T_2 is short), others are known as *split-t_1* experiments and these remove the need for shearing. Further pulse sequence modifications can be made to generally improve sensitivity. One method involves redistributing the population of the spin energy levels such that the population difference between the central $(\frac{1}{2}, -\frac{1}{2})$ pair is enhanced by a process known as rotor-assisted population transfer (RAPT). Details can be found in Further reading. High RF fields (>100 kHz) are an advantage for any MQMAS experiment and, as is the case for quadrupolar nuclei in general, so are high magnetic fields. The review article by Goldbourt & Madhu (see Further reading) provides extensive detail on the MQMAS class of experiments and includes sensitivity enhancement schemes.

Inset 6.5. Axis-labeling of MQMAS spectra

Often, it is the number of resonances (environments) in the indirectly detected (f_1) dimension of an MQMAS experiment that provides useful chemical information. Additional information on the nature of the environments can be obtained from the shape and position of the band in the directly detected (f_2) dimension. A scale for f_1 is, perhaps, not critical but, nevertheless, one should be provided. Unfortunately, there is (currently) no internationally recognized way of labeling the f_1 dimension in MQMAS spectra and, confusingly, various methods are encountered in the literature.

In figure 6.17, f_1 is a pure triple-quantum (3Q) dimension and is labeled as such. Chemical shift values can be added in the same way as for the f_2 (single-quantum or 1Q) axis (see inset 6.4, point (1)). This method is consistent with IUPAC recommendations but does mean that chemical shift values on 1Q and 3Q (or 3Q and 5Q axes, etc.) axes will be different. Here, the 1Q f_2 axis is labeled "MAS."

After shearing (figures 6.18 and 6.19), the new f_{iso} dimension is a linear combination of f_1 and f_2 frequencies and is labeled "isotropic." For a spin-$\frac{3}{2}$ nucleus, the spectral width (*SW*) in f_{iso} is related to that in f_1 through $SW_{iso} = \frac{9}{16} SW_1$. The multiplication factor is $\frac{12}{31}$, $\frac{45}{146}$, and $\frac{36}{127}$ for spin-$\frac{5}{2}$, spin-$\frac{7}{2}$, and spin-$\frac{9}{2}$ nuclei, respectively.

Shifts in the isotropic dimension are both chemical *and* quadrupolar in origin (see section 6.4.2), so the isotropic chemical shift cannot be read directly from this axis. The upshot of this is that the isotropic chemical shift can still only be determined properly by simulating the bandshape or by determining its center-of-gravity (equation 6.15)—either from the one-dimensional spectrum obtained with a small pulse angle or from the sections through an MQMAS spectrum.

6.7.5 SATELLITE TRANSITION MAGIC-ANGLE SPINNING

Up to this point, much emphasis has been placed on the central transition. However, as stated earlier, the complete static spectrum or the intensities of the satellite transition spinning sideband manifolds (in particular) may be computer-fitted to determine the quadrupolar parameters. Inset 6.6 gives the basic theory for the satellite transitions. It shows that it is possible to identify the true chemical shift by locating the various centers of gravity. This can be a simple process for spin-$\frac{5}{2}$ if the inner and outer satellite peaks can be resolved. In this case, the isotropic quadrupolar shifts for the outer, inner, and central transitions are in the ratio $28:1:-8$ (figure 6.21), with a constant factor:

$$\frac{3\chi^2}{4000\nu_0} \qquad\qquad 6.22$$

In order to achieve spectra of the type shown in figure 6.21, the magic angle must be accurately set and rotor synchronization must be used—or else all the spinning sidebands must be carefully added into the centerbands (as was done in the case of figure 6.21).

   The center of gravity of the inner satellites is actually very close to the true isotropic chemical shift. Moreover, the second-order broadening of transitions is proportional to the relevant $A^4(I, m_I)$ coefficient listed in table 6.2. For spin-$\frac{5}{2}$, these show that the widths for individual spinning sideband signals for the outer, inner, and central transitions are in the ratio:

$$-44:7:24 \qquad\qquad 6.23$$

(The negative sign implies a reversal of the shape from that shown in figure 6.9.) The inner satellite lines are appreciably sharper than the others.

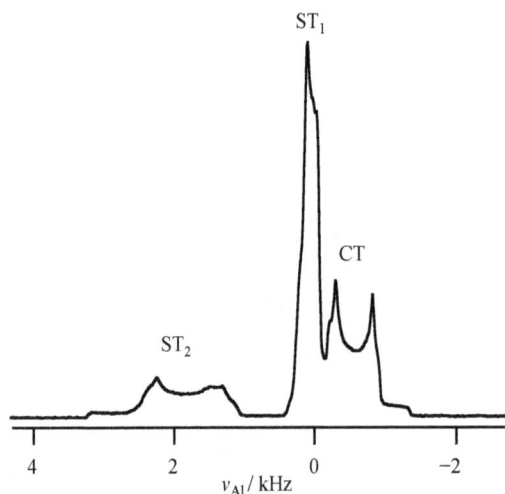

Figure 6.21. Aluminum-27 ($I = \frac{5}{2}$) magic-angle spinning (MAS) spectrum of aluminum acetylacetonate, obtained at a magnetic field of 9.4 T. All the spinning sidebands have been folded into the displayed centerbands. The outer (ST_2), inner (ST_1), and central (CT) transitions are well resolved and their differing quadrupolar isotropic shifts and bandwidths are clearly shown. Quadrupolar parameters: $\chi = 3.0$ MHz, $\eta = 0.16$.

Inset 6.6. Satellite transitions

Equation 6.14 may be expanded to the general $m_I \leftrightarrow (m_I - 1)$ case:

$$v_0^{(2)} = \frac{9}{4v_0} \left[\frac{\chi}{2I(2I-1)} \right]^2 \left[A^0(I, m_I) + A^2(I, m_I) P_2(\cos\theta) + A^4(I, m_I) P_4(\cos\theta) \right] \qquad 6.19$$

where the spin-dependent coefficients are given (for spin-$\frac{3}{2}$ and spin-$\frac{5}{2}$ nuclei) in table 6.2. The above equation assumes axial symmetry.

Table 6.2. Spin coefficients for single-quantum transitions (see equation 6.19)

	$A^0(I, m_I)$	$A^2(I, m_I)$	$A^4(I, m_I)$
Spin-$\frac{3}{2}$			
Central transition	$-\frac{2}{5}$	$-\frac{8}{7}$	$\frac{54}{35}$
Satellite transitions	$\frac{4}{5}$	$\frac{4}{7}$	$-\frac{48}{35}$
Spin-$\frac{5}{2}$			
Central transition	$-\frac{16}{15}$	$-\frac{64}{21}$	$\frac{144}{35}$
Inner satellite transitions	$\frac{2}{15}$	$-\frac{4}{3}$	$\frac{6}{5}$
Outer satellite transitions	$\frac{56}{15}$	$\frac{80}{21}$	$-\frac{264}{35}$

The analogous equation to 6.19 for sample spinning can be obtained by replacing $P_2(\cos\theta)$ and $P_4(\cos\theta)$ by $P_2(\cos\beta)P_2(\cos\Theta)$ and $P_4(\cos\beta)P_4(\cos\Theta)$, respectively (as in going from equation 6.14 to equation 6.16). Thus, under MAS conditions, equation 6.19 becomes, for the $m_I \leftrightarrow (m_I - 1)$ transition:

$$v_0^{(2)} = \frac{9}{4v_0} \left[\frac{\chi}{2I(2I-1)} \right]^2 \left[A^0(I, m_I) - \frac{7}{18} A^4(I, m_I) P_4(\cos\Theta) \right] \qquad 6.20$$

For the specific case of the $I = \frac{3}{2}$ satellite transitions, the quadrupolar shift is:

$$v_0^{(2)} = \frac{\chi^2}{60v_0} \left[3 + 2P_4(\cos\Theta) \right] \qquad 6.21$$

Since the $A^0(I, m_I)$ coefficients differ for the central and satellite transitions, so will the centers of gravity.

Clearly it would be useful to separate the different subspectra. This can be achieved by a two-dimensional experiment that correlates the satellite transitions to the central transition. This is usually referred to as the satellite transition magic-angle spinning (STMAS) experiment. To understand this experiment, one needs to realize that a nonselective (but not truly "hard") pulse will convert any particular coherence into all the other coherences (see section 6.6.1). Consider then a series of two such pulses. The first will create single-quantum coherence (for both the central transition and the satellite transitions). Such coherences will evolve during the interpulse time t_1. The second pulse will spread these coherences around further. Thus, a satellite single-quantum coherence created by the first pulse will be partially converted into central-transition single-quantum coherence by the second pulse. Detection of the eventual FID (over time t_2), followed by double Fourier transformation over t_1 and t_2, results in a two-dimensional spectrum that correlates the different coherences in the two times.

As with MQMAS, there are many versions of the relevant pulse sequence, each with its special characteristics. One possibility is the z-filtered experiment shown in figure 6.22. Again as with MQMAS (figure 6.16), this involves two nonselective pulses and a soft pulse. A major difference lies in the phase cycling, which, for STMAS, selects $\Delta p = \pm 1$ by the first pulse. The time t_1 must be accurately rotor-synchronized. The second pulse converts the magnetization into the z direction and the final pulse results in selective detection (by recording an echo) of the central transition. The second-order effect is refocused at the center of the echo, the appearance of which depends on time t_1. After Fourier transformation, the resulting two-dimensional spectrum contains separate ridge-lines for CT–CT and ST–CT correlations.

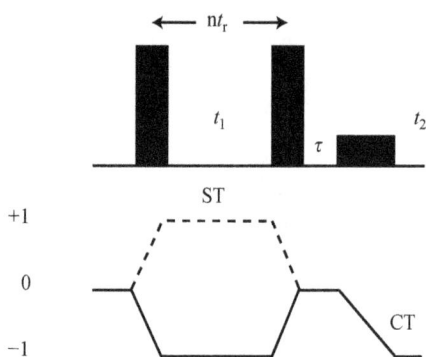

Figure 6.22. Pulse sequence and coherence transfer pathway for a z-filtered satellite transition magic-angle spinning (STMAS) experiment.

An experimental example ($I = \frac{3}{2}$), for the [87]Rb resonance of rubidium nitrate, is shown in figure 6.23. Suitable processing has resulted in horizontal lines for the ST–CT transitions along the f_2 dimension so that projection onto the f_1 dimension will give a high-resolution (isotropic) spectrum free from second-order quadrupolar effects. There are three crystallographic sites. The quadrupole coupling constants for the signals at –27.4, –28.5, and –31.3 ppm are 1.68, 1.94, and 1.72 MHz, with asymmetries 0.2, 1.0, and 0.5, respectively.[10] The two giving signals at higher

[10] D. Massiot, B. Touzo, D. Trumeau, J.P. Coutures, J. Virlet, P. Florian & P. J. Grandinetti, *Solid State NMR* **6** (1996) 73.

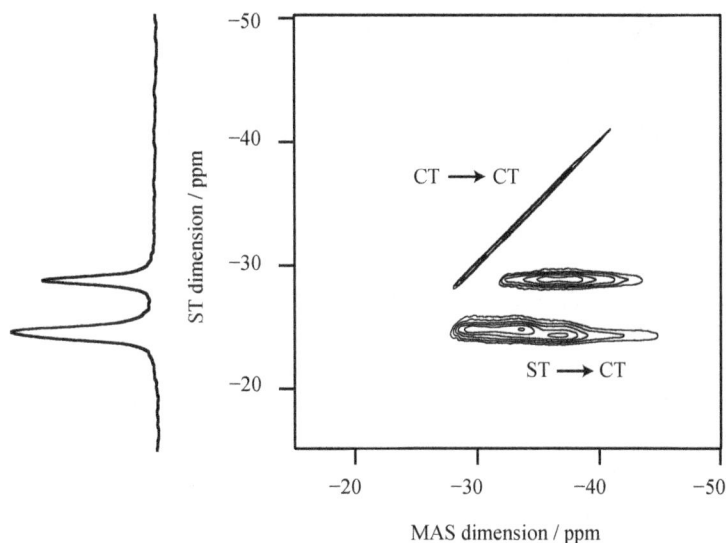

Figure 6.23. Satellite transition magic-angle spinning (STMAS) ^{87}Rb spectrum of rubidium nitrate after shearing. Operating parameters: the first two pulses were of duration 2.3 μs and angle 90° (as measured for an aqueous solution). The RF power was equivalent to 108 kHz. The *z*-filter pulse was of duration 6.0 μs at a 14 kHz power equivalent. The spin rate was 10 kHz and the spectral width was 10 kHz in both dimensions.

frequency are marginally resolved in the two-dimensional plot but not at all in the projection onto the ST dimension.

While STMAS is somewhat more sensitive than MQMAS, it is significantly more demanding to set up. The magic angle *must* be set precisely and the spin rate must be highly stable.

6.7.6 SUMMARY FOR SPECTROSCOPY OF HALF-INTEGER QUADRUPOLAR NUCLEI

Obtaining useful spectra of quadrupolar nuclei is more of a challenge than is the case for spin-$\frac{1}{2}$ nuclei. The choice of experiment must be tailored to the situation in question. When quadrupole coupling constants are small or when there is only one crystallographic site for the nucleus of interest, simple MAS operation may suffice, but when second-order effects are significant and resolution of multiple sites is required, it may be useful to employ one of the special techniques described above, though these do not work satisfactorily for substantial quadrupole coupling constants (> 8 MHz, say). The use of DOR or DAS is limited by the requirement for special hardware, so in most cases it is necessary to turn to either MQMAS or STMAS. The spectrometer manufacturers supply suitable software for these experiments, so most users will be well advised to follow the relevant instructions. However, some understanding of the principles given in section 6.7 will be essential to the interpretation of the spectra obtained. Detailed analysis will, in many cases, involve fitting bandshapes using standard computer programs. It will frequently be necessary to use both quadrupolar and anisotropic chemical shift parameters in such fitting. An ability to obtain spectra at different applied magnetic fields will be a big advantage in this process. Moreover, when large quadrupole coupling constants are involved, the use of the highest possible field will be preferred.

6.8   SPECTRA FOR INTEGRAL SPINS

Mention was made in section 2.6 that there are only four nuclides of significance for NMR with integral spin quantum numbers. Of these, ^{10}B has $I = 3$, while the others (^{2}H, ^{6}Li, and ^{14}N) are spin-1. Boron-10 NMR is unimportant because ^{11}B is the preferred boron nuclide. Lithium-6 has a low quadrupole coupling constant and can usually be treated as a "pseudo spin-$\frac{1}{2}$" nuclide—it provides a useful foil to the spin-$\frac{3}{2}$ nuclide ^{7}Li. Consequently, only deuterium and nitrogen-14 need to be considered in any detail here.

Quadrupole coupling constants for ^{2}H are generally modest ($< \sim 300$ kHz), so second-order effects are not a problem and methods analogous to those used for spin-$\frac{1}{2}$ nuclei generally suffice (though the spin dynamics is different). An example has been given as figure 6.2. Static bandshapes and spinning sideband manifolds in ^{2}H spectra frequently reflect local molecular-level mobility and are much used in that context. Figure 6.24 shows an example (a ^{2}H MAS spectrum). The spinning-sideband manifolds for finasteride hydrate THF solvate cover only ~ 35 kHz, whereas the expected quadrupole coupling constants for a static molecule are ~ 200 kHz. There must be rapid (though not isotropic) motion of the THF molecule in the solvate. However, the quadrupole parameters for the two sites are still unequal—for the high-frequency peak $\chi = 20.2$ kHz and $\eta = 0.99$, whereas for the low-frequency peak $\chi = 16.4$ kHz and $\eta = 0.93$.

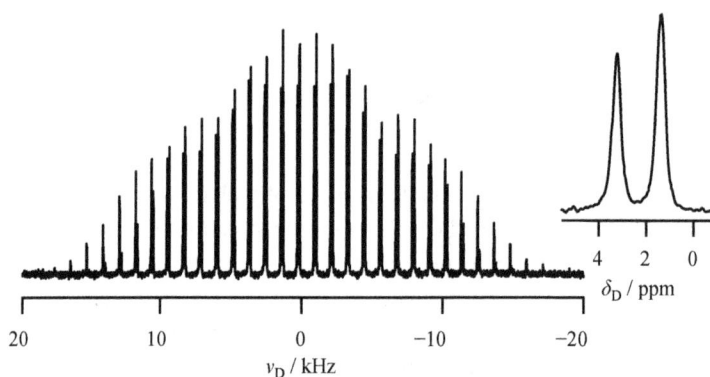

Figure 6.24. Deuterium magic-angle spinning (MAS) spectrum at 11.7 T of finasteride hydrate THF solvate with fully deuterated THF. The expansion of the centerband shows the two different chemical shifts. The spin rate was kept low (~ 1 kHz) in order to show the complete spinning sideband manifolds, which illustrate that the two chemically distinguishable sites have different quadrupole coupling constants.

Nitrogen-14 is something of a problem nucleus because there is no central transition and quadrupole coupling constants can be significant (e.g., ~3 MHz)—sufficiently high that MAS results in innumerable spinning sidebands and thus a loss of S/N. Various schemes have been mooted to alleviate this situation, but there is no widely used protocol yet.

FURTHER READING

GENERAL TEXTS ON QUADRUPOLAR NMR

"Quadrupole effects in NMR studies of solids", M.H. Cohen & F. Reif, *Solid State Physics*, **5** (1957) 321–438.
"Quadrupole effects in solid-state NMR", D. Freude & J. Haase, in *NMR basic principles and progress*, Eds. P. Diehl, E. Fluck, H. Günther, R. Kosfeld & J. Seelig, Vol. **29**, Springer (1993) 1–90.

TEXTS ON SPECIFIC ASPECTS OF QUADRUPOLAR NMR

"Nutation spectroscopy of quadrupolar nuclei", B.C. Gerstein, in *"Encyclopedia of NMR"*, Eds. D.M. Grant, R.K. Harris & R.E. Wasylishen, John Wiley & Sons Ltd. (online posting date 15 March 2007), DOI: 10.1002/9780470034590.emrstm0359.
"Solid-state NMR line narrowing methods for quadrupolar nuclei: Double rotation and dynamic-angle spinning", B.F. Chmelka & J.W. Zwanziger, in *NMR basic principles and progress*, Eds. P. Diehl, E. Fluck, H. Günther, R. Kosfeld & J. Seelig, Vol. **33**, Springer (1994) 79–124.
"Multiple-quantum magic-angle spinning: High-resolution solid-state NMR spectroscopy of half-integer quadrupolar nuclei", A. Goldbourt & P.K. Madhu, *Monat. Chem.*, **133** (2002) 1497–1534. DOI: 10.1007/s00706-002-0502-y.
"High-resolution NMR of quadrupolar nuclei in solids: The satellite-transition magic angle spinning (STMAS) experiment", S.E. Ashbrook & S. Wimperis, *Prog. NMR Spectry.*, **45** (2004) 53–108. DOI: doi:10.1016/j.pnmrs.2004.04.002.

CHAPTER 7

RELAXATION, EXCHANGE & QUANTITATION

7.1 INTRODUCTION

It is important not to picture a solid as a rigid entity but rather as one in which there *is* motion, driven by thermal energy, albeit on a more restricted scale than in solution (figure 7.1). Motion is the driving force for relaxation and it can have an impact on the appearance of the NMR spectrum.

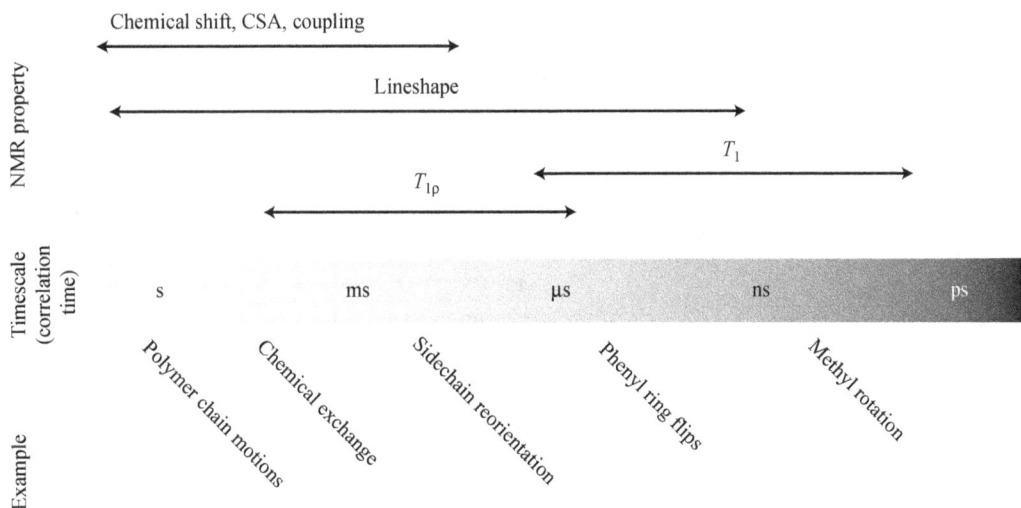

Figure 7.1. Examples of the timescales of motions in the solid state and the NMR properties that can be used to explore them.

This chapter starts by exploring the relationship between motion and relaxation. It is shown how the relaxation behavior can be used to deduce information on sample morphology and how relaxation data can be combined with high-resolution measurements to carry out spectral editing. Methods for measuring relaxation times are described.

Section 7.3 shows how motion in the form of reorientation or, for example, exchange of hydrogen, can affect the appearance of a spectrum. Measurements designed to determine rates of motion are introduced there. It is demonstrated how the anisotropic interactions important in solid-state NMR (shielding/chemical shift anisotropy (CSA), dipolar coupling, quadrupolar coupling) are modified by anisotropic motion and how experiments can be used to probe details of this motion.

As motion affects relaxation and relaxation influences the intensity of the resonances in the spectrum, section 7.4 of the chapter deals with quantitative aspects of solid-state NMR. Some of the special effects seen in paramagnetic systems are discussed in section 7.5 at the end of the chapter.

7.2 RELAXATION

Three modes of relaxation important to solid-state NMR were defined in section 2.8. Spin–lattice (or longitudinal) relaxation, characterized by the time constant T_1, describes the process of restoring equilibrium to the z component of the net magnetization for an ensemble of spins.[1] Spin–spin (or transverse) relaxation, with time constant T_2, relates to the xy component of the magnetization. Spin–lattice relaxation in the rotating frame, with a time constant $T_{1\rho}$, describes the return to equilibrium of transverse magnetization in the presence of spin-locking from an RF magnetic field.

7.2.1 THE CAUSE OF RELAXATION

Relaxation in NMR, unlike the case for most other spectroscopies, is non-radiative. That is, the dominant cause of relaxation for the sample under study is not by emission of electromagnetic photons. Instead, fluctuations in the local field at the excited nucleus allow it to relax. Molecular motion causes this fluctuation (see inset 7.1). In the solid state, the main contributions to the local *magnetic* field arise from

Inset 7.1. When solids lack motion

That relaxation is related to motion is illustrated by considering the spin–lattice relaxation time in systems where there is little or no motion. Silicon carbide exists as a rigid, 3D network of silicon and carbon atoms. The silicon T_1 in one polymorph has been measured as 35 minutes! In such systems, the only relaxation mechanism is often via paramagnetic defects. So relaxation may be extremely slow in high-purity network (3D) materials.

---

[1] When energy is lost through relaxation, it has to go somewhere. The term "spin–lattice" is indicative of where:- energy is lost to the "lattice."

shielding anisotropy and from dipolar coupling to the magnetic nuclei of surrounding atoms. Due to its magnitude, the quadrupolar interaction, where present, often gives rise to very efficient relaxation, though in this case it is a fluctuation in the local *electric* field that is effective.

A full treatment of relaxation is not something to be undertaken lightly! The analysis of anything more than an isolated spin pair soon gets complicated. A detailed understanding of the relaxation of a real system, with many interacting nuclei and a number of degrees of freedom for motion, is not generally feasible. Nevertheless, facets of the relaxation behavior can usefully be related to the properties of the system under study.

Relaxation at a frequency v (which might be the resonance frequency of the protons in a sample, for example) is most efficient when the molecular motion results in a fluctuating magnetic field of the same frequency. A quantity called the *spectral density*, $J(v)$,[2] plays a key role in the theory of relaxation (see inset 7.2). It can be written as:

$$J(v) = \frac{2\tau_c}{1 + (2\pi v \tau_c)^2}$$

7.1

Inset 7.2. Spectral density

In a solid, random motion produces fluctuations in the local magnetic field. These fluctuations can be characterized by a time correlation function, $G(t)$, which describes the correlation between the average local field at a moment in time and at a time t later. In effect, this is a measure of the persistence of the local field. An unchanging (correlated) field or one that changes very quickly (uncorrelated) provide poor relaxation pathways. Often, the value of the correlation function is large over short times and decays exponentially with a time constant τ_c, the correlation time, as time increases, so:

$$G(t) \propto e^{-t/\tau_c}$$

7.2

(The proportionality constant, omitted for simplicity here, depends on the magnitude of the local field.) The spectral density is the Fourier transform of the correlation function:

$$J(v) \propto \int_{-\infty}^{\infty} G(t)e^{-i2\pi v t} \, dt$$

7.3

so that, with $G(t)$ defined as it is here, and ignoring the proportionality constant, $J(v)$ is given by equation 7.1.

---

[2] Note that the symbol J does not relate to indirect coupling in this context!

where τ_c is called the correlation time and is the time constant that characterizes the motion in question.[3] Short correlation times imply fast motion (as for liquids and solutions) and long ones slow motion (as for rigid solids).

A plot of $J(\nu)$ against τ_c puts equation 7.1 into context—figure 7.2. The higher the value for $J(\nu)$, the more efficient the relaxation (as equations 7.4 and 7.5 will illustrate). Efficient spin–lattice relaxation requires fluctuations in the local field at frequencies of hundreds of megahertz (Larmor frequencies) and this equates to correlation times on the order of nanoseconds. On the other hand, spin–lattice relaxation in the rotating frame requires fluctuations of tens of kilohertz, equating to correlation times on the order of microseconds. Also it can seen from figure 7.2 that a fluctuation in the local field at a typical RF nutation frequency (62.5 kHz, curve (c)) can be more effective at causing relaxation than a fluctuation at a Larmor frequency (curves (a) and (b)). This is why $T_{1\rho}$ is often much shorter (usually milliseconds) than T_1 (often seconds) for the solid state.

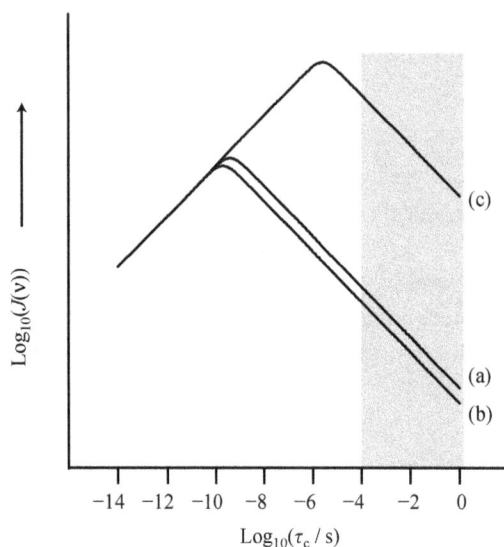

Figure 7.2. A log–log plot of the spectral density, given by equation 7.1, as a function of correlation time for frequencies of: (a) 400 MHz, (b) 800 MHz, and (c) 62.5 kHz. For a given value of ν, $J(\nu)$ reaches a maximum when $2\pi\nu\tau_c = 1$. For very slow motions (long correlation times), indicated by the gray box, "classic" relaxation theory breaks down.

For a single, simple motion, the relaxation rates encountered in NMR can be considered as being proportional to a linear combination of spectral densities. For example, suppose the dominant relaxation mechanism is a result of homonuclear dipolar coupling between spin-$\frac{1}{2}$ nuclei, as might be the case for the protons in an organic solid. Under these conditions, the relaxation times for a single spin species relate to the spectral densities as follows:

$$T_1^{-1} = K[J(\nu_0) + 4J(2\nu_0)] \tag{7.4}$$

---

[3] This spectral density is for a process with a single, well-defined correlation time. It is not so appropriate for motion with distributions of correlation times. The definition of the correlation time depends on the type of motion. For example, for rotational motion it is often defined as the average time for rotation through 1 radian.

$$T_{1\rho}^{-1} = K[\tfrac{3}{2}J(\nu_{RF}) + \tfrac{5}{2}J(\nu_0) + J(2\nu_0)] \simeq \tfrac{3}{2}K[J(\nu_{RF})] \qquad 7.5$$

The constant K depends on the magnitude of the dipolar coupling. This constant is different for other relaxation mechanisms, as are the coefficients of the individual terms. T_2^{-1} can be expressed in a similar way but this is rarely valid in the solid state so the equation is not given here and T_2 is omitted from figures 7.2 and 7.3 (it is discussed later in section 7.2.3).

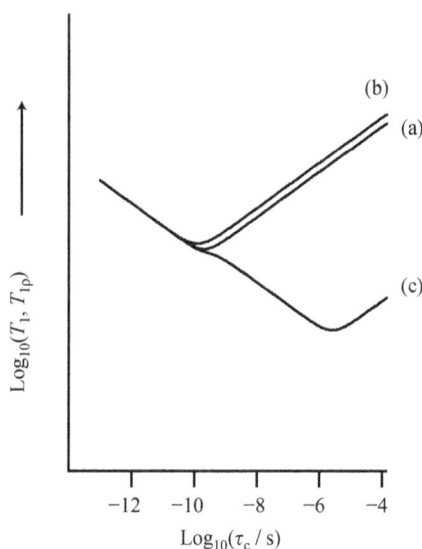

Figure 7.3. A log–log plot of the relaxation times as a function of correlation time. The curves are calculated for frequencies of: (a) 400 MHz, (b) 800 MHz, and (c) 62.5 kHz. (a) and (b) are appropriate for T_1 given by equation 7.4 and (c) is for $T_{1\rho}$ given by equation 7.5.

Two limiting conditions for $J(\nu)$ are important in NMR. If $2\pi\nu\tau_c \ll 1$ then $J(\nu) \simeq 2\tau_c$ and the relaxation *rates* are *proportional* to τ_c. This is called the *extreme narrowing condition* and applies for fast molecular motion. At this condition $T_1 = T_2 = T_{1\rho}$, which is the reason that usually $T_1 = T_2$ in solution.

The opposite end of the scale is particularly important in solid-state NMR. This is the *rigid lattice limit*. Here $2\pi\nu\tau_c \gg 1$ so that $J(\nu)$ becomes proportional to τ_c^{-1} (and so do the relaxation rates). However, for very slow motions (say, with correlation times greater than 0.1 ms) this approach to interpreting relaxation behavior is no longer appropriate.

For a given relaxation mechanism, the behavior of the relaxation times, as a function of correlation time, defined by equations 7.4 and 7.5, is shown in figure 7.3. Several observations can be made:

- T_1 in the rigid lattice limit is proportional to the magnetic field of the spectrometer.
- $T_{1\rho}$ is independent of the magnetic field of the spectrometer (because $J(\nu_{RF})$ is the dominant term in equation 7.5); however, it depends on the strength of the RF magnetic field.
- The T_1 and $T_{1\rho}$ curves go through a minimum. At the minimum $1/\tau_c = 2\pi\nu_0$ or $2\pi\nu_{RF}$ for T_1 and $T_{1\rho}$, respectively (when a single spectral density is involved).

Energy of Activation

For a thermally activated process, the activation energy, E_a, can be obtained from an Arrhenius plot of the correlation time as a function of temperature:

$$\tau_c = A \exp(E_a/RT) \qquad 7.6$$

where R is the gas constant, T is the temperature, and A is a constant (the pre-exponential factor). In the extreme narrowing limit, it follows that $\ln(T_1^{-1}) = \ln(A) - E_a/RT$, so a plot of $\ln(T_1^{-1})$ against T^{-1} will have a gradient proportional to the activation energy. The result in the rigid lattice limit is the same, apart from a change in the sign of the gradient. This applies equally well to $T_{1\rho}$. An example of such a plot is shown in figure 7.4. The correlation time can also be written in an Eyring form

$$\ln\frac{1}{\tau_c T} = -\frac{\Delta H^{\ddagger}}{RT} + \ln\frac{k_B}{h} + \frac{\Delta S^{\ddagger}}{R} \qquad 7.7$$

where $\Delta H^{\ddagger}$ is the enthalpy of activation, $\Delta S^{\ddagger}$ is the entropy of activation, k_B is the Boltzmann constant, and h is Planck's constant. Plotting $\ln(T_1/T)$ (in the extreme narrowing limit) or $\ln(1/T_1 T)$ (in the rigid lattice limit) against $1/T$ will yield the thermodynamic parameters.

In general, the relaxation behavior is much more complex, since it contains contributions from all of the motional processes present in the solid. For the case shown in figure 2.14, for example, the molecule experiences two types of motion so two minima are observed in the T_1 relaxation for the relevant temperature range. In a complex system it can, therefore, be difficult to interpret the

Figure 7.4. An Arrhenius plot of the carbon spin–lattice relaxation rate for the methyl group labeled (a) in ibuprofen. The gradient of the fitted line gives an activation energy for methyl group rotation (assuming this is the cause of the relaxation) of 7.9 kJ mol^{-1}.

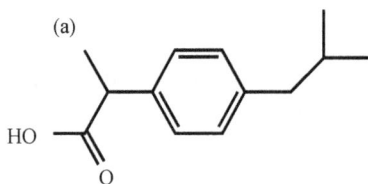

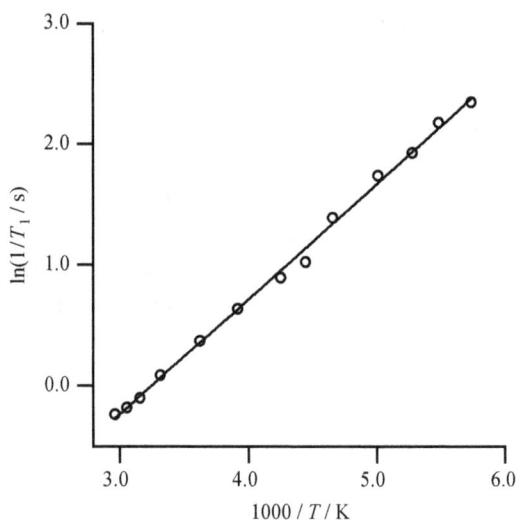

relaxation behavior, so that relating a single T_1 or $T_{1\rho}$ determination back to a particular molecular motional property is often not feasible although such information can be used to design a *filter* for more complex measurements (see section 7.2.4). Recording relaxation times as a function of temperature can give information on activation barriers and that may give some clue as to the nature of the molecular dynamics. Such an approach may also highlight physical changes to the sample under study, such as a *glass transition* (the point at which a polymer changes from a rigid glass-like state to a rubbery one, thus giving an abrupt change in T_1), even if a detailed interpretation is not forthcoming.

Which relaxation times, then, are useful in solid-state NMR? In principle, they all are but some are easier to measure than others. For spin-$\frac{1}{2}$ nuclei (other than ^1H), solid-state T_1 values tend to be long (often tens of seconds) and measuring them is time consuming. However, if they are measured through a high-resolution spectrum (see section 7.2.5) and can be associated with signals from specific nuclear species present in that spectrum, they may provide localized dynamic information. $T_{1\rho}$ values can be tens of milliseconds and, while they can provide useful information on slower motions, care has to be taken in measuring them (long, high-power spin-lock pulses are required to measure long $T_{1\rho}$ times, and these can damage probes). T_1 relaxation times for quadrupolar nuclei vary; they can be many seconds but they are frequently very short (milliseconds). Proton relaxation times are the easiest to measure and these are the subject of the next section.

7.2.2 PROTON RELAXATION TIMES

These can be measured directly, often from a non-spinning (static) sample, or they can be obtained indirectly using a high-resolution cross-polarization (CP) method (see section 7.2.5). Spin diffusion (see inset 7.3) in a tightly coupled network of protons is very efficient, so proton relaxation times in solids are rarely localized to functional groups within a molecule. Usually a molecule-wide, average relaxation time is obtained for homogeneous samples, but it may be possible to relate relaxation times to specific components in the system for heterogeneous materials (see below). Despite the limitations, proton relaxation times can yield useful information, as illustrated here with a set of examples.

Relaxation Times in Polyethylene

As polyethylene is formed of just a long chain of $-CH_2-$ units (in the absence of branching), it might be expected to show relatively simple relaxation behavior. However, samples are usually semi-crystalline (i.e., they contain crystalline and amorphous domains), so proton relaxation can exhibit complicated behavior. The usual starting point for proton relaxation studies is the measurement of T_1. This will be useful for several purposes, including determination of the recycle delay for the experiments discussed later in this section and in section 7.2.4. Sensitivity is not an issue, so the

---

Inset 7.3. Spin diffusion and heterogeneous materials

Spin diffusion is a consequence of dipolar coupling. In a network of dipolar-coupled spins, the redistribution of magnetization from a locally excited site to distant sites can be viewed as a diffusion process. However, the details are situation dependent—the 1D lamellar structures of some polymers need a different model to the 3D case of a molecular crystal, for example. Nevertheless, the conclusions are qualitatively the same. For the 3D case, the mean-square distance traversed via spin diffusion (assumed to be isotropic) in a time t, $\langle r^2 \rangle$, is given by:

$$\langle r^2 \rangle = 6Dt \qquad 7.8$$

where D is an isotropic spin diffusion constant. This parameter depends on the molecular-level mobility in the system. For the case of spin diffusion between protons in proton-rich, relatively rigid systems, it is of the order of 10^{-15} to 10^{-16} m^2 s^{-1}; this will decrease as mobility increases. The relevant time, t, over which the impact of spin diffusion needs to be considered, is the relaxation time, that is, T_1 or $T_{1\rho}$. For solids, $T_1 \gg T_{1\rho}$ so spin diffusion has an influence over greater distances for T_1 than it does for $T_{1\rho}$. In this context, it is useful to talk about *domains* within a material that are either chemically or morphologically distinct. Thus, in heterogeneous materials containing such domains, spin diffusion will lead to average T_1 values over larger domain sizes than $T_{1\rho}$.

Suppose proton measurements are made on a material where $D = 5 \times 10^{-16}$ m^2 s^{-1} such that $T_1 = 1$ s and $T_{1\rho}$ is found to have two components: 5 and 30 ms. Substituting these values into equation 7.8 gives an estimate for the limiting size of the domains. From $T_{1\rho}$ it is ~10 nm. Any smaller than this and a single average value would have been obtained. From T_1 it is ~55 nm. Any larger than this and T_1 values attributable to the two domains would have been obtained (unless they were accidentally coincident).

By generalizing, on this basis, three rules of thumb can be deduced:

- One T_1 and one $T_{1\rho}$ value is observed—any domains must be smaller than 5 nm.
- One T_1 but multiple $T_{1\rho}$ values are observed—the average domain size is between 5 and 50 nm.
- Multiple T_1 and $T_{1\rho}$ values are observed—the average domain size is larger than 50 nm.

In the case of T_2, spin diffusion is not fast enough to equilibrate magnetization in different domains so they have a separate T_2 for all sizes greater than the molecule.

---

simplest approach is to make measurements on a static sample. The result of such a measurement on a commercial polyethylene sample is shown in figure 7.5(a). Plotted on a log scale, the behavior of the signal is linear, that is, a single time constant derived from equation 7.9 describes the T_1 relaxation.

$$S_\infty - S(t) = [S_\infty - S(0)]\exp(-t/T_1) \hspace{3cm} 7.9$$

Proton $T_{1\rho}$ values often provide extra information about the system. The result of such a measurement on the same polyethylene sample is shown in figure 7.5(b). This time, the signal decay is nonlinear and the relaxation cannot be represented by one time constant. Such behavior is often an indication of physical heterogeneity within a sample. For this polyethylene sample, the T_1 relaxation is that expected of a homogeneous material but the $T_{1\rho}$ behavior suggests heterogeneity. As discussed in inset 7.3, it is likely that this sample is semicrystalline and that the average domain size

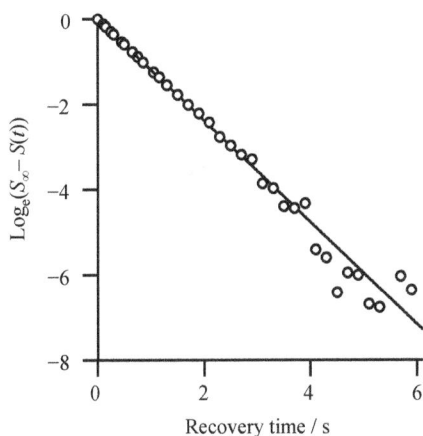

Figure 7.5. (a) A plot from a saturation-recovery T_1 measurement on a commercial polyethylene sample. $S(t)$ is the signal after a recovery time t and S_∞ is the signal at $t = \infty$ (in practice, the signal with a recovery time of $5 \times T_1$). The solid line is the least-squares fit to the experimental data (the circles) and gives a time constant of $T_1 = 0.84$ s.

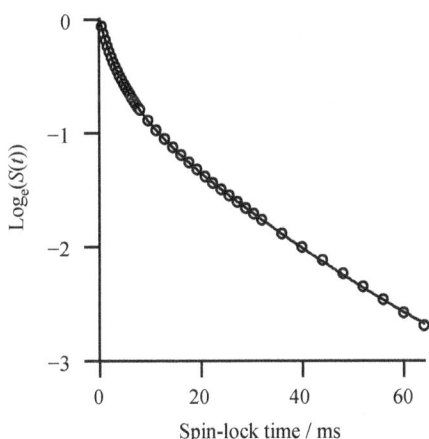

(b) The results from a $T_{1\rho}$ measurement on the same sample, showing the signal $S(t)$ as a function of spin-lock time (t). The solid line is a least-squares fit to the experimental data. The best fit is obtained from the sum of three discrete exponential decays with time constants of $T_{1\rho} = 1.2$, 4.5, and 27.1 ms. Fitting curves like this to more than two components can be contentious as the chemical/physical significance of the additional component(s) may be debatable! Careful consideration of the error in each value is essential. Both (a) and (b) were obtained with a *solid echo* (see inset 3.5) and $S(t)$ is the intensity of the echo maximum.

is between 5 and 50 nm. Analyzing relaxation data in terms of discrete exponentials is one approach but it is not the only one, as shown in inset 7.4.

Inset 7.4. Modeling relaxation behavior

Least-squares fitting of the experimental data to one or more discrete exponential functions is a common method found in the literature for analyzing relaxation measurements. However, the nature of some materials is such that *distributions* of relaxation times are likely to be encountered. Then, analyzing in these terms is more appropriate. The result of such a treatment for the decay in figure 7.5(b) is shown here, together with the values from the discrete component model (the vertical lines). The major components are treated consistently but the problem of dealing with a weak component is illustrated in the difference for the minor one. Measurement of pore-size distribution in oil-bearing rock is one example where this type of approach is used (the relaxation time of the oil is pore-size dependent).

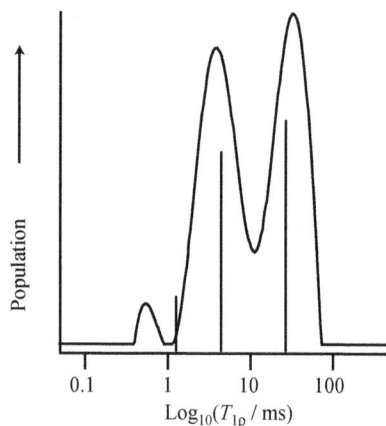

In a system with multiple relaxation time constants, each one will have a *population* (the number of ^1H nuclei relaxing with that time constant) associated with it. If there is no interaction between the domains to which the time constants relate, then the populations will be proportional to the number of ^1H atoms in the domains. However, if the domains interact so that spin diffusion can occur between them, the populations of the faster relaxing domain will be enhanced at the expense of that relaxing more slowly and the measured populations will no longer reliably relate to the relative numbers of ^1H atoms in each domain. In such cases, the lineshape is a better indicator of the composition of a sample.

7.2.3 LINESHAPES & LINEWIDTHS FOR NON-SPINNING SAMPLES

The interpretation of the relaxation results discussed in the previous section is helped by knowledge of the ^1H lineshapes observed for non-spinning samples. Homonuclear dipolar coupling is the main contributor to the linewidth and the bandshape for an isolated, dipolar-coupled spin pair has already been introduced (see section 2.5). In polyethylene the interaction between the hydrogens in each CH_2 unit might be dominant (25.1 kHz) but these protons cannot be considered as an isolated pair and there are many longer-range interactions that contribute to the bandshape, resulting in a broad, featureless line (see section 5.2).[4] Any motion in the system will tend to reduce the dipolar coupling, which results in line narrowing. This fact leads to an explanation of the ^1H bandshape for the polyethylene sample discussed in the previous section and shown in figure 7.6(a). The bandshape has two components, one of which is broad with a full width at half height, $\Delta v_{\frac{1}{2}}$, of around 50 kHz. Values of this magnitude are typical of rigid, crystalline organic materials. The second component is narrower, $\Delta v_{\frac{1}{2}} \simeq 10\,kHz$, which implies that the coupling is reduced through motion. This is typically associated with an amorphous domain (the looser structural constraints allow more freedom for motion). So the polyethylene sample consists of crystalline and amorphous domains. It is possible to show, by measuring the bandshape as a function of spin-lock time, that the broad component is associated with the longest $T_{1\rho}$ value (figure 7.6(b)). The crystallinity of the sample can be determined from the relative intensities of the broad and narrow components. This is not necessarily a trivial exercise, as the components of the bandshape may not fit well to simple (Gaussian or Lorentzian; see inset 7.9) lineshape functions.

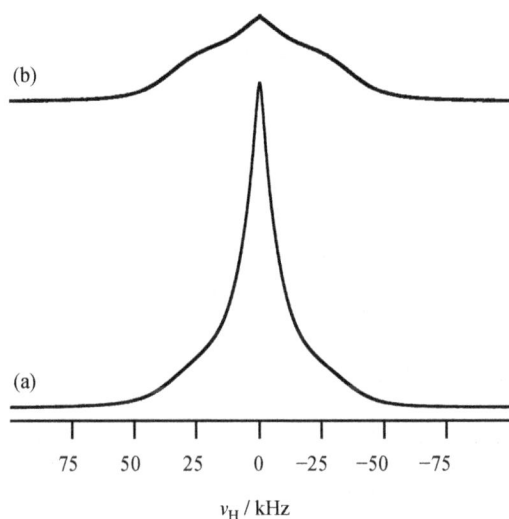

Figure 7.6. (a) The static ^1H spectrum from polyethylene. (b) The spectrum from the same sample obtained after spin locking the ^1H magnetization for 18 ms. The part of the spectrum with the narrowest line is largely lost and therefore must be associated with the shortest proton $T_{1\rho}$ value.

---

[4] Some texts introduce a term called the *second moment* (M_2) to help describe the bandshape. This is a measure of the mean, local magnetic field generated by all the pairwise dipolar interactions experienced by a particular nucleus. High values for M_2 equate to broad lines and are associated with rigid systems where the dipolar coupling is at its strongest. For a powdered sample, the second moment for heteronuclear interactions is:

$M_{2IS} = \frac{1}{5}\left(\gamma_I \gamma_S \frac{h}{2\pi}\frac{\mu_0}{4\pi}\right)^2 \sum_k r_{jk}^{-6}$ and for the homonuclear case $M_{2II} = \frac{9}{4}M_{2IS}$. r_{jk} is the distance between spins j and k.

T_2, T_2' & T_2^*

So far much has been made of the properties of T_1 and $T_{1\rho}$ but little has been said about T_2. As has been shown, T_1 and $T_{1\rho}$ are relatively easy to characterize but the same is not true for T_2. For this reason, transverse relaxation measurements tend to receive less attention in solid-state NMR studies.

For a solution, T_2 is the time constant that describes the irreversible magnetization decay in the xy plane, in the absence of the effects of magnetic field inhomogeneity. It can be measured using the Carr–Purcell–Meiboom–Gill (CPMG) pulse sequence (figure 7.7(a)), which creates a series of echoes, refocusing the effects of the magnetic field inhomogeneity. In solutions this decay is generally exponential. However, the observed decay of the NMR signal (because of the effects of magnetic field inhomogeneity) occurs on a shorter timescale than T_2. A related quantity T_2^* (the rate of loss or dephasing of the phase coherence produced by the excitation) is defined as the time taken for the signal to decay to $1/e$ of its initial value (figure 7.7(b)). For a solution, the exponential decay of the signal leads to a Lorentzian lineshape with full width at half-height given by $1/\pi T_2^*$.

For a solid the definition of a specific time T_2 is a matter for discussion. For a rigid solid, the echo decay from the CPMG experiment is not usually exponential (see figure 7.8) so a value for T_2 is not forthcoming in the same way as it is in solution. However, a value of

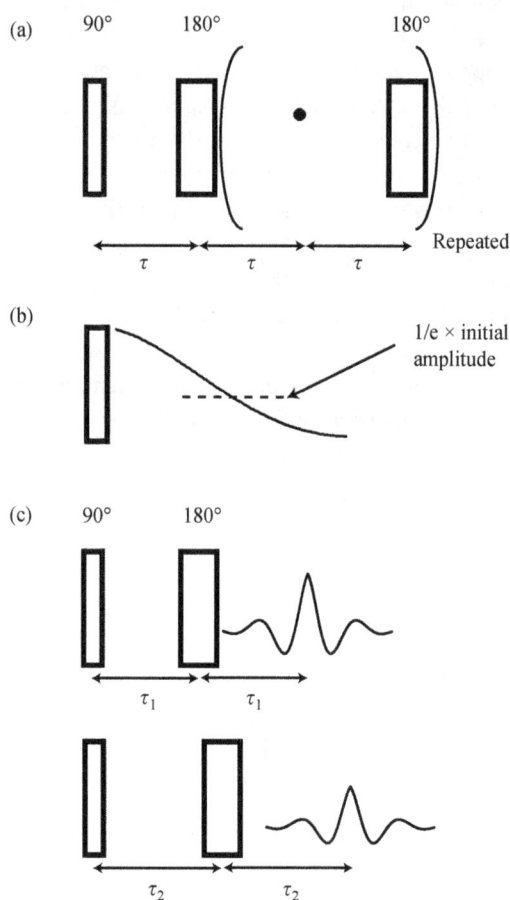

Figure 7.7. (a) Point-by-point Carr–Purcell–Meiboom–Gill (CPMG) measurement for T_2. (b) T_2^* measurement from the free-induction decay (FID). (c) T_2', measurement by varying the echo delay in a spin-echo experiment.

T_2^* can be measured as it is in solution. For a Lorentzian lineshape, it relates to the linewidth as it does for a solution, or for a Gaussian shape (see section 7.4) the full width at half-height is given by $1.67/\pi T_2^*$. So a short T_2^* (and that can mean less than a few tens of microseconds) implies a broad line (a static proton line, with Gaussian shape—often a good approximation—and a half-height width of 40 kHz, equates to $T_2^* = 13$ μs). A long T_2^* corresponds to a narrow line. In solid-state NMR, a slightly different parameter, T_2', can be obtained from the decay of the echo intensity as shown in figure 7.7(c). This parameter is related to the homogeneous linewidth (see section 5.2).

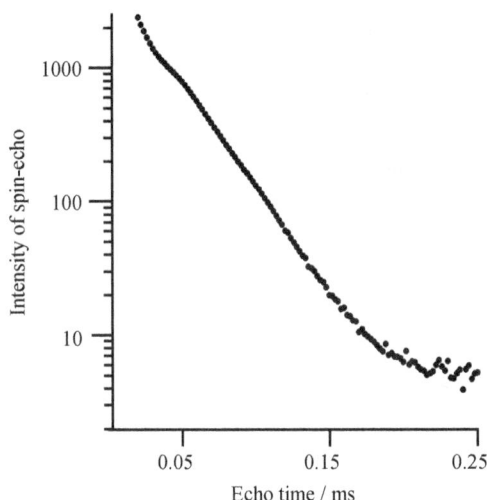

Figure 7.8. The decay of the proton spin-echo intensity for a non-spinning, polycrystalline sample of glycine.

7.2.4 RELAXATION & HIGH-RESOLUTION MEASUREMENTS COMBINED

Interrupted Decoupling

The *interrupted decoupling* experiment (also known as dipolar dephasing, non-quaternary suppression, and protonated carbon dephasing) exploits differences in T_2^* values to simplify (or *edit* or *filter*) the spectrum. It is one of the most widely used editing experiments in the solid-state NMR study of organic materials. As shown in inset 7.5, the pulse sequence includes a short window where the ^1H decoupling is turned off. Briefly (often 40–50 μs for carbon or ~200 μs for nitrogen) turning off the decoupling connects the ^{13}C (say) magnetization of protonated carbon sites to the large proton *spin bath*. This drains intensity away from the observable ^{13}C signal and suppresses the signal from those protonated carbons. The magnetization from quaternary carbons (or tertiary nitrogens) will not be suppressed to the same extent as they are only weakly coupled to protons (they have no directly bonded hydrogens).

Inset 7.5. Interrupted decoupling

The interrupted decoupling experiment is used to simplify spectra from organic materials. The delay τ is sometimes referred to as the dephasing delay. The rotor-synchronized 180° pulse on the X-channel refocuses the signal, which makes the result easier to process (it avoids the need for a large first-order phase correction). τ_r is the rotor period.

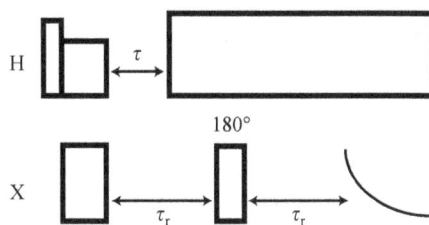

An example of such an experiment is shown in figure 7.9. Molecular motion, either of a molecule as a whole or of part of a molecule, such as methyl group "rotation", tends to reduce dipolar coupling (to ^1H) and T_2^* is not reduced to the same extent as for a rigid CH grouping, so signals from mobile species (particularly methyl groups) are rarely fully removed from the spectrum. Fortunately, methyl groups can usually be distinguished from quaternary carbons by their chemical shift. This experiment is not the only means of editing spectra on the basis of multiplicity, but it is simple and reliable.

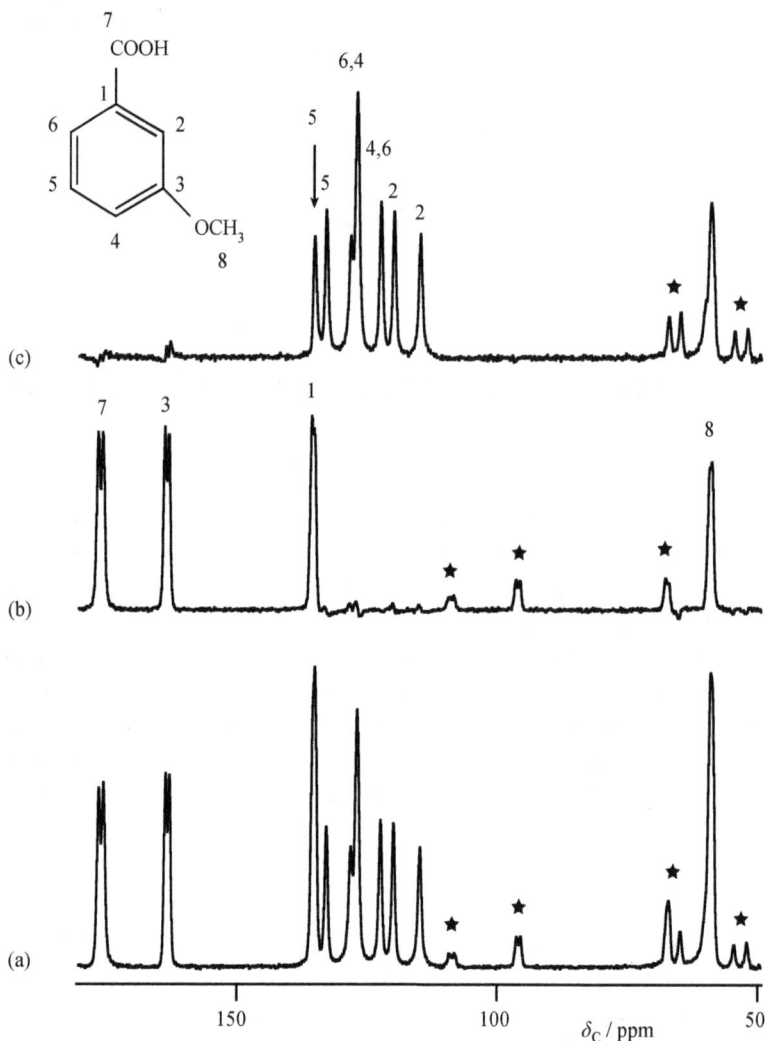

Figure 7.9. Carbon-13 cross-polarization magic-angle spinning (CPMAS) spectra from 3-methoxybenzene carboxylic acid recorded (a) without and (b) with 40 μs of interrupted decoupling. (c) A difference spectrum (showing the CH carbons) which confirms the presence of a CH carbon signal under that from carbon 1 (marked with an arrow). There are two molecules in the asymmetric unit (note the pairs of lines) and the ★ indicate spinning sidebands.

Relaxation Filters

Any difference in relaxation times in systems exhibiting multicomponent relaxation behavior can be exploited to simplify (or aid the interpretation of) spectra. Most of the pulse sequences given in section 7.2.5 potentially can be used to do this, but two specific examples are included here using the polyethylene example discussed earlier in this chapter. Figure 7.10(a) illustrates the carbon-13 cross-polarization magic-angle spinning (CPMAS) spectrum. It shows two major lines, the narrower one at 32.8 ppm and a broader one at about 31 ppm. Delaying the contact time, in a modified CP experiment (see figure 7.14(b)), allows the proton magnetization from the component of the material with the shortest $T_{1\rho}^{H}$ to decay. Subsequent CP can then occur only from that part of the material with the longer $T_{1\rho}^{H}$ (the rigid, crystalline component). As shown in figure 7.10(b), the broad line is reduced in intensity and the narrow signal from crystalline polyethylene dominates. (Note that it is not always the case that a crystalline component has the longer $T_{1\rho}^{H}$.) This is an example of a $T_{1\rho}^{H}$ filter.

A filter based on T_2^{*H} produces the opposite result (figure 7.10(c)). Now the crystalline component, which has the shorter T_2^{*H}, is the one that is reduced in intensity (the experiment used to produce this result is described in the next section).

If signals in a high-resolution spectrum can be separated in this way, then it becomes possible to explore any interactions between the components of the sample by adding a mixing time to the experiment (as in the EXchange SpectroscopY (EXSY) experiment described in the next part of this chapter).

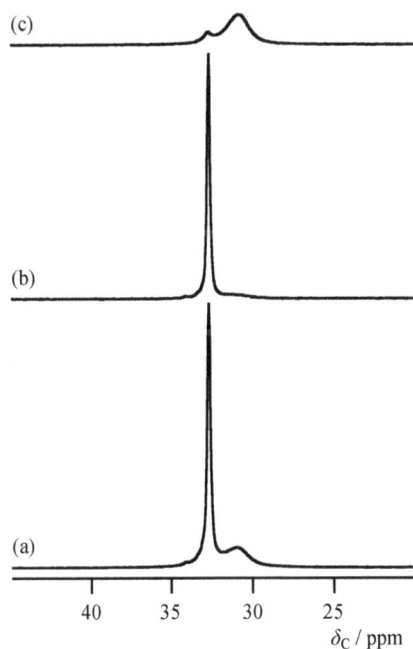

Figure 7.10. (a) The cross-polarization magic-angle spinning (CPMAS) spectrum from a commercial polyethylene sample. (b) The spectrum from the same sample but recorded with a $T_{1\rho}^{H}$ filter (18 ms delay before contact); (c) obtained from the sample but with a T_2^{*H} filter (a 16 μs delay in the sequence shown in inset 7.6).

Suppose that signal separation has been achieved with a relaxation filter and the spectrum is then monitored as a function of mixing time. If the suppressed signal(s) increase in intensity, then spin diffusion must be transferring magnetization from one component of the sample to another and there must be an interaction between them.[5] This is generally termed a Goldman–Shen experiment, and the result of one such measurement, based on the filter used for figure 7.10(c), is shown in figure 7.11. From this result, it is possible to estimate a domain size, although this is not a trivial exercise and requires knowledge (or an assumption) of the rate of spin diffusion, together with a model for the shape of the domains. Note that it is usually easier to monitor an increase in the height of a narrow line (it increases more quickly than for a broad one) so a filter that suppresses a crystalline component is preferable.

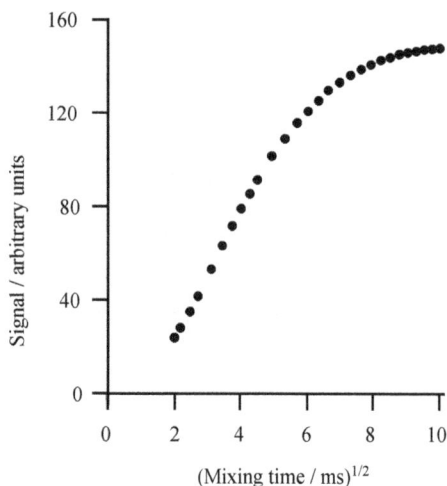

Figure 7.11. The carbon-13 signal height for the crystalline component of a polyethylene sample as a function of the mixing time following a T_2^{*H} filter. A curve of this type can be used to deduce a domain size for the amorphous component of the sample. Signal recovery on this timescale and with this shape correlates with a 3D amorphous domain size of the order of ~10 nm.

^{1}H–X Correlation

There is a simple but useful heteronuclear correlation experiment (see section 5.4.3) that can demonstrate the existence of different mobilities (relaxation behaviors) within a sample. This is the so-called WIdeline SEparation (WISE) experiment (inset 7.6). The 2D experiment correlates a high-resolution X (usually ^{13}C) spectrum with a low-resolution ^{1}H one. Low spin rates (~5 kHz) are sufficient to generate the high-resolution X spectrum but have little impact on the ^{1}H bandshape, which will resemble the static spectrum. The result of such an experiment on the polyethylene sample is shown in figure 7.12.

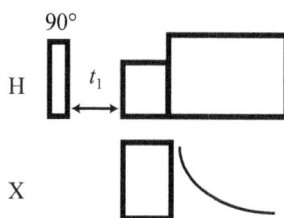

Inset 7.6. WISE

The WISE pulse sequence. Spectra are acquired in 2D fashion as a function of the delay t_1 (typically in the range 0–100 μs).

---

[5] Assuming the mixing time is much less than T_1^H (in this case).

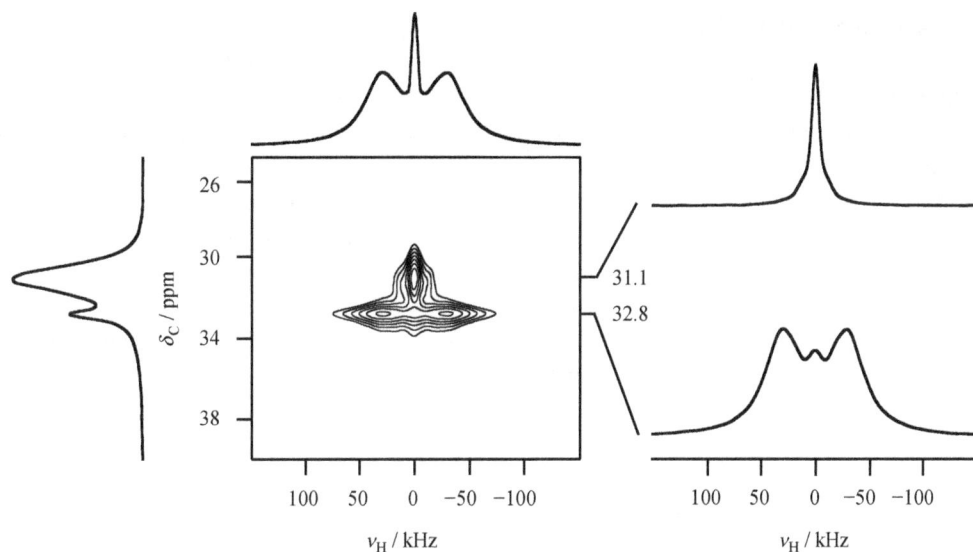

Figure 7.12. The result of a WIdeline SEparation (WISE) experiment on polyethylene. The narrow ^{13}C signal at 32.8 ppm from the crystalline domains correlates with a broad proton bandshape while the broader ^{13}C signal (31.1 ppm) from the amorphous component correlates with a narrow proton line.

7.2.5 MEASURING RELAXATION TIMES

The pulse sequences that are used to record the T_1 and $T_{1\rho}$ relaxation times discussed in this chapter have not yet been introduced. They are grouped together in this section and are summarized in table 7.1.

Table 7.1. Pulse sequences for measuring relaxation times

Pulse sequence	Relaxation time	Figure	Nucleus
Inversion recovery	T_1	7.13(a)	H or X
Saturation recovery	T_1	7.13(b)	H or X
Inversion recovery with cross polarization	T_1 (indirect)	7.13(c)	H
"Torchia" method (utilizing cross polarization)	T_1	7.13(d)	X
Variable spin-lock	$T_{1\rho}$	7.14(a)	H or X
Delayed contact	$T_{1\rho}$ (indirect)	7.14(b)	H
Variable spin-lock, cross-polarization preparation	$T_{1\rho}$	7.14(c)	X

Spin–lattice Relaxation Times

Four methods for measuring spin–lattice relaxation times are shown in figure 7.13. The inversion-recovery method (figure 7.13(a)) gives a potentially large signal range (from approximately fully inverted to fully positive) but has the disadvantage (particularly for slowly relaxing species) that the recycle delay must be at least $5T_1$ to allow the system to fully relax between repetitions. The signal recovery is fitted to the equation $S(\tau) = S_0(1 - 2e^{-\tau/T_1})$, where $S(\tau)$ is the signal intensity measured with a recovery time τ and S_0 is the signal intensity when $\tau = 0$. The time at which the signal is zero (the *null time*) is useful for a quick indication of T_1: $S(\tau) = 0$ when $\tau = \ln(2)T_1 = 0.693T_1$. The rectangular 180° pulse can be replaced by a shaped pulse for selective inversion or for the more efficient inversion of a signal from a quadrupolar nucleus.

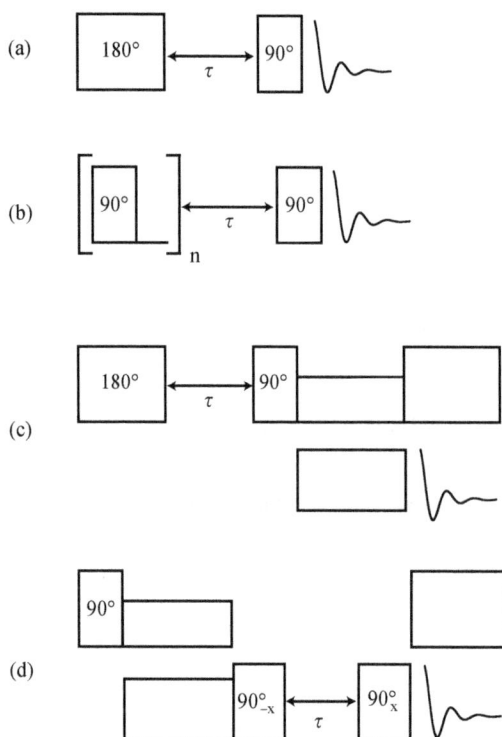

Figure 7.13. Pulse sequences for measuring spin–lattice relaxation times.

The saturation-recovery sequence (figure 7.13(b)) starts by saturating the spins (e.g., by setting n to 64), so there is no need to wait for the system to return to equilibrium after the acquisition (the recycle delay can be set to zero, though if ^1H decoupling is being used while measuring T_1^X, care must be taken to avoid overheating the probe). The signal recovery is fitted to: $S(\tau) = S_0(1 - e^{-\tau/T_1})$.

Figures 7.13(c) and (d) incorporate CP steps. The pulse sequence in figure 7.13(c) is used to measure T_1^H indirectly through an X nucleus spectrum. In a heterogeneous sample where T_1^H is not a sample-wide average (see inset 7.3), it may be possible to relate T_1^H values to chemically identifiable components within the sample (this information is not available from a wideline ^1H measurement). The sequence shown in figure 7.13(d) is for measuring T_1^X. Often, the delay τ has to extend to long times but the advantages of CP discussed in section 3.4 apply here, making this method practical for dilute and slowly relaxing nuclei. No ^1H decoupling is applied during the variable delay to prevent damage to the probe. As a consequence, relaxation behavior measured this way may be non-exponential because there is the potential for *cross relaxation* involving both ^1H and X nuclei and a simple analysis of the result may not be possible.

Spin–lattice Relaxation Times in the Rotating Frame

In the basic measurement (figure 7.14(a)), spectra (or a single point from the FID) are recorded as a function of the spin-lock time. The signal decay is fitted to an equation of the form $S(\tau) = S_0 e^{-\tau/T_{1\rho}}$. $T_{1\rho}$ depends on the RF field strength during the spin-lock (as well as on the sample) so it is important to state this when quoting results.

$T_{1\rho}^H$ can be measured indirectly through an X nucleus spectrum by incorporating a CP step (figure 7.14(b)). This is often referred to as a delayed contact experiment. As for T_1^H, this may impart some chemical significance to the result. $T_{1\rho}^H$ can also be determined by varying the contact time in a CP experiment and fitting the result to equation 3.8. However, the data analysis for the delayed contact method is simpler: $S(\tau) = S_0 e^{-\tau/T_{1\rho}}$.

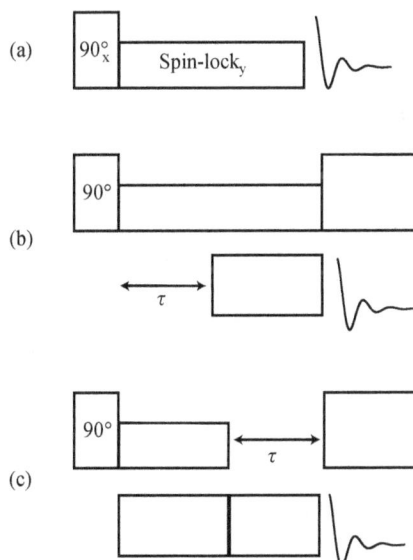

Figure 7.14. Pulse sequences for measuring spin–lattice relaxation in the rotating frame.

Figure 7.14(c) gives a way of measuring $T_{1\rho}^X$ with the advantages of the CP method. Proton decoupling is not applied during the variable delay to prevent unwanted CP. Again, cross relaxation may complicate the data analysis.

With all the spin-lock measurements, care must be taken not to damage the probe with very long spin-lock pulses.

The sequences involving CP are invariably used in conjunction with magic-angle spinning (MAS). This raises the issue of whether relaxation times measured on a non-spinning sample are the same as those from one undergoing MAS. For protons, in particular, high spin rates weaken the homonuclear dipolar coupling and it may be the case that a molecule-wide average relaxation time is no longer observed. Such behavior has been noted, for example, for hydrogen-bonded carboxylic acid protons at spin rates of 30 kHz. In this case the acid proton may have a considerably longer spin–lattice relaxation time than the other protons in the molecule. $T_{1\rho}$ depends on motion in the tens of kilohertz range—similar to those associated with MAS, so the subtle effects of interference between $T_{1\rho}$, the RF field strength, and the spin rate can complicate matters—but can provide useful information on the slower motions in a sample.

7.3   EXCHANGE

In addition to causing relaxation and influencing linewidths, motion in the form of exchange processes can also affect the appearance of a spectrum. Here, an exchange process is defined as one that *interchanges* the environment of the atoms in the system under study. The motional process is frequently an internal rotation or a change in speciation involving the movement of hydrogen. The examples in the next two sections illustrate both these types of process. There is more information on dynamic tensor averaging in section 8.2.

7.3.1   POSITIONAL EXCHANGE

An exchange process can involve two or more sites (or environments). For example, in an organometallic coordination polymer containing $-C{\equiv}N-Sn(CH_3)_3-N{\equiv}C-$, there is rotation about the N–Sn–N axis and the methyl groups interchange between three equally populated environments. The appearance of the carbon spectrum depends on the temperature of the measurement: at low temperature three methyl signals are observed but at high temperature a single signal is seen. For simplicity, exchange between only two sites (A and B, say) is considered now:

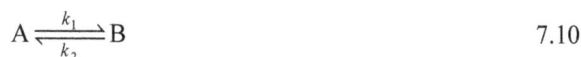

$$A \underset{k_2}{\overset{k_1}{\rightleftharpoons}} B \qquad\qquad 7.10$$

If the forward (k_1) and backward (k_2) exchange rates are the same, then A and B will be equally populated and they will contribute equally to the NMR spectrum. If the exchange rates are not equal, then the system will spend more time in one of the sites than the other (the lifetime, τ, of a particular state is given by $\tau = 1/k_1$ or $1/k_2$). In this case, sites A and B do not contribute equally to the spectrum.

The drug compound formoterol (used in the treatment of asthma) illustrates an exchange process involving 180° ring flips about a phenylene axis. Its molecular formula is shown in scheme 7.1. It can exhibit a form of positional exchange that is often observed in the solid state for molecules containing 1,4-bonded phenylene

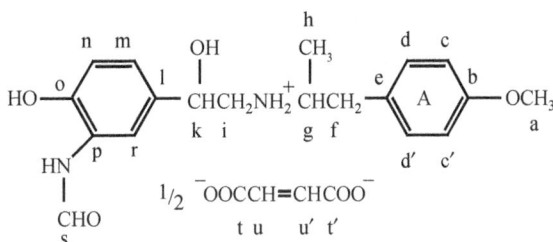

Scheme 7.1. Formoterol (fumarate salt).

rings. In solution in dichloromethane, the ring labeled A undergoes fast rotation (on the NMR timescale), which renders carbons c and c' equivalent (and similarly d and d'), so that each pair gives a single line in the carbon-13 spectrum (at 115.6 and 131.8 ppm, respectively). Part of the carbon-13 spectrum from the fumarate dihydrate form recorded in the solid state at ambient probe temperature (~25 °C) is shown in figure 7.15 and consists of a number of well-resolved lines. Comparison with the solution-state chemical shifts allows the lines to be assigned as indicated, but there are no obvious signals for carbons c, c', d, or d'. This contrasts with the situation at low temperature when four additional lines (at 107.2, 118.6, 127.5, and 130.9 ppm) are observed. These can be attributed to the missing carbons (the average shifts of these pairs, 112.9 and 129.2 ppm, are close to the solution-state values).

These observations can be explained if the phenylene ring is undergoing 180° flips about the 1,4-axis. At –50 °C, the phenylene ring is only flipping slowly on the NMR timescale (at a rate

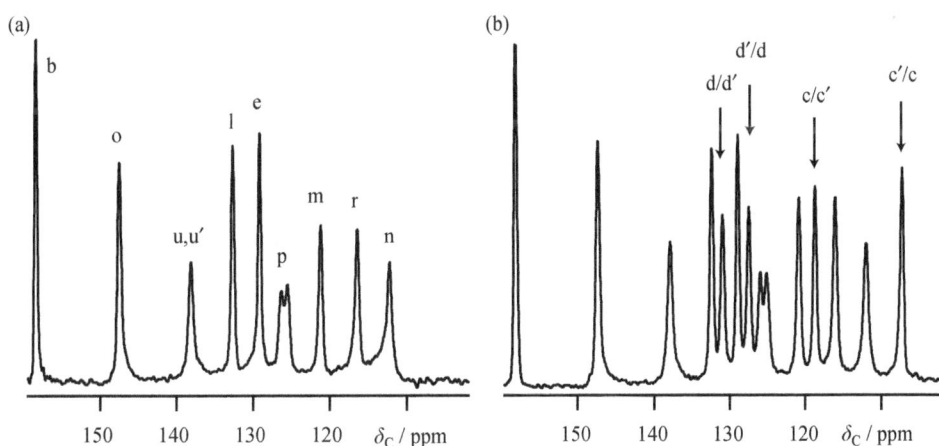

Figure 7.15. Part of the ^{13}C cross-polarization magic-angle spinning (CPMAS) spectrum from formoterol fumarate dihydrate recorded at ambient probe temperature (a) and at –50 °C (b). The lettering refers to scheme 7.1.

of less than 1000 s⁻¹) and the local environments of c and c′ and d and d′ are sufficiently different for distinct resonances to be observed. As the temperature is increased, the ring motion becomes faster and eventually one line would be observed for c and c′ and one for d and d′ (at the average of the individual chemical shifts). At intermediate temperatures, the sharp lines observed at low temperature broaden (see also section 8.8.1) and eventually each pair coalesces into one broad line before this single line narrows at higher temperatures. For formoterol fumarate dihydrate, the coalescence temperatures for the two pairs of lines are around ambient temperature and it is difficult to detect the broad lines from these carbons among the other lines from the rest of the molecule. The behavior of the signals is illustrated schematically in figure 7.16.

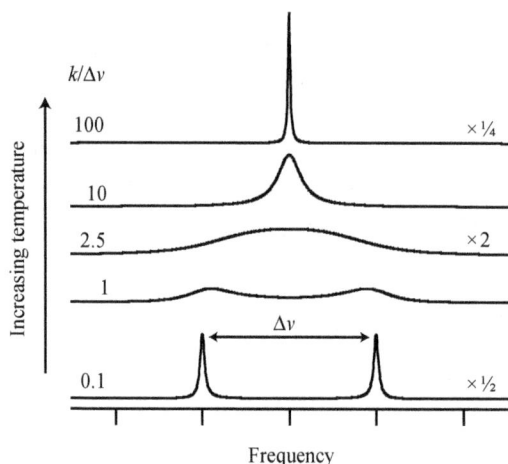

Figure 7.16. Schematic illustration of the coalescence of two lines, related by exchange. The point at which one flat-topped band is observed is known as the *coalescence temperature*. The vertical scales are varied as indicated. k is the exchange rate constant.

Coalescence Temperature

It is possible to model the lineshape to determine an exchange rate over a range of temperatures. However, it is necessary to know the shape of the lines in a non-exchanging state. Given that the general behavior of a pair of lines in the solid state is, in many cases, analogous to that in solution (i.e., they broaden, coalesce, and then narrow), then the exchange rate (k^c) at the temperature of coalescence, for a two-site exchange with equal forward and reverse exchange rates, is given by:

$$k^c = \frac{\pi}{\sqrt{2}} \Delta v \qquad\qquad 7.11$$

where $\Delta v = |v_A - v_B|$, v_A and v_B being the *frequencies* of the two lines related by exchange. For the case discussed here, the c and c′ lines coalesce at approximately 25 °C and $\Delta v = 860$ Hz, so $k^c \approx 1900$ s⁻¹.

7.3.2 HYDROGEN EXCHANGE

The exchange process for tropolone involves the transfer of a proton between the two oxygen atoms, together with a rearrangement of the double bonds (scheme 7.2). There is a subsequent rotation of the whole molecule to restore its packing in the crystal lattice so that the start and end points of the process are identical. This is not evident in an X-ray diffraction study, but the presence of an exchange process can be demonstrated using a 2D NMR experiment (called EXSY). The EXSY experiment is almost as simple as the WISE experiment; the only difficulty comes in choosing the *mixing time*, τ_{mix}. Getting that right often comes down to trial and error unless the exchange rate is already known (see below). The pulse sequence is shown in inset 7.7 and the experiment applied to tropolone gives the result shown in figure 7.17. The peaks along the diagonal are the ones that appear in the 1D spectrum. The off-diagonal peaks arise because of the exchange process. They link pairs of resonances on the diagonal that are related by exchange: 1 and 2, 3 and 7, and 4 and 6. Only carbon 5 has no partner as its environment is unaffected by the exchange.

Often, it is useful to measure the exchange rate as a function of temperature. The 2D spectrum of the quality shown in figure 7.17 took 76 hours to obtain, so recording a series of such spectra can be prohibitively time consuming. However, the EXSY experiment can be adapted to yield the exchange rate more quickly. If, instead of incrementing the evolution delay t_1 (inset 7.7), it is set to a single value determined by the frequency difference of two lines of interest, then a spectrum of the form shown in figure 7.18 will be obtained at zero mixing time. As the mixing time is increased, exchange will reduce the intensity of line A, while line B will become less negative. A plot of the intensity difference against the mixing time will yield the exchange rate, as shown in figure 7.19. In the tropolone case, at 30°C, this pro-

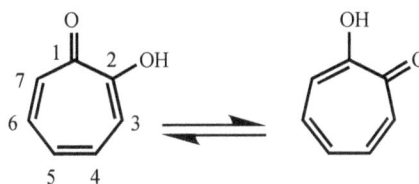

Scheme 7.2. Hydrogen exchange in tropolone.

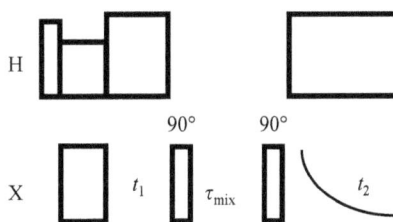

Inset 7.7. EXSY

The EXSY pulse sequence, shown with a CP preparation period and decoupling during the *evolution* time t_1 and the acquisition time (t_2). The experiment is carried out in 2D fashion by incrementing t_1. The parameter τ_{mix} is the mixing time and as this is often relatively long, it is usual to turn the decoupling off during this period. The effect of the 90° pulse at the end of the preparation period and before the mixing time is to store the (carbon) magnetization along the z axis. The result of any mixing (exchange) is then "read" by the subsequent 90° pulse. The experiment usually works best when the mixing time required to demonstrate exchange is short relative to T_1^X (and $t_1 \ll \tau_{mix}$).

cedure yields an exchange rate of 0.12 s^{-1}. The advantage with this method is that it only took a total of 2 hours to record the spectra for the 10 mixing times used to produce figure 7.19, so it is entirely feasible to obtain the exchange rate as a function of temperature.

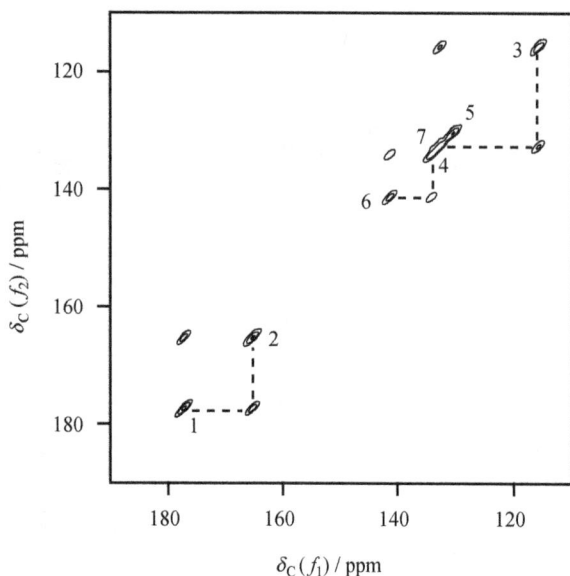

Figure 7.17. A ^{13}C EXSY experiment from tropolone obtained at 30 °C and with a mixing time of 3 s. The proton relaxation in tropolone is quite slow, so a long recycle delay is needed (60 s at 9.4 T). The numbering refers to scheme 7.2 and the dashed lines link peaks related by the exchange process.

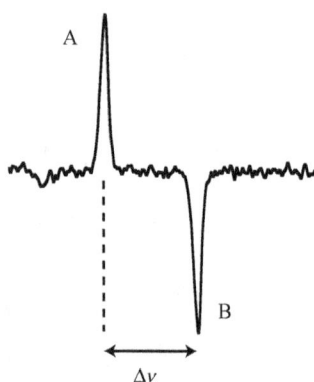

Figure 7.18. The result of a 1D form of the EXSY experiment from tropolone, obtained at 30 °C with zero mixing time. Line A is on-resonance and the t_1 delay (inset 7.7) is set to $1/2\Delta v$. This has the effect of inverting line B.

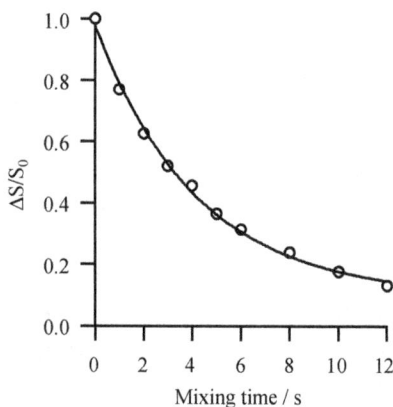

Figure 7.19. The difference in signal height as a function of mixing time for the modified EXSY experiment on tropolone at 30 °C. The time constant of the decay is (ignoring relaxation) $1/2k$.

Energy of Activation

For a thermally activated exchange process, the activation energy can be obtained from an Arrhenius plot of exchange rate as a function of temperature (see section 7.2.1). For tropolone the literature value for the activation energy is 109 kJ mol^{-1}. The activation energy for a phenylene ring flip like that in formoterol is typically lower (<50 kJ mol^{-1}).

7.3.3 REORIENTATION WITHOUT A CHANGE IN ISOTROPIC CHEMICAL SHIFT

In general, any anisotropic interaction can be modulated by molecular motion, so EXSY-based experiments (or bandshapes for quadrupolar nuclei) can often give information on a change in orientation even when there is no change in isotropic chemical shift. This contrasts with the solution state, where exchange effects are limited to jumps between inequivalent sites because anisotropic interactions are averaged out.

Deuterium

It is the high natural abundance of ^{1}H and the myriad dipolar couplings that are responsible for the relatively featureless static ^{1}H bandshapes discussed in section 7.2.3. By contrast, for ^{2}H, homonuclear dipolar coupling can be neglected so the ^{2}H resonance from a non-spinning sample is dominated by quadrupolar coupling (see chapter 6) and this may be affected by anisotropic molecular motion. This will translate into the shape of the spinning sideband manifold in a spectrum acquired under MAS.

Deuterium has a low natural abundance (0.0115%) and to observe a signal isotopic labeling is often necessary. However, if this can be done in a site-specific fashion, deuterium can be introduced into a molecule or macromolecular system as a localized probe of motion. In the simplest cases, the motion results in a scaling of the quadrupolar bandshape that would be observed when there is no motion. This is illustrated in figure 7.20. The examples here show the slow-exchange limit (figure 7.20(a)), where $\tau_c \gg 1/\chi$, and the fast exchange limit (figure 7.20(b)), where $\tau_c \ll 1/\chi$. Like the spin-$\frac{1}{2}$ case, intermediate correlation times (typically 10^{-5} to 10^{-6} s) produce distorted (and, for MAS, broadened) bandshapes.

Other types of motion give rise to different bandshapes, as illustrated in figure 7.21 (and figure 6.24). So, the appearance of the deuterium bandshape (or spinning sideband manifold in

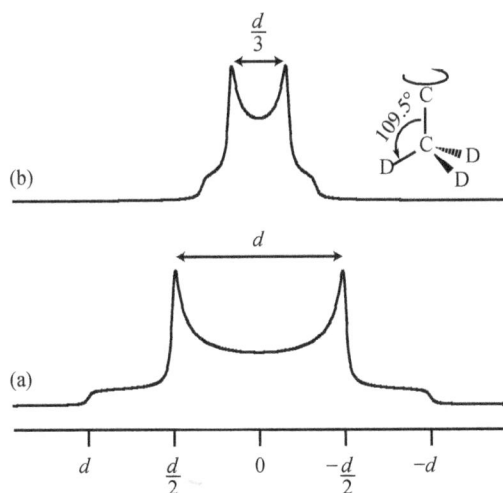

Figure 7.20. Schematic simulated ^{2}H bandshapes. For: (a) an immobile C–D bond, $d = \frac{3}{4}\chi \approx 128$ kHz and (b) a deuteromethyl group rotating about the C–C bond. For (b) the motion is a hypothetical "free" rotation with $\tau_c < 100$ ns (i.e., $\tau_c \ll 1/\chi$) and the quadrupole coupling is scaled by a factor $\frac{1}{2}(3\cos^2\psi - 1)$, where ψ is the angle between the C–D bond and the axis of rotation (109.5° here).

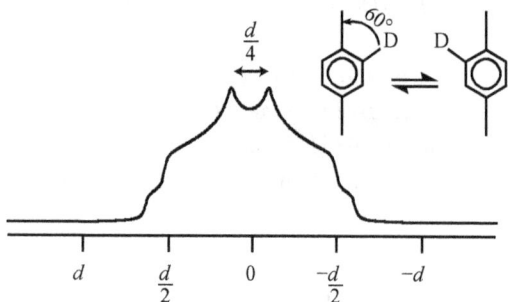

Figure 7.21. Simulated ^{2}H bandshape for a fast ($\tau_c < 100$ ns) 180° phenyl ring flip, $d = \frac{3}{4}\chi$.

a MAS spectrum) can give information on the angle of reorientation involved in the motional process. However, in practice, things might not be as straightforward as these simulated spectra imply. In polymeric systems, for example, a well-defined local motion is likely to be superimposed on motion(s) of the polymer chain. This will complicate the ^{2}H bandshape and make it more difficult to extract a picture of the motion. In some cases, synthesizing a system with ^{2}H localized to a single environment may not always be practical and then it may be necessary to unravel complex overlapping bandshapes from multiple deuterium environments. Double-quantum/single-quantum or heteronuclear correlation experiments may be necessary to achieve the required ^{2}H spectral resolution. Analogous effects also can be observed for other quadrupolar nuclei (including those where $I \neq 1$).

A non-spinning EXSY experiment can also be applied to deuterium (although in this case the 90° storage pulses are replaced by 54.7° ones). The result is a 2D spectrum with off-diagonal elliptical ridges, the precise nature of which gives information on the angle of molecular reorientation involved in the exchange process[6] (see section 8.7.2).

For spin-$\frac{1}{2}$ nuclei, the CSA can also be used to probe motion. Two-dimensional experiments can distinguish between different models of the reorientational process as the pattern of off-diagonal intensity depends on the nature of that process. For ^{13}C, MAS is necessary for sensitivity, so care is needed to ensure that the intensities of the off-diagonal spinning sidebands relate to the exchange process (and are not modulated by the sample spinning). Although 2D experiments can distinguish between different motional models, acquiring a series of them to extract exchange rates can be prohibitively time consuming. For this reason, a number of experiments, which rely only on the measurement of centerband intensities in 1D spectra, have been developed. One such is CODEX (Centerband Only Detection of EXchange), and from two short series of experiments, it is possible to obtain a correlation time and information on the type of motion present (in principle for all resolved lines in the spectrum) (see figure 7.22 and Further reading).

---

[6] The spectrum from deuterated dimethylsulphone is a classic example of this: C. Schmidt, B. Blümich and H.W. Spiess *J. Magn. Reson.* **79** (1988) 269.

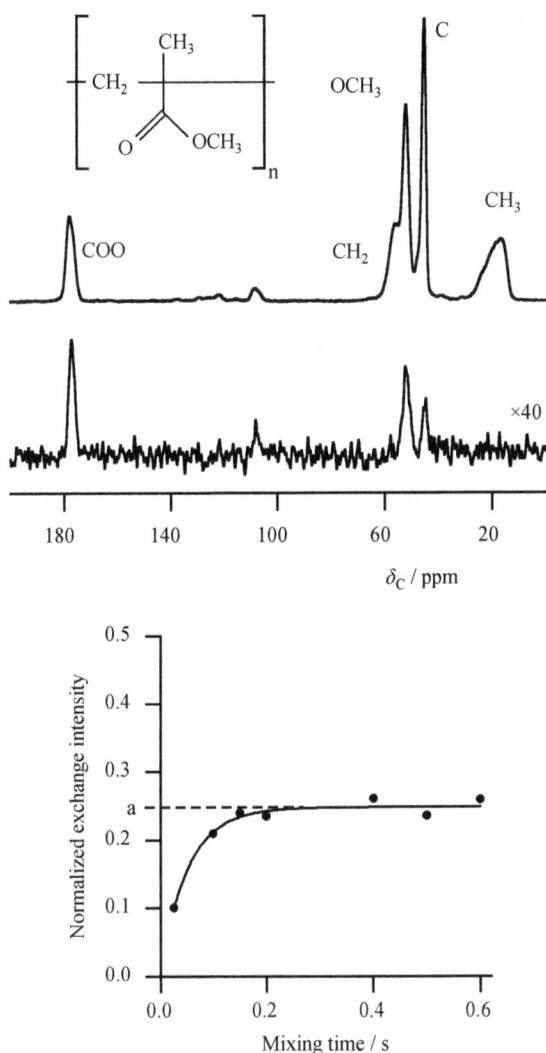

Figure 7.22. Carbon-13 cross-polarization magic-angle spinning (CPMAS) spectrum from (top) polymethylmethacrylate (PMMA) together with (middle) its pure-exchange Centerband Only Detection of Exchange (CODEX) spectrum (for which the signal from immobile sites has been subtracted). The graph at the bottom is the normalized exchange intensity as a function of mixing time for the COO carbon. The low value of the final intensity (marked "a") implies that less than 50% of the COO groups are undergoing large-amplitude motion. The correlation time is ~50 ms (determined by fitting the exponential build-up of the CODEX signal).

7.3.4 DIFFUSIVE MOTION

In most single-component molecular systems in the solid state, the motion that affects the NMR spectrum is restricted to (hindered) rotations or hydrogen exchange. In composite systems, however, translational, diffusive motion is possible. The techniques discussed earlier in sections 7.2.2 and 7.2.3 can be used to study this. For example, in materials designed for lithium-based batteries, the ^{7}Li linewidth and relaxation properties are strongly affected by the ability of the lithium ions to diffuse. In one such study, the static ^{7}Li spectrum at low temperature showed a broad and featureless line consistent with immobile lithium ions in a distribution of environments within the host matrix. As the temperature was increased, the linewidth decreased until a single narrow and symmetric line was observed. This was attributed to lithium ions undergoing rapid isotropic reorientation during diffusion within channels in the host matrix. Finally, at high temperature, a scaled quadrupole bandshape was observed, which was interpreted as arising from lithium ions in an aggregate too large for isotropic motion but undergoing anisotropic diffusion within the matrix.

Valuable information can be obtained by studying the guest in solid host–guest structures, such as water in membranes or hydrocarbons in porous zeolitic structures. However, the study of diffusion is more suitable for pulsed field gradient or imaging techniques, both of which introduce some sense of position within the sample to the NMR experiment and are therefore sensitive to translational motion.

7.3.5 "SOFT" SOLIDS

In all the cases introduced so far in this chapter, the sample would be described as a rigid solid. The general molecular motion in a "soft" solid has a major impact on the NMR spectrum. For example, in a high-resolution ^{13}C spectrum from such an organic material, the partial averaging of the $^{13}C,^{1}H$ (and $^{1}H,^{1}H$) dipolar coupling means that the spectrum is to some extent *self-decoupled* and requires relatively little RF decoupling power to reduce the linewidths to a small number of hertz (see figure 7.23 and inset 7.8). The CSA is fully averaged, so spinning sidebands are not observed

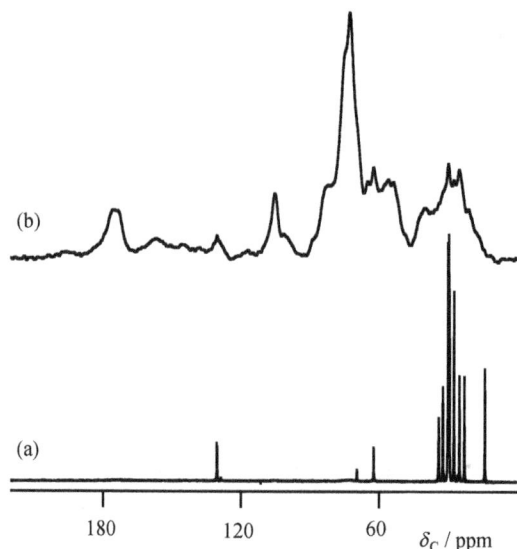

Figure 7.23. (a) Carbon-13 cross-polarization magic-angle spinning (CPMAS) spectrum from a whole hazelnut. The narrow lines and absence of spinning sidebands are indicative of a soft solid. The unsaturated fatty acid component of the nut is preferentially detected with a long (10 ms) contact time. With a short contact time (0.1 ms), (b) the "hard" component of the nut is detected.

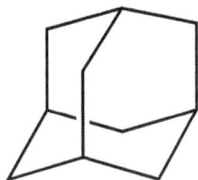

Inset 7.8. Adamantane

Tricyclo[3,3,1,1]decane, $C_{10}H_{16}$, or adamantane. Its melting point is 270 °C (although it readily sublimes, even at room temperature). At room temperature, it has a disordered structure but undergoes fast reorientation ($\sim 1.6 \times 10^{11}$ jumps per second). This pseudo-isotropic motion removes all *intra*molecular dipolar interactions but *inter*molecular ones are only partially averaged due to the lack of translational motion.

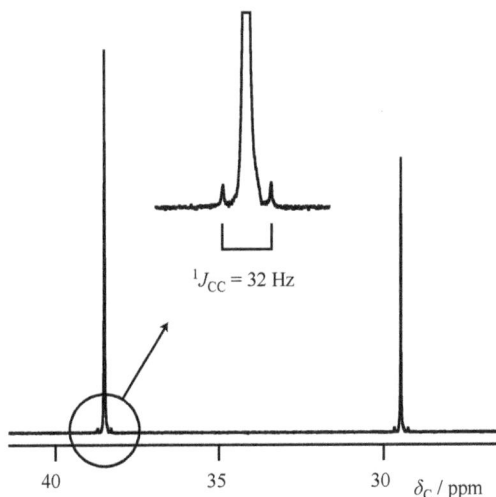

$^1J_{CC} = 32$ Hz

Adamantane is a soft solid that gives very narrow lines at low (~ 2 kHz) spin rates and low (~ 30 kHz) decoupling fields (left). It is an excellent test of shimming— recorded at 75.43 MHz, the full half-height linewidth here is less than 2 Hz and satellites from $^1J_{CC}$ can be observed. The spectrum was obtained with CP.

even at relatively low spin rates (<5 kHz).[7] Even though such materials do not necessarily possess the long-range order of a crystalline solid, the motional averaging of each environment (akin to that in solution) means that the broad lines associated with a rigid amorphous solid are not observed. In consequence, as illustrated in figure 7.23, highly resolved spectra can be obtained that begin to approach the appearance of those from a solution.

7.3.6 INTERFERENCE

Molecular mobility can produce some unexpected effects in high-resolution spectra when the rate of motion approaches that of MAS or the nutation rate of ^1H decoupling. In both cases, the effect can be thought of as a destructive interference so that the interactions which the spinning or decoupling

---

[7] Low-intensity spinning sidebands can be observed but are probably attributable to other factors such as bulk susceptibility.

should be removing from spectra are being reintroduced. Without going into the mathematics of the process here, it can be said that the result is a broadening of the affected lines. This broadening will be most extreme when the spinning or nutation and motion rates are matched, but lines narrow again as they diverge. An example is given in figure 7.24.

Interference can also arise between the timing of some pulse sequences and, for example, the sample spin rate as noted in section 5.5.1.

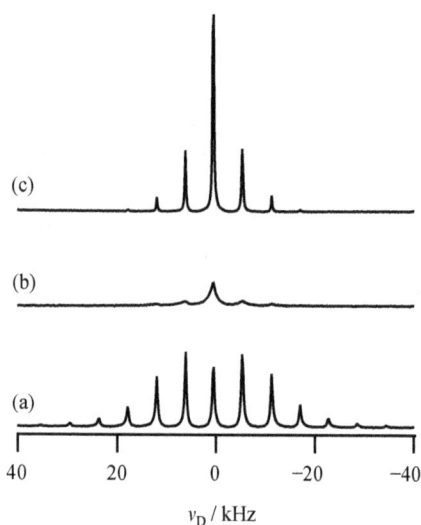

Figure 7.24. Deuterium magic-angle spinning (MAS) spectra from finasteride d_8-dioxane solvate hydrate at (a) −70 °C, (b) −5 °C, and (c) +40 °C. The dioxane undergoes a chair–chair inversion (slow at −70 °C and fast at 40 °C). At −5 °C the exchange rate is close to that of the spin rate (6 kHz) and the interference between the two results in line broadening. At +40 °C, the motion also causes a narrowing of the spinning sideband manifold.

7.4   QUANTITATIVE NMR

Determining the signal intensity is key to many of the experiments discussed in this chapter. Measuring the intensity of a single resolved (not overlapping) line in a high-resolution spectrum from a spin-$\frac{1}{2}$ nucleus is straightforward and is normally done using numerical integration. The precision of the result is determined by the signal-to-noise ratio in the spectrum.

For the integration of any signal, it is important that the spectral baseline is flat and has no offset (so the value of the integral truly represents the area under the peak and not the shape of the baseline as well). If signals from different samples are being compared, the signal intensity needs to be corrected for the amount of sample in the rotor (so the sample mass should be recorded). For a heterogeneous sample, it is important that the components are uniformly distributed through the rotor (as the sample is not uniformly excited/detected along the whole length of the rotor). When signals overlap, spectral deconvolution (see inset 7.9) may be required. For a resonance with spinning sidebands, the intensity of the latter should be added to that of the centerband.

Inset 7.9. Deconvolution

Deconvolution is the process of obtaining information about the components of a bandshape that contains overlapping lines. It is commonly applied to the spectra from spin-$\frac{1}{2}$ nuclei; software for doing this is widespread. Deconvolution of overlapping quadrupolar bandshapes is an altogether more complicated process (due to the number of variables involved).

   Gaussian or Lorentzian lineshapes (see figure 7.25), or a linear combination of the two, often can be used to represent the lines in a solid-state NMR spectrum (from spin-$\frac{1}{2}$ nuclei). The result of the deconvolution of a ^{29}Si bandshape is shown in figure 7.26. Here Gaussian shapes have been used and the position, width, and height of the lines in a three-line model have been allowed to vary in an iterative fit to the observed spectrum. The observed and simulated spectra should only differ at noise level, as in this example, but such precision, using these simple lineshapes, can be hard to achieve—at least if the number of components is restricted to being chemically meaningful.

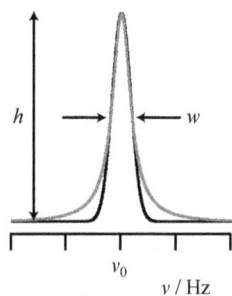

Figure 7.25. Common lineshapes.

In figure 7.25 the Lorentzian (gray line) is characterized by the broad base:

$$f(v) = \frac{2wh}{w^2 + \left(4\pi(v - v_0)\right)^2} \qquad 7.12$$

Gaussian (black line) lineshapes are used in figure 7.26:

$$f(v) = h\exp\left(\frac{-a(v - v_0)}{w}\right)^2 \qquad 7.13$$

where $a = 4\log_e(2)$. In figure 7.25 the full width at half-height (FWHH), w, is the same for both shapes. The area under each line (of most interest in a deconvolution) is:

$$\frac{\pi}{2}wh \quad \text{and} \quad \sqrt{\frac{\pi}{4\log_e(2)}}wh \qquad 7.14$$

for the Lorentzian and Gaussian cases, respectively.

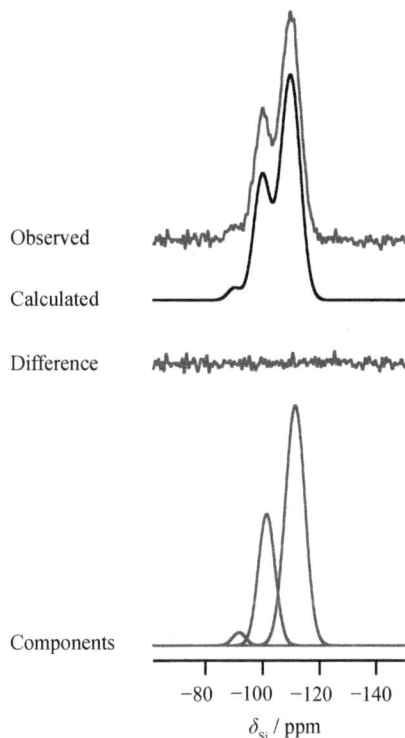

Figure 7.26. Deconvolution of a ^{29}Si bandshape into three lines with relative intensity 66%, 31%, and 3% (low to high frequency).

Additional problems arise for quadrupolar nuclei. Resonances from species with different χ values will have different excitation profiles and hence the response to a single pulse may be biased. The amount of signal located in the (possibly undetectable) satellites might also need to be considered. Furthermore, if one type of environment has a particularly large value of χ, second-order effects may broaden the signal to the point where it is unobservable. Such problems can be encountered when quantifying ^{27}Al spectra, for example, from aluminosilicate materials.[8]

7.4.1 RELATIVE INTENSITY

When measuring the relative intensities within one spectrum, it is essential to take proper account of relaxation times and to set the recycle delay accordingly. This is particularly relevant when direct excitation is used, since T_1 values for different nuclei may vary substantially and the recycle delay must be set at a value significantly higher than the longest relaxation time involved (see section 3.3.1/figure 3.7). This is not generally a problem for a homogeneous system when CP is used because recycle delays depend on T_1^H, which is uniform when spin diffusion is efficient. However, a different criterion then becomes important (see section 3.4.2 and, particularly, figure 3.20), since different carbons, for instance, will cross polarize at different rates. In particular, quaternary carbons generally show slow CP rates, as illustrated in figure 7.27, because the protons are relatively remote. In such a case, long contact times may be necessary for successful quantitation, which can be a problem if significant loss of carbon magnetization, due to $T_{1\rho}^H$ relaxation, has occurred by that stage. Ideally, signal intensity should be plotted against contact time for all resonances, but this is not normally feasible.

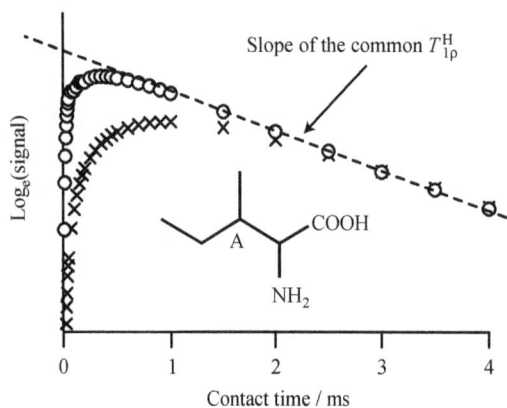

Figure 7.27. Signal intensity vs. contact time for the CH labeled A (circles) and carboxylic acid carbon (crosses) resonances in L-isoleucine.

[8] High-symmetry (small χ value) aluminum environments octahedrally or tetrahedrally coordinated by oxygen are easy to detect and are likely to be excited in a similar fashion, so measuring their relative intensity can be done reasonably accurately. However, if the environments are distorted and if their χ values diverge, this is no longer the case.

More substantial problems arise when it is important to determine relative intensities for a heterogeneous system using CP, because the recycle delay now has a significant effect if T_1^H differs for different components. Moreover, differences in $T_{1\rho}^H$ may now also need to be taken into account. Suppose a system has two components, A and B. If $T_{1\rho}^H(A) \gg T_{1\rho}^H(B)$, then a CP plot such as is shown in figure 7.28 may result, which means that signals for component B will be rapidly lost as the contact time increases. Such a case occurs for styrene/butadiene copolymers. The example here is a block copolymer, with each domain sufficiently large that it retains its own value of $T_{1\rho}^H$. At ambient temperature, $T_{1\rho}^H$ for the butadiene domains is an order of magnitude greater than that for the styrene domains. Therefore, at long contact times, the signals for styrene are lost, as shown in figure 7.29(a) and (b). However, the butadiene is slow to cross polarize and so is underrepresented in the spectrum recorded with a short contact time. Spectra recorded only at high contact times would indicate that little styrene was present, with the opposite conclusion at low contact times![9] Values of $T_{1\rho}^H$ are temperature dependent and at 200 K the situation is reversed, with styrene having the longer $T_{1\rho}^H$ (see figure 7.29(c)). Thus, at the lower temperature and long contact time, only styrene signals appear.

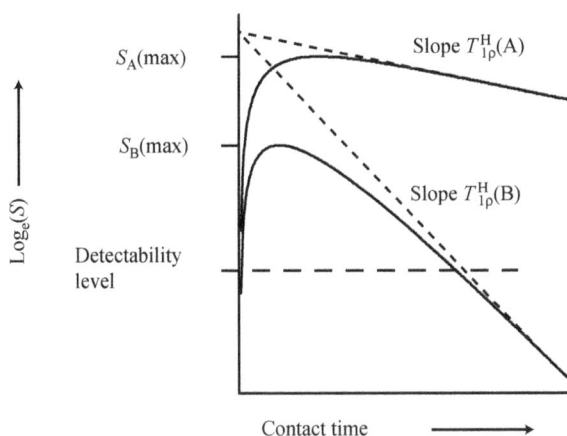

Figure 7.28. Schematic plot of signal intensity vs. contact time for a heterogeneous system of two components with *equal* concentrations but widely differing values of $T_{1\rho}^H$. $S_A(max)$ and $S_B(max)$ is the maximum amount of signal that can be observed for component A and B, respectively.

7.4.2 ABSOLUTE INTENSITY

If a spectrum is used to quantify the amount of sample, or a component of it, in the rotor on an absolute scale (i.e., in moles per rotor), an extended strategy is necessary. There are two ways of approaching the problem of referencing such a measurement.

A known quantity (as a %) of a reference compound can be added to provide a reference signal. The nature and amount of the additive need to be carefully controlled, since a clear NMR signal with a similar intensity (with respect to the signals from the compound to be measured) is required. Ideally, a calibration curve should be constructed from samples of known content and this can then

---

[9] The pair of spectra shown in figures 7.23(a) and (b) illustrate a similar selectivity.

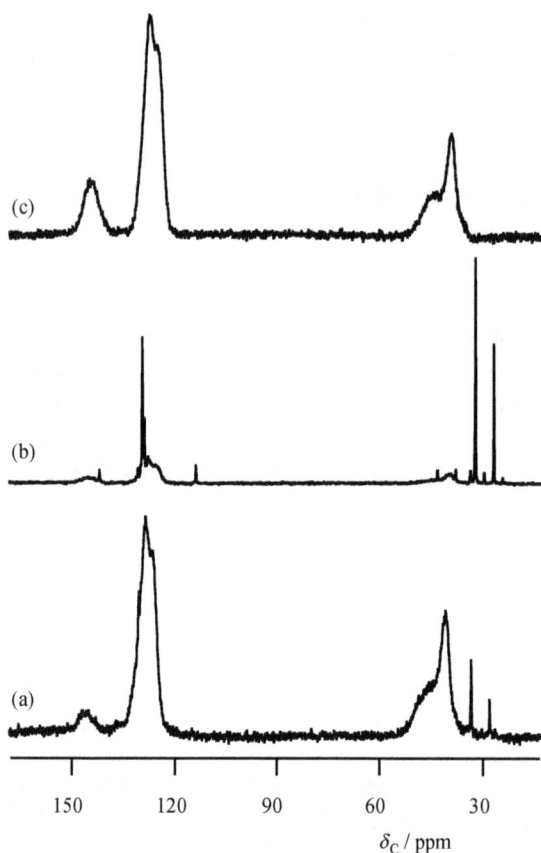

Figure 7.29. Carbon-13 spectra from a styrene/butadiene copolymer under various conditions: (a) ambient temperature, contact time 0.4 ms; (b) ambient temperature, contact time 10 ms; (c) 200 K, contact time 10 ms. The polymer contains 30 wt.% styrene.

be used for a sample of unknown content. If this is not possible, any differences in relaxation, CP behavior, or excitation efficiency must be taken into account.

Figure 7.30 shows a test case for a pharmaceutical formulation containing the drug substances bambuterol hydrochloride (BHC) and terbutaline sulfate (both used in asthma therapy) in α lactose monohydrate as the majority component, together with a small amount (1%) of magnesium stearate. The methylene ^{13}C signal for magnesium stearate was used as the reference. A calibration curve was employed to show that the amount of BHC is linearly related to the ratio (from deconvoluted peak areas) of the BHC signal at 147.3 ppm to that of the reference. This plot was used when measurements were made on samples of unknown BHC content. It was shown that the limit of detection (in 3 hours of spectrometer time) was 0.5 mol % and the limit of quantification was 1 mol %—important information when comparing the value of NMR with other techniques for such measurements.

The second method of referencing signal intensity (as an alternative to using a chemical additive to the sample) is to create a signal electronically—the so-called ERETIC method.[10] The spectrum is

---

[10] The acronym stands for "Electronic REference To access In vivo Concentrations," but the method is now used for nonmedical spectra.

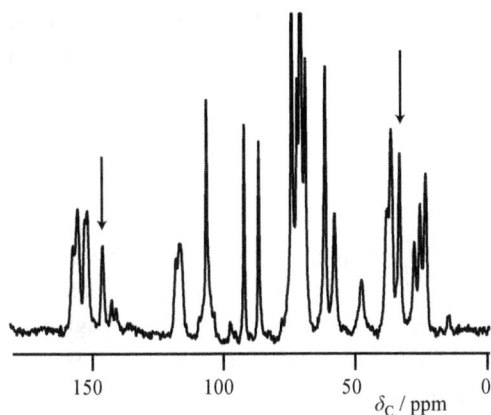

Figure 7.30. Carbon-13 cross-polarization magic-angle spinning (CPMAS) spectrum, accumulated in 3 hours, of the pharmaceutical formulation (see the text) containing 5% of bamburterol hydrochloride (BHC) and 1% magnesium stearate. The signals used for quantifying the content of BHC are indicated by arrows BHC to high frequency and magnesium stearate to low frequency). The BHC signal at 147.3 ppm arises from two carbamic acid carbons, that from the stearate ($C_{36}H_{70}MgO_4$) is from ~30 coincident mid-chain CH_2 carbons (hence the apparent mismatch in intensity given the respective concentrations of these components). Crystalline polysaccharides typically have very long T_1^H (often hundreds of seconds). With the recycle delay optimized for the drug components, the signals from the lactose saturate and are therefore severely underrepresented in this spectrum.

"spiked" with a resonance using a signal generated electronically in a spare RF channel. This signal can be shaped so that it produces a realistic lineshape when it is Fourier transformed. The signal is fed to the probe and detected in the NMR coil along with the signal from the sample. Because the two signals are simultaneously detected, they must be coherent in phase. A Fourier transformation will yield a spectrum containing a "synthetic" peak. Calibration of the area under this peak against standards of known concentration allows this signal to be used to quantify spectra of unknown concentration. The advantages of using the ERETIC method are that the sample is not contaminated, the ERETIC peak can easily be placed at a position in the spectrum where it does not interfere with signals from the sample, there is no dependence on the relaxation behavior of the reference signal, and the amplitude of the reference can be adjusted (using the pulse amplitude control of the spectrometer) to be similar to that of the spectrum of interest.

7.5   PARAMAGNETIC SYSTEMS

The presence of unpaired electrons has a profound effect on NMR. The gyromagnetic ratio of the electron is ~660 times larger than that for ^1H so the coupling between the unpaired electron spins and the nuclear spins (*hyperfine* coupling) is large and must be taken into account.[11]

Unpaired electron density is an efficient driver of nuclear spin relaxation, and so relaxation times for paramagnetic systems tend to be very much shorter than those for diamagnetic ones. Shortened T_1s can be used to advantage (see section 7.5.1), but the hyperfine coupling of sites close

---

[11] This requires an additional term to be added to equation 1.2 (see Further reading for more details on the form of this term).

to unpaired electron density leads to significant line broadening. It is usually impractical to obtain useful NMR spectra from paramagnetic atoms themselves;[12] however, it is often possible to obtain useable NMR signals from sites with weaker hyperfine couplings, that is, those at greater distances from the paramagnetic center and/or with less delocalized electron density from the unpaired electron at the nuclear site.

7.5.1 RELAXATION EFFECTS

Short T_1s in paramagnetic systems can be exploited, that is, paramagnetic species can be used to shorten recycle delays for systems with prohibitively long T_1s. For example, it has been observed that oxygen in the pores of zeolites acts as a *relaxation agent* (molecular oxygen is paramagnetic), and the oxygen partial pressure can be deliberately increased to reduce the ^{29}Si T_1. Similarly, spectra from coordination compounds, organometallics and ceramics can be enhanced in intensity if the system is doped with a small concentration of paramagnetic metal ions, such as Mn^{2+}, to form a solid solution. The magnitude of this effect is strongly dependent on the distance between the unpaired electron and the nucleus and the concentration of the relaxation agent, and so additional care is needed when choosing the parameters for quantitative measurements (see section 7.4.1).

7.5.2 SHIFT EFFECTS

The nature of the hyperfine coupling is complex and its magnitude (and sign) varies from nucleus to nucleus. The coupling has essentially three components: a Fermi contact interaction (due to unpaired electron density at the nucleus), a pseudocontact term that is equivalent to the nuclear dipole–dipole coupling, and a *spin-orbit* term (see Further reading for details). Because the electron spins generally relax very quickly, the multiplet resulting from the coupling is collapsed into a single line by processes analogous to those described in section 7.3.[13] However, the position of this single line may be significantly different from that observed from an analogous diamagnetic system. This *paramagnetic shift* arises because the difference in energy of the *electron* spin states is not negligible in comparison with kT. Consequently, the non-negligible population difference for the electron spin states means that the *population-weighted average* frequency is not at the center of the putative doublet (in the NMR spectrum).

The dependence of the population distribution on kT means that the paramagnetic shift is strongly temperature dependent, which distinguishes it from other contributions to the chemical shift. The shift can potentially provide useful information on the distribution of electron density and,

---

[12] The nucleus at the site of a localized unpaired electron will have a very large hyperfine coupling constant (potentially tens of MHz), resulting in multiplets that are far too broad to be observed by conventional NMR.
[13] The hyperfine coupling of the paramagnetic center itself is so large that even the very quick electron spin relaxation is not efficient enough to collapse the multiplet from the nucleus at the site of a localized unpaired electron.

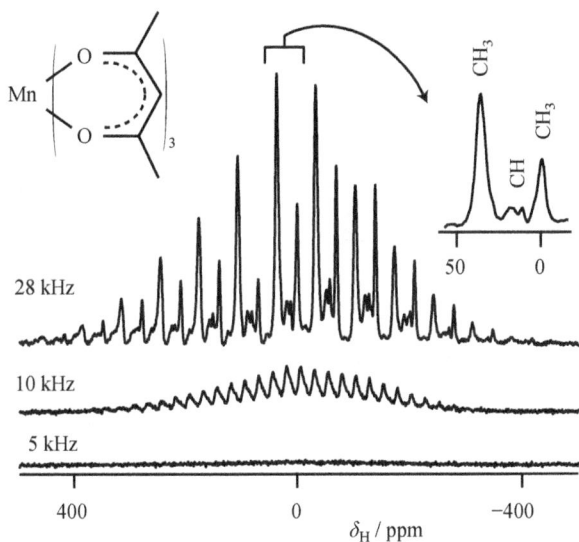

Figure 7.31. Proton magic-angle spinning (MAS) spectra from paramagnetic Mn(acac)$_3$ as a function of spin rate. (Figure adapted from data published in N.P. Wickramasinghe *et al., J. Am. Chem. Soc.* **127** (2005) 5796.)

in favorable cases, it can be experimentally measured by comparing the spectrum of the paramagnetic compound with that of an analogous diamagnetic compound.

The hyperfine coupling is anisotropic, and the anisotropy can be very large (thousands of parts per million even for ^{1}H spectra), so although it has the usual $P_2(\cos\theta)$ dependency, it can be difficult to handle with MAS. Thus the MAS NMR spectra of paramagnetic materials are often largely unresolved and contain extensive manifolds of spinning sidebands. As shown in figure 7.31, however, very fast MAS (>25 kHz) may allow useful resolution to be obtained.

FURTHER READING

"Multidimensional solid-state NMR and polymers", K. Schmidt-Rohr & H.W. Spiess, Academic Press (1994), ISBN 0 12 626630 1.

"Intramolecular motion in crystalline organic solids", P. Hodgkinson, in *"NMR Crystallography"*, Eds. R.K. Harris, R.E. Wasylishen & M.J. Duer, John Wiley & Sons, Ltd. (2009), 375–386 (Chapter 25), DOI: 10.1002/9780470034590.emrstm1002, ISBN 978 0 470 69961 4.

"Centreband-only detection of exchange (CODEX)", K. Schmidt-Rohr, E.R. deAvezado & T.J. Bonagamba, in *"Encyclopaedia of Nuclear Magnetic Resonance"*, Volume 9, Eds. D.M. Grant & R.K. Harris, John Wiley & Sons, Ltd. (2002), 633–642.

"Combining NMR spectroscopy and quantum chemistry as tools to quantify spin density distributions in molecular magnetic compounds", M. Kaupp & F.H. Köhler, *Coord. Chem. Rev.*, **253** (2009), 2376–2386. DOI: 10.1016/j.ccr.2008.12.020.

"Strategies for solid-state NMR studies of materials: From diamagnetic to paramagnetic porous solids", V.I. Bakhmutov, *Chem. Rev.*, **111** (2011), 530–562. DOI: doi:10.1021/cr100144r.

"Electron paramagnetic resonance: Elementary theory and practical applications", J.A. Weil, J.R. Bolton & J.E. Wertz, John Wiley & Sons Ltd. (1994), ISBN 0 471 57234 9.

CHAPTER 8

ANALYSIS & INTERPRETATION

8.1 INTRODUCTION

This chapter takes the reader through a variety of topics related to the interpretation of spectra and the extraction of detailed information about the NMR interactions. The first section describes the analysis of one-dimensional spectra to obtain, for example, chemical shift information and so does not involve special experimental techniques. However, obtaining reliable estimates of other parameters, such as the strength of dipolar couplings, may require specialist experiments and/or analysis, so most of the rest of the chapter deals with somewhat more sophisticated aspects of solid-state NMR, especially involving tensors and their evaluation. The final section looks to the value of NMR in crystallography, where it complements diffraction-based techniques (*diffraction crystallography*).

8.2 QUANTITATIVE MEASUREMENT OF ANISOTROPIES

8.2.1 GENERAL

As discussed in earlier chapters, all NMR interactions are essentially anisotropic. It is often necessary to reduce the effect of these anisotropies on the NMR spectrum by techniques such as magic-angle spinning (MAS) in order to improve resolution. However, the anisotropies[1] contain valuable information about the local environment. This section considers how the anisotropic components of NMR interactions, such as the shielding and quadrupole coupling tensors, can be measured and exploited. Measurements of the dipolar couplings are particularly significant, since their strength directly depends on the distance between the nuclei involved. As a result, a wide range of experimental techniques have been derived to measure dipolar couplings (see section 8.3), and hence determine structural information from solid-state NMR studies.

---

[1] Including in this term at this point asymmetries which are a form of anisotropy by a different name.

In order to measure anisotropy information, two conditions have to be met. Firstly the anisotropy has to be sufficiently large in order to be measured accurately, and secondly the different sites present need to be resolved in order to obtain anisotropy information for each distinct resonance. These conditions may conflict. For example, anisotropy information is often characterized most accurately at very low MAS speeds or by using static samples, since the overall profile of the bandshape can be fully modeled. But if multiple sites are present, spectral overlap will often prevent the resolution of the different sites. Using fast MAS rates may lead to clean, well-resolved spectra, but when spinning sidebands are completely absent, information about the anisotropies is lost.

In many cases it is possible to use intermediate spinning rates to obtain resolved spectra that still contain sufficient spinning sidebands for analysis. The fitting of such spectra to determine anisotropic information is discussed in the following section. In more difficult cases, alternative strategies are required. If the interactions involved are inhomogeneous (see section 2.7 and inset 4.5), for example the shielding anisotropy, and so are associated with sharp spinning sidebands, it is common to use two-dimensional spectra which contain isotropic information in one dimension and anisotropic information in the other. These are discussed in section 8.2.5. In other cases, particularly when the interaction is homogeneous (such as is common for the homonuclear dipolar coupling), the usual strategy is to use moderate to fast MAS to suppress the anisotropies and obtain high-resolution spectra, and use recoupling sequences to selectively reintroduce anisotropic information. The qualitative application of recoupling was discussed in section 5.4.2. The quantitative use of recoupling, particularly in the context of dipolar interactions, is discussed in section 8.3.

Finally, it is important to be aware of potential interactions between dynamics and NMR anisotropies; indeed the modulation of NMR anisotropies by motional processes is a powerful tool for the characterization of dynamic processes in solids, as previously discussed in section 7.3. If the frequency of the dynamic process is significantly faster than the anisotropies involved,[2] then an averaged tensor will be observed. In the case, for example, of a two-site exchange between sites 1 and 2 with fractional populations p_1 and $p_2 = 1 - p_1$, the average of a tensor A will be:

$$A_{\text{average}}^{\text{M}} = p_1 A_1^{\text{M}} + p_2 A_2^{\text{M}} \qquad 8.1$$

Note that a *common* reference frame must be used for the tensors; here M denotes a molecular frame of reference, rather than the unrelated principal axis systems of the individual tensors. The parameters of the averaged tensor can be fitted in the same way as "normal" unaveraged tensors, although it is important to note that averaged tensors do not necessarily have the same symmetry properties as the original tensors. For example, averaged dipolar tensors may have non-zero asymmetries even though the original dipolar tensors are axially symmetric by definition.

If the dynamic process is slow compared with the anisotropies (the slow exchange limit), then tensor information can be determined independently for the different sites (assuming that they can be resolved). In the intermediate case, the form of the spectrum will be changed by the dynamic process, with the exact shape depending on the tensors involved and the nature of the dynamic process.

---

[2] More correctly, the relevant parameter is the *modulation* of the tensor by the dynamics, rather than the magnitude itself. For example, there will be no effect on the observed spectrum if the tensor involved is unchanged by the dynamic exchange process.

Fitting such exchange-modulated spectra is a powerful means of characterizing the dynamic process. The focus here is on spectra that are unaffected by intermediate timescale dynamics.

8.2.2 QUANTITATION OF POWDER LINESHAPES

In favorable cases, it may be possible to read off tensor information from turning points in powder spectra. For instance, the separation of the horns of the Pake doublet from an isolated dipolar coupling (see figure 2.4) can be accurately measured and provides the dipolar coupling constant directly. When the edges of the powder pattern are defined sharply enough, the principal components of the shielding tensor can also be read off from the turning points, as in figure 2.2.

In more difficult cases, and particularly when estimates of errors are required, it is necessary to fit the powder pattern to simulated spectra; the NMR frequencies are calculated as a function of crystallite orientation, and the individual spectra summed to give the overall pattern. Note that a large number of orientations may need to be accumulated in order to obtain a smooth lineshape if the ratio of the anisotropy to the intrinsic linewidth is very large.

Figure 8.1(a) shows an example of fitting the central-transition bandshape of a MAS powder pattern (obtained at 104.20 MHz) for a ^{27}Al spectrum resulting from second-order quadrupolar effects (the small feature around 7 ppm is due to the satellite transitions). Exact modeling of powder

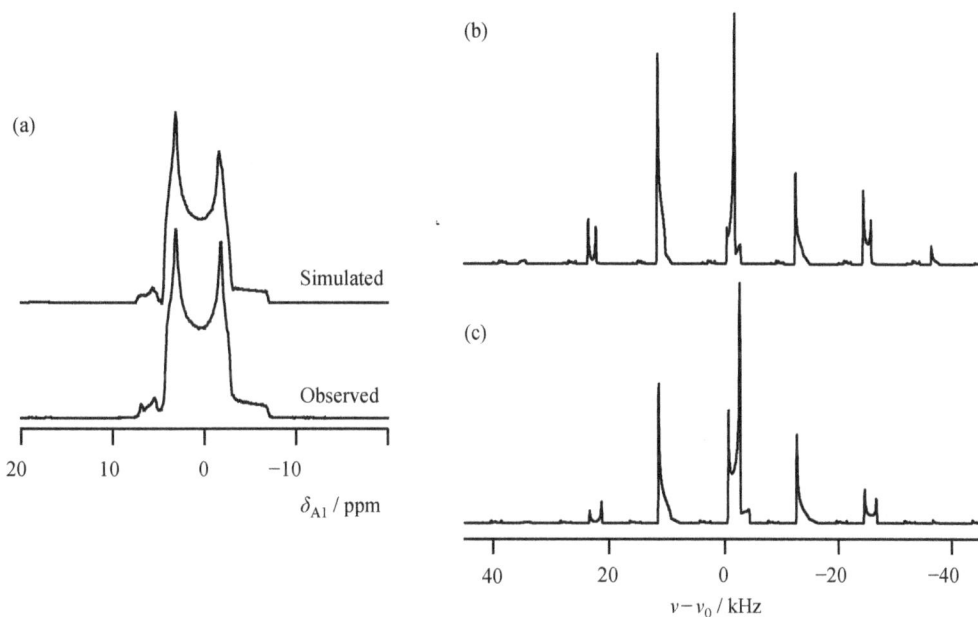

Figure 8.1. (a) Fitting of a MAS powder lineshape due to second-order quadrupolar effects on the central transition of ^{27}Al in aluminum acetylacetonate. Quadrupolar parameters of $\chi = 3.0$ MHz and $\eta = 0.16$ match the turning points of the simulated and experimental spectra. (b) and (c) Simulated spectra for a spin-$\frac{3}{2}$ nucleus influenced by the quadrupolar interaction ($\chi = 3$ MHz) and shielding anisotropy ($\zeta = 20$ kHz), both with zero asymmetry and with coaxial tensors, at (b) 14.1 T and (c) at 9.4 T.

lineshapes is often difficult. For example, the underlying (homogeneous) lineshape cannot easily be modeled as it is swamped by the inhomogeneous lineshape, and broad patterns are often not excited uniformly by the NMR experiment. In figure 8.1(a), however, the turning points of the powder pattern are well defined, and MAS has suppressed other orientation-dependent interactions, such as the shielding anisotropy. This allows the principal components of the quadrupole coupling tensor to be estimated with confidence.

In other cases, it may be necessary to obtain complementary data sets in order to determine individual tensor contributions with confidence. For example, if the central-transition spectrum of figure 8.1 had been obtained under non-spinning conditions (see the upper spectrum in figure 6.6), the powder pattern would contain a contribution from the shielding anisotropy, which would distort the fitting of the quadrupole coupling parameters. Obtaining spectra at multiple magnetic fields is the best solution for distinguishing shift interactions (which scale in frequency units with B_0), from (first-order) coupling effects (independent of field) and/or from second-order effects (which scale inversely in frequency units with B_0; see figures 8.1(b) and (c) and also section 8.7.1). A single set of shielding and quadrupolar parameters[3] that fits spectra at different fields simultaneously is likely to be robust.

8.2.3   QUANTITATION OF SPINNING SIDEBAND MANIFOLDS

Fitting of anisotropy information is commonly performed using data from MAS experiments, where the anisotropy parameters are encoded in the pattern of intensities formed by the spinning sidebands. In comparison with fitting spectra from static samples, MAS spectra have the advantage of being better able to discriminate multiple sites. It is also generally easier to phase spectra and correct baselines where peaks are separated by clear regions of baseline; flat baselines and well-phased spectra are very important when using intensity information quantitatively (see section 7.4).

Note that the number of spinning sidebands required depends to some extent on the information wanted; anisotropies (i.e., essentially the "width" of the pattern) can be quantified with good accuracy with only four to five sidebands of measurable intensity. By contrast, determination of the asymmetry parameter (which loosely determines the *shape* of the pattern) steadily improves as more spinning sidebands are included. Unless the signal-to-noise ratio is a major problem or slower spinning rates cause centerbands and sidebands to overlap, it is better to err on the side of using slower spinning rates to increase the number of spinning sidebands above the minimum number required to quantify the anisotropy parameter effectively. This makes it easier to identify any systematic deviations between the model and the data.

If the sidebands of interest are well resolved from other peaks, then the most straightforward approach is to determine the peak heights (or, better, peak areas) of the observed sidebands.

---

[3] Note that the spectra would also be dependent on the relative orientations of the tensors involved, which highlights the difficulties of quantifying spectra when multiple interactions are present. Section 8.7 discusses the question of relative tensor orientations in more detail.

The intensity pattern of the spinning sideband manifold is then fitted numerically to obtain the anisotropy parameters. The Further reading has information about some of the various software tools available.

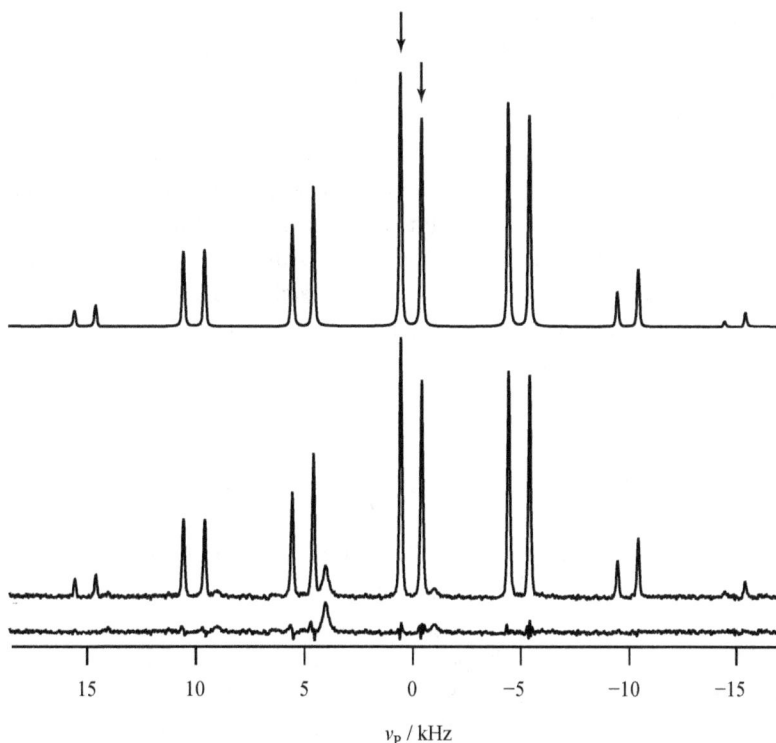

Figure 8.2. Fitting of the ^{31}P spinning sideband manifolds of the two phosphorus sites in γ-(MoO$_2$)$_2$P$_2$O$_7$. Although obtaining the shielding parameters from the pattern of peak heights would be quite feasible in this case, fitting the complete spectrum (middle trace) makes the most use of the available data. The fitted spectrum is the top trace and the lowest trace shows the difference between the fitting and experiment. The experimental spectrum was obtained at 202.28 MHz with a spin rate of 5 kHz. The zero of the frequency scale is arbitrary. (This original figure has been adapted for publication in the Supplementary Information to Lister et al., *Inorg. Chem.* **49** (2010) 2290.)

In cases where signal overlap prevents the determination of accurate intensity information for the individual sidebands, it is necessary to fit the complete spinning sideband pattern directly. This is a little more involved since additional parameters are required to model the lineshape of the spinning sidebands, but fitting the complete manifold in one step is more robust than using deconvolution to estimate individual sideband integrals and then fitting the resulting intensity pattern. Figure 8.2 gives an example of such a fit where complete sideband manifolds were modeled (see also figure 2.13, which shows an example with a single chemical shift). The difference between fitted and experimental data sets (lowest trace) shows features due to an impurity peak (and its associated spinning sidebands). The residual plot also shows small features indicating that the modeled lineshape (a mixed Lorentzian/Gaussian; see inset 7.9) does not perfectly match the experimental shape. Such minor deviations are to be expected and will have negligible impact on the accuracy of the overall fitting, which here gave ^{31}P shielding anisotropy (asymmetry) parameters of 65.5 ppm (0.28) and 69.3 ppm (0.69) for the high- and low-frequency resonances, respectively. The fitting also confirms that the overall intensity ratio for the two sites is exactly 1:1 (within the fitting error), which is not immediately apparent from the spectrum.

As with any fitting of experimental data, it is important to note the associated confidence bounds on the fitted parameters. Although these may not be entirely reliable (particularly if there are significant systematic deviations between fitted and experimental data), they will give a good indication of the precision that can be reasonably quoted. This is particularly important for asymmetry parameters, since they are often difficult to fit accurately—for example, if there are too few spinning sidebands to determine the "shape" of sideband manifold and/or the symmetry is close to axial. Small changes in the asymmetry parameter when it is less than about 0.1 have very little effect on the sideband manifold, meaning that the asymmetry is not well constrained in this case.

8.2.4 SINGLE-CRYSTAL VS. POLYCRYSTALLINE SAMPLES

The drawback of using polycrystalline ("powder") samples, as described above, is that information about the orientation of the tensor with respect to the crystal axes cannot generally be obtained; it is normally only possible to determine the magnitudes of the principal components of the tensor. In favorable cases, discussed below, it may be possible to determine the tensor orientation with respect to another tensor, but because polycrystalline samples contain a random distribution of all possible crystal orientations, it is impossible to orient the tensor with respect to the crystal axes.

In principle, tensor orientation can be obtained by working with single-crystal samples. However, there are a number of major practical obstacles to single-crystal NMR, not least the difficulties of obtaining crystals of sufficient volume, and the need for a specialist NMR probe containing a goniometer (required to orient the crystal within the magnetic field). Full characterization of NMR tensors via single-crystal experiments is much more demanding than the experiments on polycrystalline samples described above.

8.2.5 RESOLVING ANISOTROPY INFORMATION BY ISOTROPIC SHIFT

Fitting anisotropy information becomes difficult when multiple tensors are involved. The case of multiple interactions involving the same site is discussed in sections 8.5 and 8.6, but another important case is when spectral overlap involving distinct sites prevents the measurement of anisotropies.

Fortunately, a variety of two-dimensional NMR experiments can be used to resolve the problem of overlap. For example, *magic-angle turning* (MAT) experiments for shielding anisotropy measurements involve very slow MAS and result in two-dimensional spectra with anisotropy information in one dimension and isotropic frequencies in the other; slices through the 2D spectrum at the isotropic frequencies provide shielding anisotropy powder patterns for the different sites. A relatively straightforward alternative for shielding anisotropies that works under moderate to fast MAS is the phase adjustment of spinning sidebands (PASS) experiment. This uses rotor-synchronized π-pulses to manipulate phases of the spinning sidebands as in the TOSS experiment for the suppression of spinning sidebands (see section 3.3.6), except that, rather than just providing a center-band-only ($m = 0$) spectrum in which sidebands have cancelled, the result is a series of spectra containing just the peaks from a single sideband order ($m = -1$, $m = 0$, $m = 1$, etc.) (see figure 8.3). Collating the intensity information gives the spinning sideband manifold patterns for the different resolved sites.

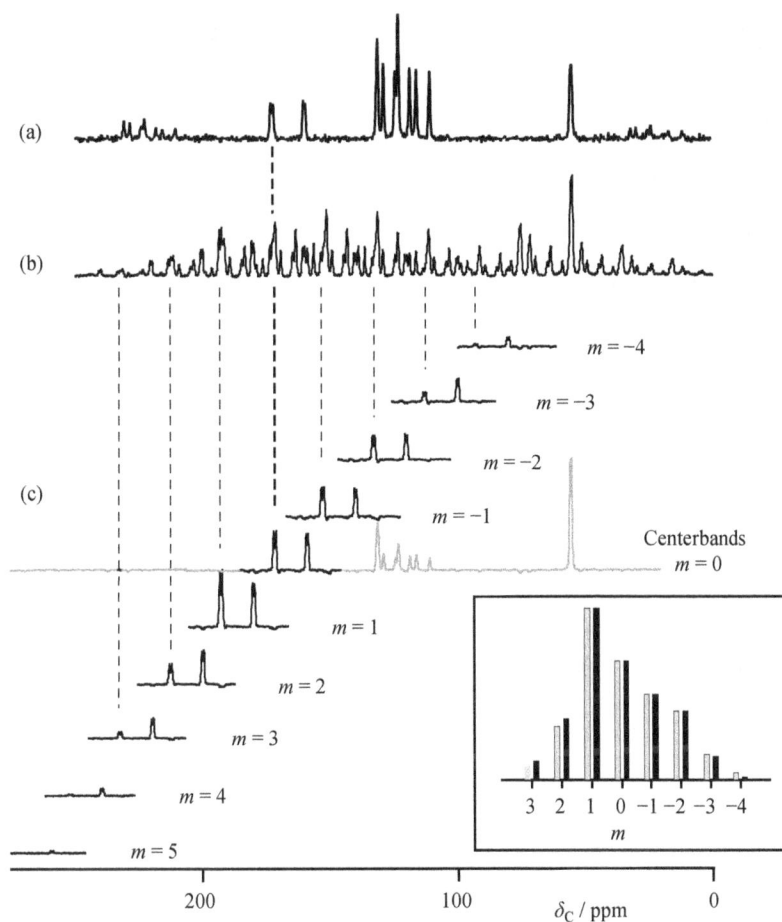

Figure 8.3. Results (obtained at 100.56 MHz) of a ^{13}C PASS experiment on 3-methoxybenzoic acid: (a) spectrum using a spin rate of 10.0 kHz, showing only weak sideband signals; (b) spectrum using a spin rate of 2.0 kHz, showing many sideband signals; (c) separate sideband signals for the acid and ring C–O carbons, the former with a centerband at $\delta_C = 172.8$ ppm, and sidebands indicated by the vertical dashed lines. The PASS centerband is shown for all signals. The inset shows the experimental (black) and simulated (gray) sideband manifolds for the 172.8 ppm centerband. The shielding parameters used for the simulation were anisotropy $\zeta = 65$ ppm and asymmetry $\eta = 0.55$.

The separation of anisotropy information by isotropic shift is especially straightforward when the anisotropy involved is a coupling rather than a shift anisotropy, since the two types of interaction can be readily distinguished, for example, by a spin echo. Such *separated local field* (SLF) experiments are often used to measure the dipolar *local field* at different sites. For instance, the cross-polarization (CP) experiment discussed in section 3.4 provides CP build-up curves for different sites resolved by their ^{13}C chemical shift, allowing $^{1}H,^{13}C$ dipolar couplings to be measured for each resolved site.

8.3 MEASUREMENT OF DIPOLAR COUPLINGS

The direct relationship between dipolar coupling constants and internuclear distance (equation 2.24) makes them a particularly attractive target for quantification. In contrast to measurements of the shielding anisotropy, however, it is rare to encounter a case where the powder spectrum or spinning sideband manifold is dominated by an isolated dipolar coupling (except where specific sites have been isotopically enriched). In general, specialized experiments are required to isolate the effects of the desired dipolar coupling from other NMR interactions. Readers are referred to the literature (see Further reading) for information on quantitative measurement of dipolar couplings involving quadrupolar nuclei using sequences such as TRAPDOR and REAPDOR.

The other inherent problem with dipolar couplings (in contrast to single-spin interactions such as the shielding and quadrupole coupling) is the promiscuous nature of through-space couplings; in a typical organic solid, for example, a given ^{1}H spin will have significant dipolar couplings to a number of nearby ^{1}H spins, resulting in a broad overall spectrum that cannot be fitted in terms of individual couplings (see figure 2.5). As mentioned above, selective labeling is often necessary to isolate an individual pair of spins. Moreover, it cannot be assumed that an observed dipolar coupling corresponds to an internuclear distance within the same molecule—it could correspond to a short intermolecular distance. Particularly in small-molecule systems and/or when accurate quantitative results are required, it is common to recrystallize specifically labeled molecules with an excess of the corresponding unlabeled molecule in order to reduce the impact of intermolecular couplings.

8.3.1 MEASUREMENT OF HETERONUCLEAR COUPLINGS

Figure 8.4 illustrates two possible pulse sequences that can be used to measure the heteronuclear dipolar couplings between a pair of spins, I and S. The simplest approach (which works well if the coupling between I and S is relatively strong) involves exploiting the CP experiment. If spin diffusion among the I (and the S) spins is slow (compared with the timescale for CP), then the CP build-up, obtained by measuring the intensity of the S spin signal as a function of τ_{CP}, is oscillatory. This is illustrated in figure 8.5 for $^{1}H \rightarrow ^{31}P$ CP in $SnHPO_4$. A strong oscillation due to the $^{1}H,^{31}P$ dipolar coupling is observed. As discussed in section 5.1.3, normal Hartmann–Hahn matching is inefficient in cases where MAS is effectively suppressing ^{1}H spin diffusion; hence it is necessary to match on

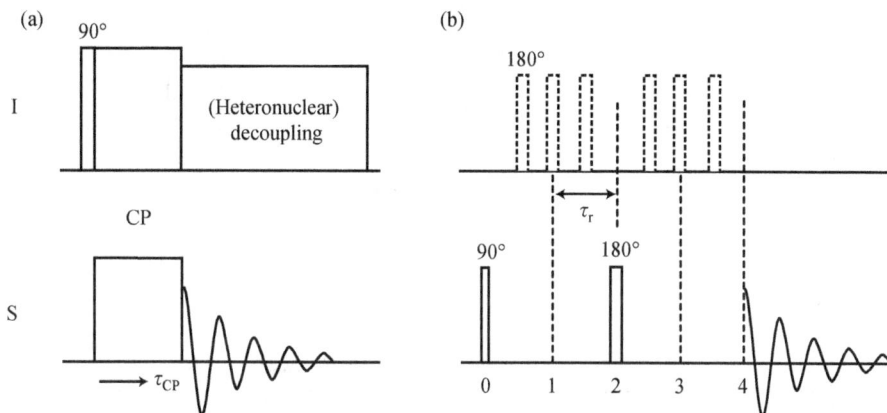

Figure 8.4. Two pulse sequences that can be used to measure heteronuclear dipolar couplings: (a) cross-polarization (CP) signal acquired as a function of CP contact time, τ_{CP}, (b) a rotational-echo double-resonance REDOR experiment, in which the S spin signal is measured after each spin echo, with and without a train of rotor-synchronized 180° pulses applied to the I spins. Note that phase cycling of the I spin pulses using so-called XY schemes such as XY8 is widely used to minimize the effects of RF imperfections.

a sideband, here $n = 1$. The "spectrum" associated with this build-up curve then takes the form of a Pake-like pattern[4] with the separation of the horns given by $\Delta = D_{PH}/\sqrt{2}$.

Note that the chemical shift (including its anisotropy) is suppressed by the spin-locking during the contact time and so the oscillation is purely a function of the dipolar couplings. This experiment is then in effect a separated local field experiment, since Fourier transformation with respect to both τ_{CP} and t_2 provides a spectrum with the dipolar coupling information in one dimension and S spin chemical shifts in the other (the IS coupling being suppressed in t_2 by the heteronuclear decoupling). Additional weak IS couplings and spin diffusion will tend to broaden the lines in the dipolar spectrum. As discussed in section 5.5.1, *Lee-Goldburg CP* can be used to suppress the effects of spin diffusion to first order, although this is unlikely to significantly sharpen the dipolar dimension for powder samples.[5]

The second relevant pulse sequence shown in figure 8.4(b) is the rotational echo double resonance (REDOR) experiment, which uses recoupling (as discussed in section 5.4.2) to reintroduce the effects of dipolar coupling that are otherwise suppressed by MAS. In terms of the observed S spins, the pulse sequence involves a spin echo, with the echo detected after an even number of complete rotor periods (see Further reading). In the absence of additional pulses on the I spins, the

---

[4] The pattern is not the same as the normal Pake pattern illustrated in figure 2.4 for a heteronuclear spin pair ($\Delta=2D$) due to the spin-lock conditions and the $n=\pm 1$ sideband matching. The splitting is even smaller ($\Delta=D/2$) for the $n=\pm 2$ sidebands, making them a less desirable option.

[5] More sophisticated schemes for achieving polarization transfer while suppressing spin diffusion are sometimes used if the dipolar spectrum is intrinsically sharp (e.g., the PISEMA experiment for biomolecules oriented in membranes). If multiple IS couplings are present, it is advisable to use a variant experiment, such as the *proton-detected local field* (PDLF) experiment, which allows different IS couplings to be resolved.

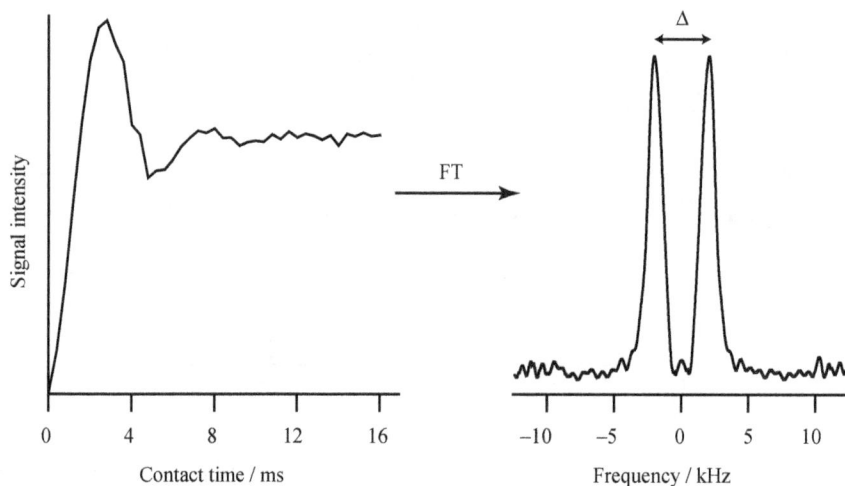

Figure 8.5. Illustration of the use of *transient oscillations* in ^{1}H to ^{31}P cross-polarization build-up curves for SnHPO$_4$ to measure dipolar couplings. The separation of the horns, Δ, of the resulting "spectrum" gives the dipolar coupling ($D_{PH} = \sqrt{2}\Delta = 5.80$ kHz), and hence the ^{31}P,^{1}H distance (here 0.203 nm) via equation 2.24. (This original figure was adapted for publication in Amornsakchai et al., *Mol. Phys.* **102** (2004) 877).

evolution due to the IS coupling refocuses over a rotor period, while the chemical shift evolution is refocused by the spin echo (see sections 4.6.1 and 4.4.2). Hence the only evolution of the "reference" signal is due to homogeneous factors (including relaxation) that are not refocused by the spin echo. If, however, 180° pulses are applied to the I spins twice per rotor period, then the refocusing of the IS coupling is disrupted, and the observed signal is reduced relative to the corresponding reference signal. Plotting the (normalized) difference between the reference signal and the signal with the REDOR pulses as a function of the dephasing time gives a build-up curve which can be fitted to obtain the IS dipolar coupling. Figure 8.6 shows an example where NMR gives crystallographic information complementing that from diffraction measurements. The aluminophosphate sieve in question has cages containing a fluorine atom. Diffraction results suggest this is centrally located in the cages, whereas the Al–F distance determined by REDOR indicates that it is in a bonding position close to one aluminum atom. The XRD and NMR results can be reconciled if the fluorines are disordered over the multiple bonding sites in each cage. The diffraction studies (erroneously) locate the fluorines at the cage centers.

Unlike the CP-based experiment described earlier, the REDOR experiment does not involve observing magnetization transferred from, say, I to S. This may lead to problems from "background" signals from any S spins that are not coupled to I. Variant pulse sequences, such as *TEDOR*, have been devised which observe magnetization transferred between I and S to avoid this problem. Because they involve polarization transfer, TEDOR-type sequences can be readily incorporated into HETCOR experiments. 3D-TEDOR experiments have, for example, been used to quantify the ^{13}C,^{15}N couplings responsible for the cross-peaks in a ^{13}C/^{15}N HETCOR experiment, although such experiments are only viable in isotopically enriched samples.

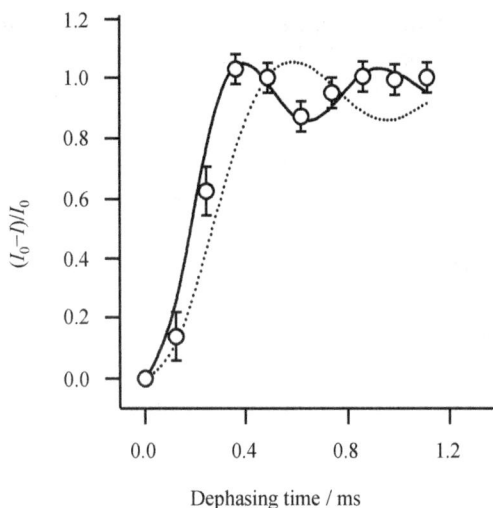

Figure 8.6. ^{27}Al$\{^{19}$F$\}$ REDOR results for the Al signal (at 10 ppm) of the aluminophosphate molecular sieve AlPO$_4$-5. The open circles are the experimental data. Calculated curves correspond to Al–F dipolar coupling constants of (dotted line) 2.8 kHz ($r_{\mathrm{AlF}} = 0.219$ nm) and (continuous line) 4.15 kHz ($r_{\mathrm{AlF}} = 0.192$ nm). (This original figure was adapted for publication in R.D. Gougeon et al., *J. Phys. Chem. B* **105** (2001) 12249.)

The REDOR experiment and its variants have been widely used to measure dipolar couplings between a variety of nuclei. It can even be applied to couplings between spin-$\frac{1}{2}$ and half-integer quadrupolar nuclei, with the quadrupolar nucleus as the S spin, since refocusing the chemical shift evolution at the midpoint of the recoupling sequence can be achieved relatively cleanly via a single 180° pulse.

As the signal is only measured at (even) multiples of the rotor period, the REDOR experiment is not suitable for strong dipolar couplings (of the order of the spinning speed or greater) since not enough data points can be obtained to characterize the dipolar oscillation. The CP-based experiment of figure 8.4(a) is better suited to these cases. In the other direction, the S spin T_2' sets the ultimate limit on the smallest couplings that can be measured. It is important to avoid overloading the probe when using long recoupling times to measure weak couplings. For example, if ^{1}H decoupling is required, it will need to be applied continuously during *both* recoupling and signal acquisition periods. It may be necessary to truncate the acquisition (and sacrifice some resolution of the S spin spectrum) to keep the duty cycle within acceptable limits.

8.3.2 MEASUREMENT OF HOMONUCLEAR COUPLINGS

Extracting accurate estimates of dipolar coupling constants is often more difficult for homonuclear spins, in large part because isolated spin pairs are relatively uncommon. Moreover, the spin dynamics are often complex if more than one coupling is involved, making it difficult, or impossible, to interpret the results in terms of individual couplings.

In the case of ^{1}H spins in typical organic solids, there are multiple strong homonuclear dipolar couplings between nearby H atoms, making it impossible to isolate the effect of an individual coupling. The dynamics of magnetization transfer between the spins are best described in terms of an overall spin diffusion rather than the kind of oscillatory build-up curves observed for isolated couplings seen in figure 8.5. That said, the rate of spin diffusion between a pair of spins will be directly related to the coupling strength and can provide a qualitative indication of the proximity of the spins involved.

The principal difficulty with estimating dipolar couplings between ^{1}H nuclei is the strength of the couplings and the resulting limited spectral resolution; fast MAS or strong homonuclear decoupling is required to suppress the homonuclear couplings sufficiently to resolve the different sites. The dipolar couplings between most other spins are much weaker, and so, by contrast, it is often necessary to use recoupling techniques (as described in section 5.4.2) to prevent the couplings being eliminated by MAS. In this case, there is the option of using experiments that selectively recouple individual spin pairs. For example, the rotational resonance experiment involves matching the difference in NMR frequencies of two spins to an integer multiple of the spinning frequency, that is deliberately overlapping a spinning sideband of one peak with the centerband of the other. Under these resonance conditions, magnetization can be exchanged between the spins (in a similar way that magnetization is exchanged between heteronuclei under Hartmann–Hahn matching), and the dipolar coupling extracted from the build-up curve. Figure 8.7 shows an example.

Where the magnetization exchange is not unduly affected by multiple couplings, the broadband recoupling schemes described in section 5.4.2 (and related pulse sequences) can be used to good effect. Sequences such as POST-C7 have the advantage that the recoupling efficiency is independent (to first order) of the chemical shift Hamiltonian. The exact spin dynamics under other recoupling

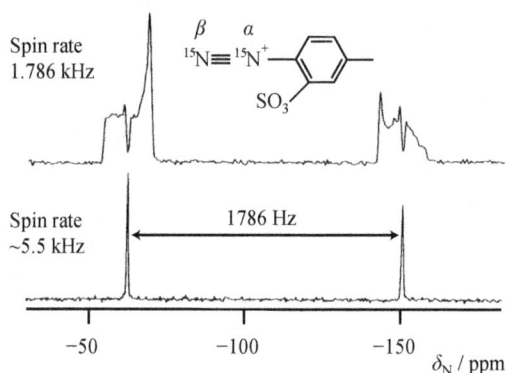

Figure 8.7. Nitrogen-15 spectra, obtained at 20.28 MHz, of doubly ^{15}N-labeled 5-methyl-2-diazobenzene-sulfonic acid hydrochloride. Bottom: Fast magic-angle spinning case, showing sharp peaks separated by 1786 Hz. The peak to low frequency is assigned to the a nitrogen while the high-frequency signal is for the β one. Top: The centerband region of the spectrum obtained with a spin rate to equal the frequency difference between the two signals ($n = 1$ rotational resonance). Detailed analysis of the full rotational resonance spectrum yields 0.1110 ± 0.0004 nm for the N–N distance (see R. Challoner & R.K. Harris, *Chem. Phys. Lett.* **228** (1994) 589.)

schemes, such as radiofrequency driven recoupling (RFDR), often depends on the chemical shift anisotropies and/or the isotropic chemical shifts of the spins involved. This is rarely a problem in qualitative applications (such as those described in chapter 5), but complicates quantitative analysis. An example of a recoupling experiment for the 1,10-difluorodecane/urea inclusion compound (using the MELODRAMA pulse sequence) is shown in figure 8.8. Analysis yields a motionally averaged intermolecular (F,F) dipolar coupling constant, $\langle D \rangle$, of 995 Hz, from which deductions have been made regarding the conformation of the guest molecule in the urea tunnels.

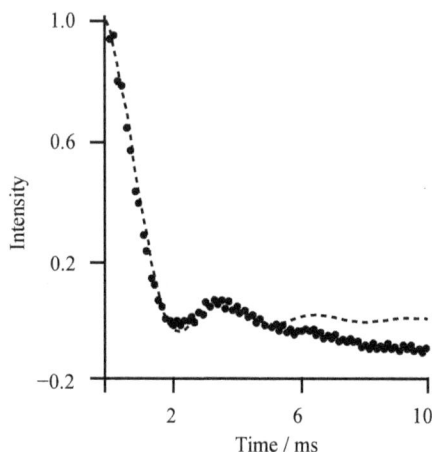

Figure 8.8. $^{19}F\{^1H\}$ MELODRAMA plot for the 1,10-difluorodecane/urea inclusion compound. The $n = 4$ matching condition was used. The experimental points are shown, together with a dashed line for the optimized simulated plot. The horizontal axis represents a variable time interval in the MELODRAMA pulse sequence. (This original figure was adapted for publication in "Recent advances in solid-state NMR", R.K. Harris, in "*Magnetic resonance in food science: A view to the future*", Eds. G.A. Webb, P.S. Belton, A.M. Gil & I. Delgadillo, Royal Society of Chemistry (2001), 3–16.)

8.4   QUANTIFYING INDIRECT (J) COUPLINGS

Scalar (J) couplings are measured straightforwardly using spin-echo experiments in which the echo delay periods are multiples of the rotation period. The spin echo refocuses evolution under the isotropic shift, and when the rotor angle is set sufficiently accurately that the evolution due to anisotropic interactions is refocused over the rotation cycle, then the intensity of the spin-echo signal is determined only by scalar couplings that are not refocused.

As discussed in section 5.3, the spin-echo signal as a function of the time τ when a single coupling is active will fit to:[6]

$$S(\tau) = A\cos(2\pi J\tau)\exp(-2\tau/T_2')$$
8.2

where 2τ is the total spin-echo period (see sections 4.4.2 and 7.2.3), J is the coupling constant, and T_2' is the time constant used to fit the signal decay. Figure 8.9 shows an example of fitting

---

[6] The circumstances in which this expression is valid are examined in detail by Duma *et al.*, *ChemPhysChem* **5** (2004) 833. Their conclusion was that this fitting is generally valid even in the presence of much stronger dipolar and shielding interactions.

J couplings between pairs of ^{31}P nuclei. Couplings between different NMR nuclides are fitted in the same way, except that the refocusing 180° is applied to both nuclei involved to select for heteronuclear rather than homonuclear couplings.

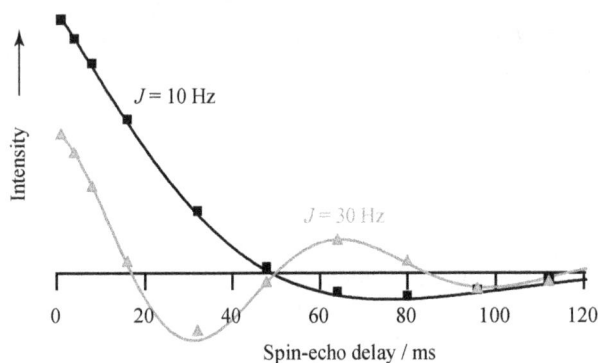

Figure 8.9. Example of fitting $^2J_{PP}$ couplings from ^{31}P spin-echo data. See figure 5.6 for full details.

Very small J couplings can be fitted in this way with good accuracy, with the limit set by the T_2' decay. In this case, $T_2' \approx 40$ ms, corresponding to a homogeneous "linewidth" of $1/\pi T_2' = 8.0$ Hz, that is, the oscillation due to the 10 Hz coupling can still be resolved, but errors on fitting couplings smaller than 8 Hz will increase sharply.

8.5 TENSOR INTERPLAY

So far, it has mostly been assumed that only a single type of tensor affects the spectrum, but clearly, in many (if not most) cases, several NMR interactions contribute to defining it. Thus one must consider interplay between tensors. Two such cases will be considered in some detail, one in this section and a second one in section 8.6. Here the case of interplay between the chemical shift and heteronuclear dipolar coupling will be discussed. This will also involve indirect coupling. The situation can be understood relatively simply because the tensors involved commute and therefore affect the spectrum in an additive fashion. Spectra obtained at more than one applied magnetic field can assist in understanding the situation (since J and D are field independent to first order, whereas chemical shifts scale linearly with the applied field), but the treatment of an individual spectrum is considered here.

The simplest case, which illustrates the essential features, occurs for isolated heteronuclear two-spin (AX) systems with $I_A = I_X = \frac{1}{2}$, in which axial symmetry prevails so that asymmetry in σ_A, σ_X, and J_{AX} can be ignored. Moreover, all three tensors for, say, the A spins, will then be coaxial (with the major principal axis along the internuclear A, X distance, r_{AX}). The anisotropy in J_{AX} has exactly the same form as the dipolar coupling, D_{AX}, and therefore an effective parameter D_{AX}' may be defined as:

$$D_{AX}' = D_{AX} - \frac{\Delta J_{AX}}{3} \qquad\qquad 8.3$$

where D_{AX} is the dipolar coupling constant in frequency units, $\left(\mu_0/4\pi\right)\gamma_A\gamma_X h/4\pi^2 r_{AX}^3$ (equation 2.24), and ΔJ_{AX} is the anisotropy in J_{AX} expressed as $J_\parallel - J_\perp$.[7] The theory for this situation is outlined in inset 8.1.

Inset 8.1. Interplay between σ, D, and J for a heteronuclear two-spin system

The NMR Hamiltonian may be written as

$$h^{-1}\hat{H} = -\nu_{NMR}^A\left[1-\left\{\sigma_A^{iso}+\zeta_A P_2\left(\cos\theta\right)\right\}\right]\hat{I}_{zA}$$
$$-\nu_{NMR}^X\left[1-\left\{\sigma_X^{iso}+\zeta_X P_2\left(\cos\theta\right)\right\}\right]\hat{I}_{zX} \qquad 8.4$$
$$+\left[J_{AX}^{iso}-2D'_{AX}P_2\left(\cos\theta\right)\right]\hat{I}_{zA}\hat{I}_{zX}$$

where $\nu_{NMR}^A = \gamma_A B_0/2\pi$ and $\nu_{NMR}^X = \gamma_X B_0/2\pi$ are the A and X Larmor frequencies in the absence of shielding, σ_A^{iso} and σ_X^{iso} are the isotropic A and X shielding constants, ζ_A and ζ_X are the shielding anisotropies (see inset 2.2) expressed as $\sigma_A^\parallel - \sigma_A^{iso}$ and $\sigma_X^\parallel - \sigma_X^{iso}$, respectively, J_{AX} is the isotropic (A, X) coupling constant, and θ is the angle between r_{AX} and B_0. The transitions of the A nucleus are therefore described by:

$$\nu^A = \nu_{NMR}^A\left[1-\left\{\sigma_A^{iso}+\zeta_A P_2\left(\cos\theta\right)\right\}\right] - J_{AX}^{iso}m_X - 2D'_{AX}P_2\left(\cos\theta\right)m_X \qquad 8.5$$

where m_X is the appropriate spin component quantum number for the X spin. This equation reduces to:

$$\nu^A = \nu_{iso}^A - \nu_{NMR}^A\zeta_A^{eff}P_2\left(\cos\theta\right) - J_{AX}m_X \qquad 8.6$$

where ζ_A^{eff} is an effective tensor anisotropy given by:

$$\zeta_A^{eff} = \zeta_A - \frac{2D'_{AX}m_X}{\nu_{NMR}^A} \qquad 8.7$$

The equations above are given under the assumption that γ_A and γ_X are both positive. Careful consideration needs to be given to signs if that is not the case.

---

[7] This implies that it is not experimentally possible to separate D_{AX} and ΔJ_{AX} *in this case*. However, when **D** and **J** are not coaxial, equation 8.3 is no longer valid. In principle, **J** generally has an asymmetry, which, in the absence of motion, **D** does not, but this factor is usually ignored.

Equations 8.5 and 8.6 show that the A transitions form two subspectra ($m_X = \pm \frac{1}{2}$), in one of which the shielding and dipolar tensors reinforce one another, whereas in the other they partially cancel each other. Thus, one static powder subspectrum is "stretched" whereas the other is "squeezed", as indicated schematically in figure 8.10. This hypothetical case is drawn for the case when J_{AX}, D_{AX}, and ζ_A are all positive[8] (involving $\sigma_{\parallel} > \sigma_{\perp}$), with $\Delta J_{AX} = 0$ and $|D_{AX}| > |\zeta_A|$,

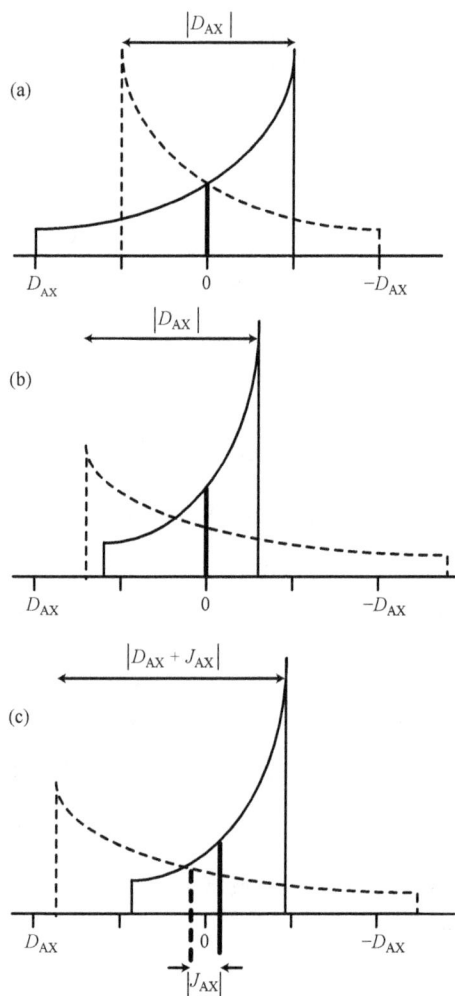

Figure 8.10. Schematic powder pattern for the A nucleus, say, of a heteronuclear two-spin (AX) system subject to (a) dipolar coupling only, (b) both dipolar coupling and shielding anisotropy, and (c) both dipolar coupling and shielding anisotropy, with the subspectra separated by an isotropic indirect coupling constant. The isotropic positions of the subspectra are indicated in bold; for (a) and (b) they are at the same position. Under the conditions as described in the text, the subspectrum when m_X is $+\frac{1}{2}$ is "squeezed" (indicated by the solid lines), while that when m_X is $-\frac{1}{2}$ (indicated by the dashed lines) is "stretched." The plot parameters used were: $\zeta_A = 2D_{AX}/5$ and $J_{AX} = D_{AX}/6$.

---

[8] The way in which the signs of J_{AX}, D'_{AX}, and ζ_A affect the spectrum can be assessed from the equations in inset 8.1. There are eight possible cases, but they give only two different spectral appearances.

so that it is the low-frequency subspectrum that is "squeezed". Figure 8.10(a) shows a case with only dipolar coupling, while figure 8.10(b) illustrates the effect when both dipolar coupling and shielding anisotropy affect the spectrum. The "horns" of the total pattern for figure 8.10(b) remain separated by $|D_{AX}|$. A static powder pattern could be analyzed to give information on both D_{AX} and ζ_A. Figure 8.10(c) shows that the subspectra are distinguished when indirect coupling is present. The "horns" are now separated by $|D_{AX} + J_{AX}|$.

Under slow MAS conditions, the influence of the angle-dependent term $3\cos^2\theta - 1$ is translated into a distribution of intensities among the spinning sidebands, which extend over a larger frequency range for the stretched subspectrum than for the squeezed subspectrum. For figure 8.10(b) (as for figure 8.10(a)), the isotropic shift positions for the two subspectra are the same, so while MAS would give rise to two spinning sideband manifolds, these would overlap and so be difficult to disentangle. The isotropic indirect coupling serves to resolve the subspectra, as shown in figure 8.10(c), enabling the spinning sideband manifolds to be analyzed separately and thus yielding the two values of ζ_A^{eff}. One consequence is that the centerbands (split by J_{AX}) of the two spinning sideband manifolds have different intensities, as can be seen for an experimental spectrum in figure 8.11. The magnitudes of ζ_A, D'_{AX}, and J_{AX} are readily obtained by the analysis. Since the sign of ζ_A will normally be obvious (and can be obtained separately by spinning sideband analysis of an A spectrum obtained with X-decoupling) sign information on J_{AX} and D'_{AX} are obtained, albeit with some ambiguity. Only the relative signs of D'_{AX} and J_{AX} are determined, the rule being as follows: If the high-frequency spinning sideband manifold has the larger magnitude of ζ_A^{eff}, then (i) for positive ζ_A, D'_{AX}, and J_{AX} have the same sign, whereas (ii) for negative ζ_A, D'_{AX},

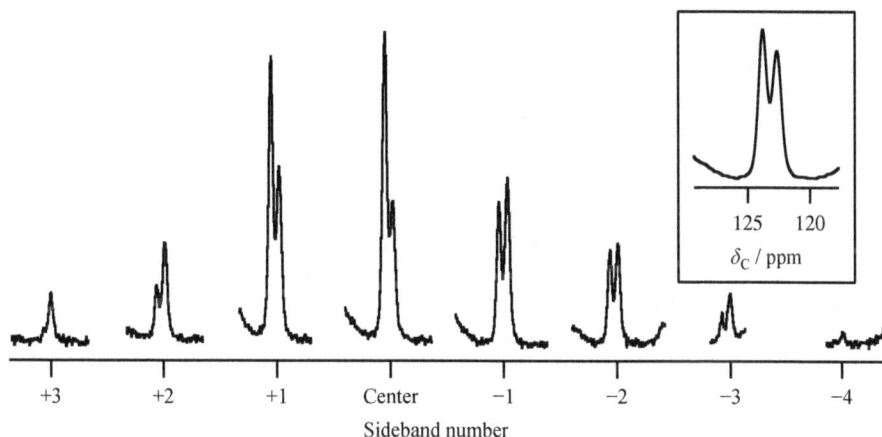

Figure 8.11. Spinning sidebands of the proton-decoupled ^{13}C cross-polarization magic-angle spinning (CPMAS) spectrum for the *ipso* phenyl carbon of solid trimethylphenylphosphonium iodide at 75 MHz. The centerband doublet shows a splitting arising from isotropic indirect coupling to the ^{31}P nucleus of 84 Hz in magnitude, but the intensities of the two centerband peaks are markedly different. The intensities of the doublet peaks in the spinning sidebands are mostly biased in the opposite direction, giving equal total intensities in the two subspectra, as shown by a summation of all the sidebands (see the inset).

and J_{AX} have opposite signs. The inverse is true if the larger magnitude of ζ_A^{eff} belongs to the low-frequency subspectrum. If r_{AX} is known (e.g., from diffraction studies), D_{AX} can be calculated, so ΔJ can be derived, though with an ambiguity unless the sign of J_{AX} is known. On the other hand, if the relative magnitudes of D_{AX} and $\Delta J/3$ can be assumed, the absolute sign of J_{AX} can be determined.

Extension to a linear AX_2 case is straightforward. There will now be three spinning sideband manifolds in the A spectrum, with effective tensor anisotropies given by:

$$\zeta_A^{eff} = \zeta_A - \frac{2 D_{AX}'}{v_A^0} \sum m_X \qquad 8.8$$

where $\sum m_X = 1$, 0, or -1, associated with the three centerbands at $v_A^{iso} - J_{AX} \sum m_X$. The central sideband manifold gives ζ_A directly, because it is independent of D'. Clearly, this is an advantage, particularly if X-decoupled A spectra cannot be obtained because of spectrometer limitations.

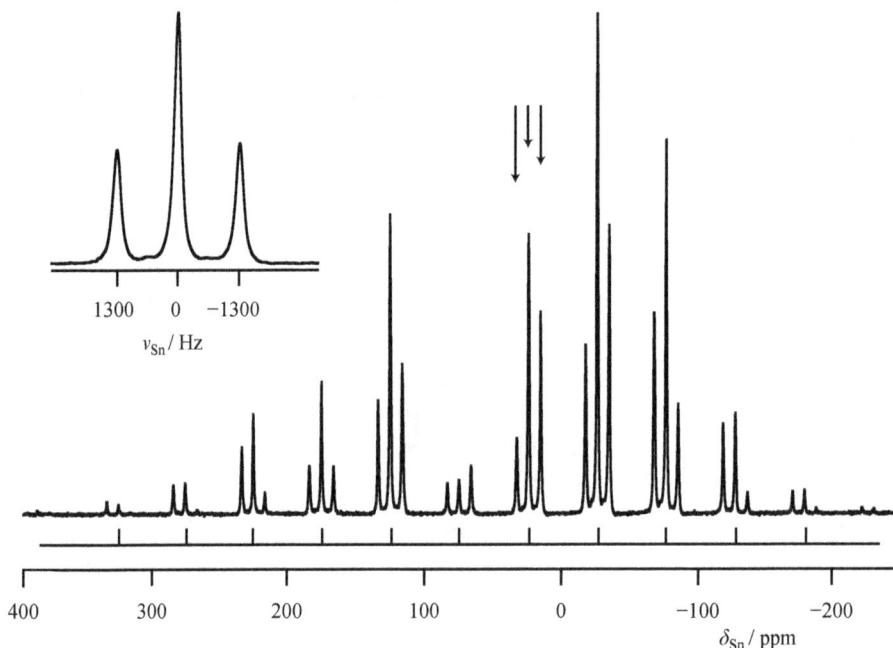

Figure 8.12. Tin-119 cross-polarization magic-angle spinning (CPMAS) proton-decoupled spectrum at 149.12 MHz of Me_3SnF. The spin rate was 7.5 kHz. The centerbands of the three subspectra are indicated with arrows. The linked marks under the spectrum indicate the spinning sideband manifold for the central transition of the triplet. The inset shows a summation over the whole spinning sideband manifold. (See H. Bai, R.K. Harris & H. Reuter, *J. Organomet. Chem.* **408** (1991) 167.)

Figure 8.12 illustrates this case, for the ^{119}Sn spectrum (proton decoupled) of Me$_3$SnF, which has a structure with F...Sn-F...Sn chains in which each tin is effectively equally coupled to two fluorine nuclei. The arrowed lines form the centerband of the J-coupled triplet. The spinning sideband manifold of the high-frequency triplet line is stretched, and that of the low-frequency triplet line is squeezed, as expected from the theory mentioned above. However, when the respective spinning sidebands are summed, the expected 1:2:1 triplet is obtained. Analysis of the central spinning sideband manifold gives the Sn shielding anisotropy ζ_{Sn} as –221 ppm and D'_{SnF} as –4.02 kHz (negative because the magnetogyric ratio for ^{119}Sn is negative). The isotropic $^1J(^{119}$Sn,^{19}F) coupling constant is 1290 Hz, known to be positive.

The situation is more complicated when the tensors involved are not coaxial. This will be discussed qualitatively in section 8.7.

8.6 EFFECTS OF QUADRUPOLAR NUCLEI ON SPIN-$\frac{1}{2}$ SPECTRA

Another case of tensor interplay arises for NMR spectra of quadrupolar nuclei because of the magnitude of the quadrupolar interaction (see chapter 6), the effect of which can be "transferred" to spectra of spin-$\frac{1}{2}$ nuclei by dipolar coupling. Consider a two-spin (I,S) system consisting of a spin-$\frac{1}{2}$ nucleus (I) and a quadrupolar nucleus (S). If the spin wavefunctions are unperturbed from their Zeeman forms, then only the secular $\left(\hat{I}_z\hat{S}_z\right)$ term of the I,S dipolar interaction needs to be considered for the energies of the spin states. Since this depends on $P_2\left(\cos\theta\right)$, its effects are eliminated by MAS. However, any substantial quadrupolar coupling will mix the Zeeman wavefunctions of the S spin states, as mentioned in section 6.3. In this case, dipolar coupling terms such as $\hat{I}_z\hat{S}_+$ (for the notation see section 4.2.1) will now affect the S spin in a way which will have a different angle dependence and so will not be eliminated by MAS. However, the value of this effect will depend on the spin state of the I spin, and so I-spin spectra will also be affected. The case of a two-spin system comprising a ^{13}C nucleus and a ^{14}N (spin-1) nucleus will be considered in inset 8.2 in more detail for the simplified situation where (a) the quadrupolar and dipolar tensors are coaxial, (b) the EFG tensor is axial, and (c) indirect (J) coupling can be ignored.

The effect on the energy levels is shown schematically in figure 8.13. The $m_C = \frac{1}{2}$ manifold of levels is affected by the second-order terms in the opposite way to that of the $m_C = -\frac{1}{2}$ levels. The energy changes on the $m_N = \pm1$ levels are equal whereas the effect on the $m_N = 0$ levels is opposite in sign and twice the magnitude.

This means that the three allowed ^{13}C transitions are moved away from the isotropic chemical shift, though the center of gravity remains unchanged. The result is a 1:2 or 2:1 doublet, giving what is termed a *residual dipolar splitting*. For the same reason that MAS does not eliminate second-order effects on a quadrupolar spectrum, it cannot remove them from the residual dipolar splitting of a spin-$\frac{1}{2}$ spectrum, though it does cause significant averaging of the orientation-dependent factors. Therefore, the resonance frequencies depend on the angles between $\boldsymbol{B}_0$ and the dipolar and quadrupolar axes, so that the observed MAS spectrum for a microcrystalline sample will be a powder pattern. Figure 8.14 illustrates the spectrum for a ^{14}N,^{15}N example, which clearly shows the splitting

---

Inset 8.2. Second-order quadrupolar effect on the energy level of a ^{13}C, ^{14}N spin system

Perturbation theory shows that the mixing of the nitrogen spin states by second-order terms results in (un-normalized) wavefunctions:

$$|\psi'_{-1}\rangle = a_1 |1\rangle - a_2 |0\rangle + |-1\rangle$$
$$|\psi'_0\rangle = a_2 |1\rangle + |0\rangle + a_2 |-1\rangle \qquad \text{8.9}$$
$$|\psi'_1\rangle = |1\rangle - a_2 |0\rangle + a_1 |-1\rangle$$

in which $a_1 = \dfrac{\langle 1| \hat{H}_Q^N |-1\rangle}{2\gamma_N \hbar B_0}$ and $a_2 = \dfrac{\langle 1| \hat{H}_Q^N |0\rangle}{2\gamma_N \hbar B_0}$.

In these circumstances, the dipolar coupling cannot be truncated to its secular part, but other terms need to be included, in particular those involving ^{14}N step-up and step-down operators, $\hat{S}_+^N$ and $\hat{S}_-^N$. Such terms have non-zero secular parts. For example, the nitrogen level $m_N = 1$ now includes an energy contribution proportional to

$$\left\langle \hat{\psi}'_1 \hat{\psi}^C \left| \hat{I}_z^C \hat{S}_+^N \right| \hat{\psi}'_1 \hat{\psi}^C \right\rangle = \langle \hat{\psi}'_1 | \hat{S}_+^N | \hat{\psi}'_1 \rangle \langle \hat{\psi}^C | \hat{I}_z^C | \hat{\psi}^C \rangle$$
$$= \left(\left(\langle 1| - a_2 \langle 0| + a_1 \langle -1| \right) \hat{S}_+^N \left(|1\rangle - a_2 |0\rangle + a_1 |-1\rangle \right) \right) m_C \qquad \text{8.10}$$
$$= -a_2 \langle 1| \hat{S}_+^N |0\rangle m_C = -\sqrt{2} a_2 m_C$$

An equal term is required for the level $m_N = -1$, whereas for the level $m_N = 0$, the analogous quantity is $2\sqrt{2} a_2 m_C$. The contribution to the energy levels also involves the dipolar coupling constant and the polar angles of the dipole direction in the magnetic field, but the net effect will be proportional to a_2 and $-2a_2$ for the ^{14}N levels $m_N = \pm 1$ and $m_N = 0$ respectively. The sign of the second-order terms clearly depends on m_C. (However, terms in a_1, involving the double step-up operator $\left(\hat{S}_+^N \right)^2$ and its partner, $\left(\hat{S}_-^N \right)^2$, are independent of m_C and therefore do not affect the I-spin transitions.)

---

and the powder-pattern nature of the two components. However, for the ^{13}C,^{14}N case details of ^{13}C patterns are only usually seen at very low magnetic fields; usually a broadened 2:1 or 1:2 doublet is recorded. For such cases (under the assumptions already mentioned), the doublet splitting is:

$$\frac{9\chi D}{10\nu_S} \qquad \text{8.11}$$

Since ν_S is the Larmor frequency of the quadrupolar spin, its value is known. Therefore, if χ is available (e.g., from NQR measurements), D can be obtained (and therefore the internuclear distance

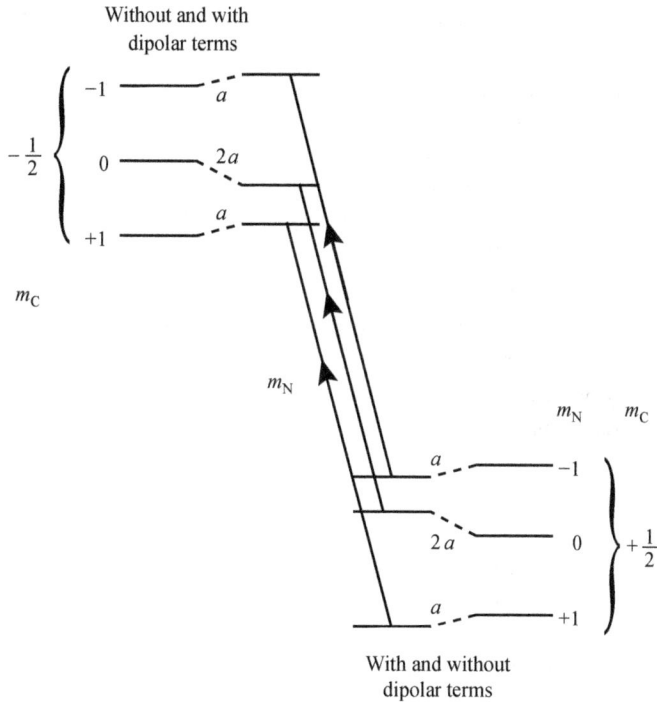

Figure 8.13. Energy-level diagram for a two-spin system involving one spin-$\frac{1}{2}$ (^{13}C) nucleus and one spin-1 (^{14}N) nucleus. The parameter a is proportional to the a_2 factor mentioned in inset 8.2.

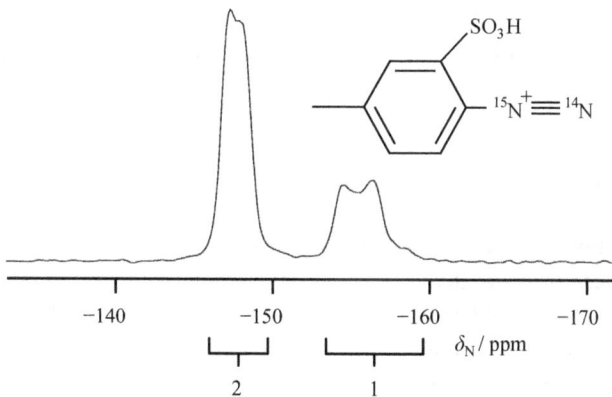

Figure 8.14. Nitrogen-15 magic-angle spinning (MAS) spectrum at 20.3 MHz of 5-methyl-2-diazobenzenesulfonic acid hydrochloride, enriched in ^{15}N at the α position, showing the residual dipolar splitting arising from $^{15}N,^{14}N$ interactions. The splitting between the two major components is 155 Hz. (This original figure was adapted for publication in R. Challoner & R.K. Harris, *Solid State NMR* **4** (1995) 65.)

between the two spins derived). Conversely, if D is obtained from diffraction measurements, χ can be estimated.

Of course, life is rarely that simple and a number of additional features need to be understood:

- In principle, the anisotropy in the indirect coupling tensor, ΔJ, behaves exactly like $\boldsymbol{D}$, and the dipolar parameter D_{AX} in figure 8.10 must be replaced by equation 8.3 (though this is only valid if $\boldsymbol{D}$ and $\boldsymbol{J}$ are coaxial), thus complicating analysis (though if both χ and D are known, ΔJ can then be obtained). This problem is only likely to be significant for situations involving heavy nuclides. It can be (and is) generally ignored for $^{13}C, ^{14}N$ cases.
- The isotropic indirect spin–spin coupling constant, J, may also need to be taken into account, but this is easy since simple splitting is involved.
- Whereas axiality of the EFG and its coaxiality with $\boldsymbol{D}$ are reasonable approximations for cases with single bonds linking I and S, as when univalent atoms such as chlorine are involved, that will rarely be true for examples with ^{14}N. Thus, more complicated expressions are frequently required for residual dipolar coupling. Strategies for determining the angular relationships between tensors will be discussed in the next section.
- The sign of the effect, for example whether a 2:1 or 1:2 pattern is obtained for the $^{13}C, ^{14}N$ case, depends on the sign of χ and the angle between the dipolar and quadrupolar tensors. However, sign matters may be further complicated if D and D' are opposite in sign (i.e., for $\frac{1}{3}\Delta J > D$) and if the quadrupolar asymmetry is non-zero.

Nuclei with large quadrupole moments (e.g., Cl or Br) cause analogous effects even at very high fields such that second-order perturbation theory may be inadequate. The reader is referred to Further reading for details.

The effects described in this section can, of course, be decreased if the applied magnetic field is increased. However, since information can be derived from the tensor interplay in question, this may not be desirable and occasionally it may be helpful to reduce the applied field so as to enhance the second-order effects. Analyzing spectra obtained at two or more applied magnetic fields increases confidence in the resulting parameters.

8.7 QUANTIFYING RELATIONSHIPS BETWEEN TENSORS

The two preceding sections have discussed how NMR spectra are affected when a given resonance is significantly influenced by more than one tensor interaction, for example, a strong dipolar coupling combined with a large shielding anisotropy. However, only cases with coaxial tensors were examined. Frequently, this is not the situation in practice. Fitting NMR parameters to spectra that are determined not only by the parameters of the individual interactions but also by their relative orientations can be difficult. However, the *relative* orientations of tensors form a potential useful source of additional information, particularly in the context of powder samples where it is impossible to establish the *absolute* orientations of individual tensors. A wide variety of experiments has been proposed for different combinations of dipolar coupling, quadrupolar coupling, shielding tensors, etc., and this section simply summarizes the different broad strategies in qualitative terms. Details can be found in the Further reading.

8.7.1 FROM ONE-DIMENSIONAL SPECTRA

As mentioned in sections 8.5 and 8.6, if a given spin is affected by more than one anisotropic interaction, then the overall spectrum is a function not only of both the interactions involved but also of their relative orientations (see figure 8.15). Hence, the relative orientation of tensors can be determined in favorable cases by careful fitting of conventional NMR spectra. Such a fitting is not necessarily straightforward, however, as the spectrum may not be particularly sensitive to the relative orientation—often only the key β Euler angle (see section 4.6) can be fitted with any confidence. Simultaneous fitting of complementary data sets (e.g., for spectra obtained at different magnetic fields) is often necessary to obtain robust quantification of the individual interactions (see figures 8.1(b) and (c)), and hence their relative orientations. Analysis of such spectra yields accurate parameters for the anisotropic interactions, which would not be obtained robustly from a spectrum obtained at a single $\boldsymbol{B}_0$ field.

Static vs. MAS spectra can be useful when dealing with large quadrupole couplings. The other anisotropic interactions are averaged out by (sufficiently fast) MAS, leaving just a second-order quadrupole-broadened lineshape (see section 6.4) that can provide accurate estimates for the quadrupolar interactions, allowing the other interactions (and relative orientations) to be determined from the static spectrum.

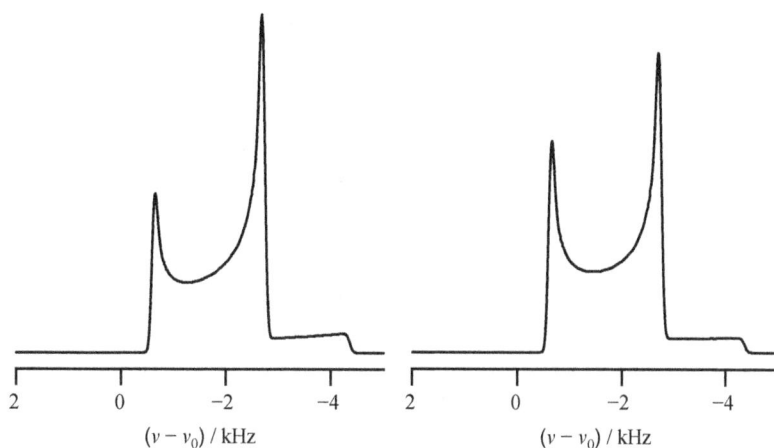

Figure 8.15. Simulated spectrum (central transition only) for a spin-$\frac{3}{2}$ nucleus influenced by the quadrupolar interaction ($\chi = 3$ MHz) and shielding anisotropy ($\zeta = 20$ kHz), both with zero asymmetry. The tensors are coaxial for the left trace and unaligned (with the β Euler angle set to 20°) for the right trace.

8.7.2 FROM CORRELATION SPECTRA

The principal drawback of determining relative tensor orientations from one-dimensional spectra is that they are principally determined by the individual interactions, and the relative orientation has only a minor effect on the overall spectrum.

By contrast, appropriate 2D correlation spectra can be directly dependent on relative orientations. If one tensor is "active" in the indirect dimension and the other in the direct dimension, then the form of the correlation peak will depend directly on the relative orientation; if the principal axes of the tensors are coincident, then there will be a perfect correlation between the frequencies in the two dimensions, leading to a diagonal correlation peak. If, however, the principal axes are not the same, then off-diagonal intensity will be observed, with the form of the pattern being directly related to the relative orientation. These patterns are most easily obtained and interpreted when powder (continuous) bandshapes are involved. It is, however, possible to correlate spinning sideband patterns in two dimensions, although specialist experiments are required to obtain clean in-phase peaks (see Further reading).

Such 2D techniques are most commonly used in the context of correlation spectra involving the transfer of magnetization from one site to another. For example, correlation peaks in EXSY-type spectra (see section 7.3.2) are often simply used to determine which sites are connected by exchange, but provided the anisotropy information has not been suppressed by fast MAS, the patterns of the cross-peaks can be used to determine the relative orientations of the tensors at the different sites.[9] Figure 8.16 shows an example: Comparison of the observed and simulated spectra reveals a substantial difference between the α and β forms of methanol. The α form has four different molecular orientations in the unit cell, with relatively large angles between the C–O bond vectors, whereas the β form has only two such orientations, with a small (only 12°) angle between C–O bonds in adjacent molecules.

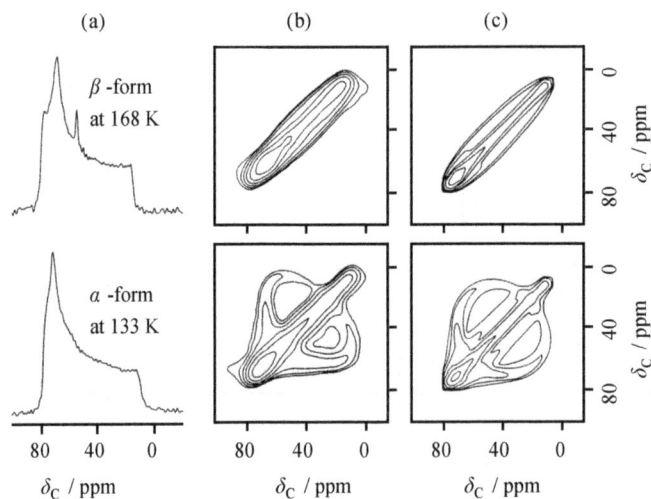

Figure 8.16. Carbon-13 spectra of static samples for two forms of methanol: (a) one-dimensional spectra; (b) experimental EXSY-type spectra (involving spin diffusion during the mixing time); (c) simulated spectra. (Figure supplied courtesy of K. Zilm. It has been partly adapted for publication in "Crystallography and NMR: An overview", R.K. Harris, Chapter 1 of "*NMR Crystallography*", Eds. R.K. Harris, R.E Wasylishen & M.J. Duer, John Wiley & Sons Ltd. (2009). ISBN 978-0-470-69961-4.)

[9] There is a close interaction between chemical exchange and relative tensor orientation. Chemical exchange can only be observed if the NMR frequencies are modulated by changes in isotropic frequency and/or tensor orientation. As a result, experiments such as CODEX (see section 7.3.3) provide information about relative orientation as well as rates of exchange.

Other correlation experiments can also be used to probe relative tensor orientation, for example, an MQMAS experiment (section 6.7.4) can be readily modified to insert an exchange period between indirect (multiple quantum) and direct (single quantum) dimensions. Magnetization transfer during this period will lead to additional correlation peaks, the shapes of which are sensitive to the relative orientation of the quadrupolar coupling tensors of the sites involved.

8.7.3 SPECIALIZED EXPERIMENTS

A number of experiments have been developed to probe the relative orientations of dipolar tensors. For instance, the relative orientation of the two $^{13}C,^{1}H$ dipolar coupling tensors in a $^{1}H-^{13}C-^{13}C-^{1}H$ molecular fragment provides direct information on the H–C–C–H torsion angle. This makes it unnecessary to acquire a full two-dimensional spectrum, and such experiments are usually run in a quasi-one-dimensional fashion; double-quantum coherence between the ^{13}C spins is created (using a recoupling sequence) and allowed to evolve under the $^{13}C,^{1}H$ dipolar couplings before being reconverted and measured. The dynamics of the evolution under the CH couplings is dependent on the magnitude of the couplings and the H–C–C–H torsion angle. The major drawback of such experiments is that specific isotopic labeling is required in order to isolate a suitable spin system.

8.8   NMR CRYSTALLOGRAPHY

Throughout this book, a number of ways in which NMR reveals details of crystallography have been mentioned. Section 1.3.3, for instance, describes how the number of resonances seen in MAS spectra gives information about the crystallographic asymmetric unit. At this point, it is appropriate to enlarge on the emerging subject of NMR crystallography. While this has only recently become generally recognized as a subject area in its own right, NMR has given crystallographic information from the earliest days. Thus, Pake was able, in 1948, to report (by measurement of dipolar coupling) the distance between hydrogen atoms in water of hydration in gypsum ($CaSO_4 \cdot 2H_2O$) crystals as 0.158 nm—even now locating hydrogen by X-ray diffraction techniques is difficult.

Each of the parameters measured by NMR, including full tensor information when relevant, is affected by crystal structure and can therefore give information about such structure. Thus, for instance, dipolar coupling constants give direct (through-space) distances between nuclei, indirect coupling reveals interactions between nuclei mediated by electrons, chemical shifts indicate both intramolecular chemical structure and intermolecular effects, quadrupolar coupling depends on the local electronic structure, and relaxation times are sensitive to molecular-level motion. Moreover, different timescales are effective for NMR and diffraction. That of diffraction experiments is the time taken to record the diffracted beam, which may be from seconds to hours. The NMR timescale, on the other hand, is related to the inverse of the frequency difference between the exchanging signals, which may be of the order of milliseconds. NMR therefore probes faster motions than diffraction.

8.8.1 CHEMICAL EXAMPLES

The simplest use of solid-state NMR in crystallography lies in its ability to determine the chemical structure present. This is valuable when there are structural differences between a solid and its solution. For example, 2-aminobenzoic acid can exist as either a neutral molecule or as a zwitterion. Carbon-13 MAS NMR shows quite clearly (figure 8.17) that form I contains equal numbers of neutral molecules and zwitterions in the unit cell (the chemical shifts for the *ipso* carbons being significantly different between the two forms). In similar fashion, NMR can distinguish between a co-crystal and a salt—the difference lying in the migration of a hydrogen atom. NMR can also readily identify the resonances of "solvent" molecules in a solvate and can quantify the host:solvent molecular ratio. Figure 8.18 shows the ^{13}C spectrum of a 1:1 solvate of phenobarbital and acetonitrile. The signals indicated by arrows arise from the acetonitrile and occur at chemical shifts close to those for acetonitrile in solution.

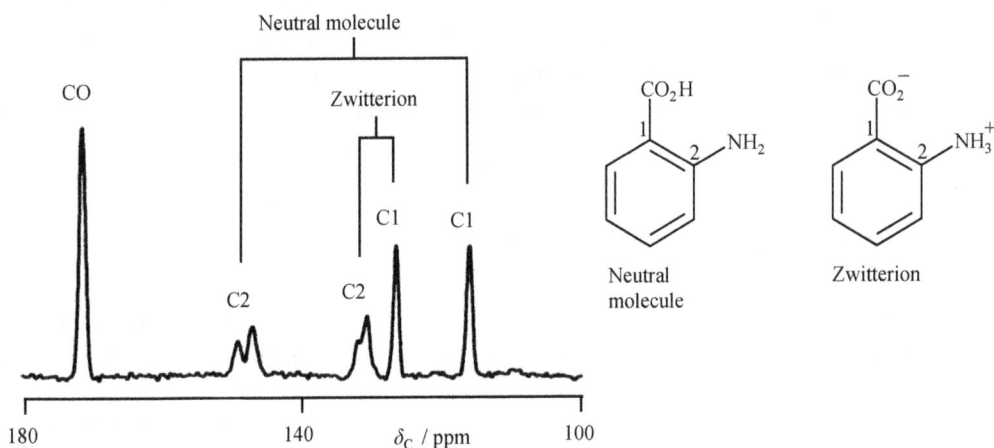

Figure 8.17. Carbon-13 cross-polarization magic-angle spinning (CPMAS) spectrum of form I of 2-aminobenzoic acid. This was obtained using the interrupted decoupling pulse sequence (see subsection 7.2.4) so that only signals arising from quaternary carbons were observed. The unit cells of this polymorph contain both neutral molecules and zwitterions, the resonances of which are separately indicated. The two signals for the carboxyl carbons overlap.

Of course, diffraction methods are generally regarded as the "gold standard" for crystal structure determination, and this situation will undoubtedly continue to be true. However, such methods have limitations. For instance, any mobility at the molecular level is hard to assess. Moreover, departure from strict three-dimensional repetition (which is implicit in the concept of a unit cell) presents problems. Therefore, systems with non-stoichiometric formulae or with disorder (either static or dynamic) are difficult to tackle. Fortunately, NMR gives both supplementary and complementary data to that supplied by diffraction, as is increasingly being recognized by the diffraction community. Thus, NMR can supply information to assist structure determination from diffraction data, and it can also provide data missed by diffraction (especially for molecular mobility). Moreover, NMR information can be obtained from polycrystalline, amorphous, and even heterogeneous samples. A number of examples are given here which demonstrate the power of NMR crystallography.

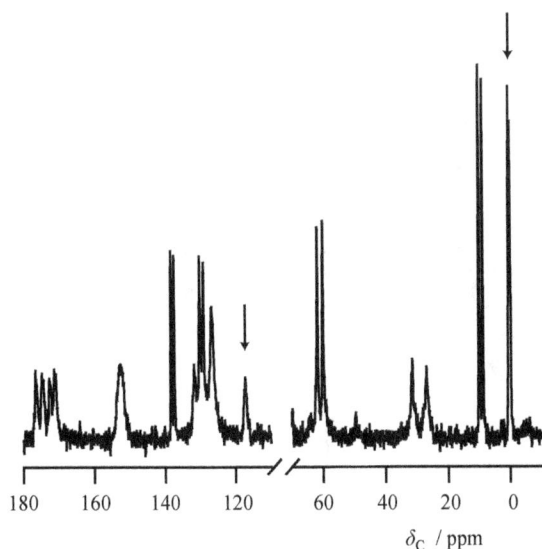

Figure 8.18. Carbon-13 cross-polarization magic-angle spinning (CPMAS) NMR spectrum of the acetonitrile solvate of phenobarbital. The arrows indicate the "solvent" peaks.

The multinuclear facet of NMR can allow deductions that go beyond the determination of the crystallographic asymmetric unit to obtain molecular symmetry information in the solid state. For example, the cyclic trimeric organotin chalcogenides $(Me_2SnX)_3$, X = S or Se, give rise to two ^{119}Sn signals (and two ^{77}Se signals in the selenide case) with intensity ratio 1:2, showing that the molecule has symmetry. However, this could consist of either a mirror plane or a twofold axis. In the former case, four ^{13}C signals in intensity ratio 2:2:1:1 are predicted, whereas the existence of a twofold axis would lead to only three signals (with equal intensities). The ^{13}C spectrum of the sulfide shows unequivocally that the latter case is the correct one. NMR can go further and determine the space group of a crystalline solid. Thus, for the low-temperature phase of zirconium phosphate, $Zr_2P_2O_7$, which was known to have a superstructure of the cubic arrangement of the high-temperature phase (limiting the number of compatible space groups), advanced NMR techniques (such as those discussed in section 5.4.4) applied to the ^{31}P spectrum (figure 8.19) shows the existence of 27 peaks arising from 14 diphosphate ions, one (and one only) of which contained phosphorus atoms related by symmetry. This information established that the space group is *Pbca*, enabling the powder diffraction pattern to be solved to give the full crystal structure.

This synergy between NMR and diffraction can be taken further. For instance, solving structures from powder diffraction data by *simulated annealing* procedures can be done using computer programs that introduce constraints based on NMR information (interatomic distances from dipolar coupling constants or knowledge of intermolecular hydrogen bonding from chemical shifts). In the

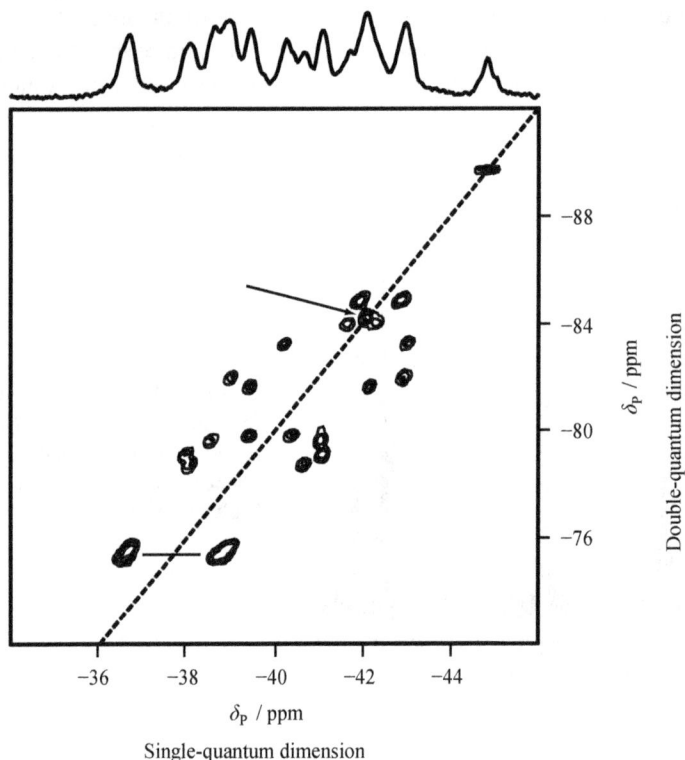

Figure 8.19. Phosphorus-31 POST-C7 two-dimensional spectrum of $Zr_2P_2O_7$, obtained at 121.48 MHz. Horizontal links show that there are 13 dipolar coupled ^{31}P pairs in nonequivalent sites (once the triple intensity of the pair at highest double-quantum frequency—indicated by a horizontal connection—is taken into account). There is also one peak (indicated by the arrow) arising from a dipolar-coupled pair of equivalent ^{31}P nuclei. The one-dimensional projection spectrum is shown at the top. (An adapted version of this figure appears in I.J. King, F. Fayon, D. Massiot, R.K. Harris & J.S.O. Evans, *Chem. Commun.* (2001) 1766.)

future, it may prove to be possible to simultaneously refine structures to fit diffraction information and NMR spectra. Already, NMR can be used to refine structures obtained from diffraction data, as in the case of zeolite ZSM-12, where a relatively poor structure obtained by powder X-ray diffraction was improved to give good agreement between the principal components of the ^{29}Si shielding tensors while retaining acceptable computed diffraction patterns. It has also been shown that ^{13}C tensor components for crystalline naphthalene reveal subtle distinctions between the geometry at different carbon atoms that are consistent with the diffraction-determined space group, whereas the diffraction-determined geometry gives no such distinctions outside experimental error.

An example of the ability of NMR to probe disorder is provided by the 3:2 phenol-triphenylphosphine (TPPO) complex, as mentioned in section 1.3.4. The diffraction work was unable to determine whether the disorder is spatial (static) or temporal (dynamic). The reason for the success of NMR and the failure of diffraction to distinguish the nature of the disorder lies in the different effective timescales of the two techniques, as mentioned earlier in this section. Analysis of the

^{31}P CPMAS NMR spectra as a function of temperature, as illustrated in figure 8.20, enables the barrier to the reorientation process to be determined: $\Delta H^{\ddagger} = 38$ kJ mol^{-1}; $\Delta S^{\ddagger} = -23$ J mol^{-1} K^{-1}. Other examples of the measurement of barriers to molecular-level motions in solids by NMR have been given in chapter 7.

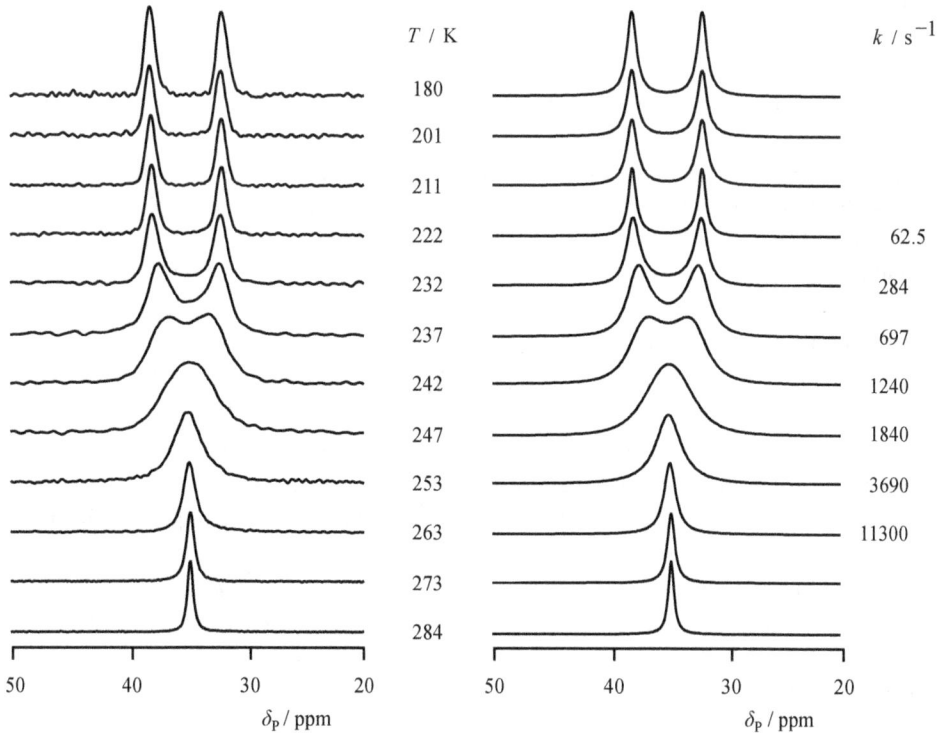

Figure 8.20. Phosphorus-31 cross-polarization magic-angle spinning (CPMAS) spectra of the 3:2 phenol:TPPO complex, obtained at 81.02 MHz as a function of temperature. Left-hand side: experimental. Right-hand side: simulated (with the rate constants to site exchange given at the right-hand side). (Figure adapted, with permission, from D.C. Apperley, P.A. Chaloner, L.A. Crowe, R.K. Harris, R.M. Harrison, P.B. Hitchcock & C.M. Lagier, *Phys. Chem. Chem. Phys.* **2** (2000) 3511.)

Such a case, where the extreme positions are distinguishable, should be contrasted with situations where molecular-level motion involves mutual exchange (i.e., where the two or more extremes are indistinguishable). In these circumstances, diffraction techniques may not detect any mobility at all. Common examples are internal rotation of a C–C(CH$_3$)$_3$ group about the C–C direction and internal rotation of a phenyl or phenylene group about the *ipso*-carbon-to-*para*-carbon axis. If the relevant carbon sites are nonequivalent, then rapid motion will reduce the number of NMR signals observed, that is, coalescence phenomena may occur. The example of formoterol fumarate has been given in section 7.3.1. Several other NMR techniques are available to quantify the dynamics in such situations. For instance, if the motional rate is of the same order as the rotation rate of MAS or of the frequency equivalent of the decoupling power, the resonance will broaden. Either variable temperature or variable spinning rate/decoupling power will reveal this phenomenon and can lead to quantitative evaluation of the barrier to the molecular motion under consideration. Relaxation techniques also allow measurement of motional rates in crystalline materials, as already illustrated in figure 7.3.

In summary, NMR has a significant and increasing role to play in crystallography, in terms of both static and dynamic structure.

8.8.2 COMPUTATION OF NMR PARAMETERS

The value of NMR in crystallography is greatly increased by the use of computation to estimate chemical shifts. Such computations have been carried out for many years on molecules in solution. Advances in the last decade have enabled use to be made of the spatial repetition inherent in crystalline solids so as to include the effects of the environment of molecules or of long-range bonding effects for network systems. For instance, the question of identifying resonances from the same independent molecule in α testosterone, which has two molecules in the asymmetric unit, has been discussed in chapter 5 (see figure 5.19). The difficulty was resolved by use of the INADEQUATE experiment. The major problem with this technique, at least for ^{13}C, is that it requires observation of signals from two bonded ^{13}C nuclei. With natural isotopic abundance, this involves only ~ 0.01% of the molecules for each atom pair. Therefore, while setting up the experiment is relatively easy, implementation takes a lot of spectrometer time (often several days!) and requires efficient proton decoupling.

This is an issue that computation can help to solve. The *gauge-invariant projected augmented wave* (GIPAW) method for computations of crystalline materials involves *density functional theory*, plane waves to cope with the unit-cell repetition, and so-called *pseudopotentials* to mimic the effect of the core electrons of atoms. The limitations imposed by the different conditions involved in the computations and the experiment must be borne in mind. This implies that the GIPAW results are particularly useful for comparisons between shifts for the same chemical atom in different situations (e.g., for two resonances for the same carbon in different crystallographically independent molecules). For the testosterone case discussed in section 5.6, the computations reveal which resonances correspond to the two different molecules in the asymmetric unit. In consequence, they predict that the separations between the pairs of signals for C8 and C9 are of opposite sign, which was confirmed experimentally by INADEQUATE spectra (as mentioned above), though the absolute splitting magnitudes are not well fitted by computation. Moreover, the GIPAW computations give extra information since they relate the signals for the two independent molecules to their sites in the crystal structure.

GIPAW computations can also assist in assignments more generally. Thus, for form I of finasteride (scheme 8.1), a dispute over the assignment of signals for C3 and C20

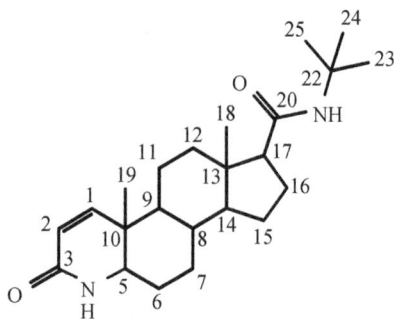

Scheme 8.1. Finasteride.

was solved by GIPAW computations (see A. Othman et al., *J Pharm. Sci.* **96** (2007) 1380). The latter gave C3 at 161.7 ppm and C20 at 166.6 ppm (a difference of 4.9 ppm), while experiment showed signals at 164.3 and 169.3 ppm (a difference of 5 ppm in magnitude). Thus computation showed that the high-frequency signal could be assigned to C20. This was confirmed by an INADEQUATE experiment—but 6 days of spectrometer time were needed!

Computations can also establish the assignment of resonances in ambiguous cases. For instance, form III of phenobarbital (scheme 8.2) contains two nitrogen atoms which are made equivalent in solution by molecular-level motion, but the symmetry is lost in the crystalline state (see figure 8.21). Thus, N1 and N3 now have different environments. There are two observed chemical shifts, at –232.6 and –226.7 ppm. Computation shows that the latter can be assigned to N3 since the computed shifts are at –232.7 and –226.6 ppm for N1 and N3 respectively.

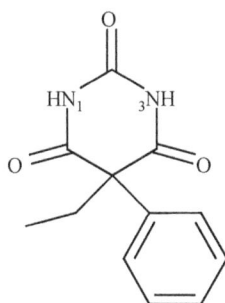

Scheme 8.2. Phenobarbital.

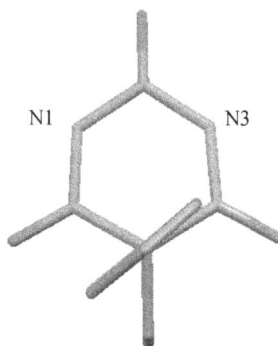

Figure 8.21. View along the axis of the aromatic group in phenobarbital (scheme 8.2) form III to illustrate how the nitrogen atoms N1 and N3 are differentiated in the crystal structure.

FURTHER READING AND SOFTWARE

"Quadrupolar effects transferred to spin-$\frac{1}{2}$ magic-angle spinning spectra of solids", R.K. Harris & A.C. Olivieri, *Prog. NMR Spectry.*, **24** (1992), 435–456. DOI: 10.1016/0079-6565(92)80004-Y.

"NMR crystallography", Eds. R.K. Harris, R.E. Wasylishen & M.J. Duer, John Wiley & Sons Ltd. (2009), ISBN 978-0-470-69961-4.

"NMR crystallography: The use of chemical shifts", R.K. Harris, *Solid State Sciences*, **6** (2004), 1025–1037. DOI: 10.1016/j.solidstatesciences.2004.03.040.

MEASUREMENT OF DIPOLAR COUPLINGS

"Dipolar coupling: Its measurement and uses", M.J. Duer, Chapter 3, pp. 111–178 (DOI: 10.1002/9780470999394. ch3) and "Molecular structure determination: Applications to biology", O.N. Antzutkin, Chapter 7, pp. 280–390 (DOI: 10.1002/9780470999394.ch7) of *Solid-state NMR spectroscopy: Principles and applications*, Ed. M.J. Duer, Blackwell (2002), ISBN 0.632-05351-8.

"Probing proton–proton proximities in the solid state", S.P. Brown, *Prog. NMR Spectry.*, **50** (2007), 199–251. DOI: 10.1016/j.pnmrs.2006.10.002.

"Measuring heteronuclear dipolar couplings for $I = 1/2$, $S > 1/2$ spin pairs by REDOR and REAPDOR NMR", T. Gullion & A.J. Vega, *Prog. NMR Spectry.*, **47** (2005), 123–136. DOI: 10.1016/j.pnmrs.2005.08.004.

SOFTWARE TOOLS FOR THE FITTING OF NMR TENSOR INFORMATION

Programs such as WSOLIDS, DMFIT include various models for fitting anisotropy parameters. General simulation programs such as SIMPSON and pNMRsim (see appendix D) can be used to fit anisotropy information including, specialized cases that are not handled by fixed-purpose programs.

THE SPIN PROPERTIES OF SPIN-$\frac{1}{2}$ NUCLIDES[a]

Isotope	Natural abundance $x/\%$	Magnetogyric ratio $\gamma/10^7$ rad s^{-1} T^{-1}	Frequency ratio $\Xi/\%$	Reference compound	Relative receptivity[b]	
					D^p	D^C
^{1}H	99.9885	26.752 2128	100.000 000	Me$_4$Si	1.000	5.87×10^3
^{3}H	–	28.534 9779	106.663 974	Me$_4$Si-t_1	–	–
^{3}He	1.37×10^{-4}	−20.380 1587	76.178 976^c	He	6.06×10^{-7}	3.56×10^{-3}
^{13}C	1.07	6.728 284	25.145 020	Me$_4$Si	1.70×10^{-4}	1.00
^{15}N	0.368	−2.712 618 04	10.136 767	MeNO$_2$	3.84×10^{-6}	2.25×10^{-2}
^{19}F	100	25.181 48	94.094 011	CCl$_3$F	0.834	4.90×10^3
^{29}Si	4.6832	−5.3190	19.867 187	Me$_4$Si	3.68×10^{-4}	2.16
^{31}P	100	10.8394	40.480 742	H$_3$PO$_4$	6.65×10^{-2}	3.91×10^2
^{57}Fe	2.119	0.868 0624	3.237 778	Fe(CO)$_5$	7.24×10^{-7}	4.25×10^{-3}
^{77}Se	7.63	5.125 3857	19.071 513	Me$_2$Se	5.37×10^{-4}	3.15
^{89}Y	100	−1.316 2791	4.900 198	Y(NO$_3$)$_3$	1.19×10^{-4}	0.700
^{103}Rh	100	−0.8468	3.186 447	Rh(acac)$_3$	3.17×10^{-5}	0.186
(^{107}Ag)	51.839	−1.088 9181	4.047 819	AgNO$_3$	3.50×10^{-5}	0.205
^{109}Ag	48.161	−1.251 8634	4.653 533	AgNO$_3$	4.94×10^{-5}	0.290
(^{111}Cd)	12.80	−5.698 3131	21.215 480	Me$_2$Cd	1.24×10^{-3}	7.27
^{113}Cd	12.22	−5.960 9155	22.193 175	Me$_2$Cd	1.35×10^{-3}	7.94
(^{115}Sn)	0.34	−8.8013	32.718 749	Me$_4$Sn	1.21×10^{-4}	0.711

(*continued on following page*)

(*Continued*)

Isotope	Natural abundance $x/\%$	Magnetogyric ratio $\gamma/10^7$ rad s^{-1} T^{-1}	Frequency ratio $\Xi/\%$	Reference compound	Relative receptivity[b]	
					D^p	D^C
(^{117}Sn)	7.68	−9.588 79	35.632 259	Me$_4$Sn	3.54×10^{-3}	20.8
^{119}Sn	8.59	−10.0317	37.290 632	Me$_4$Sn	4.53×10^{-3}	26.6
(^{123}Te)	0.89	−7.059 098	26.169 742	Me$_2$Te	1.64×10^{-4}	0.961
^{125}Te	7.07	−8.510 8404	31.549 769	Me$_2$Te	2.28×10^{-3}	13.4
^{129}Xe	26.44	−7.452 103	27.810 186	XeOF$_4$	5.72×10^{-3}	33.6
^{183}W	14.31	1.128 2403	4.166 387	Na$_2$WO$_4$	1.07×10^{-5}	6.31×10^{-2}
^{187}Os	1.96	0.619 2895	2.282 331	OsO$_4$	2.43×10^{-7}	1.43×10^{-3}
^{195}Pt	33.832	5.8385	21.496 784	Na$_2$PtCl$_6$	3.51×10^{-3}	20.7
^{199}Hg	16.87	4.845 7916	17.910 822	Me$_2$Hg[d]	1.00×10^{-3}	5.89
(^{203}Tl)	29.524	15.539 3338	57.123 200	Tl(NO$_3$)$_3$	5.79×10^{-2}	3.40×10^2
^{205}Tl	70.476	15.692 1808	57.683 838	Tl(NO$_3$)$_3$	0.142	8.36×10^2
^{207}Pb	22.1	5.676 25[e]	20.920 599	Me$_4$Pb	$2.11^e \times 10^{-3}$	12.4[e]

[a] The table is taken from *Pure Appl. Chem.* **73** (2001) 1795, which should be consulted for the explanatory footnotes and for the precise conditions for the references. The full text of that article is freely available at http://pac.iupac.org/publications/pac/pdf/2001/pdf/7311x1795.pdf. The copyright, however, remains with IUPAC. Note that all the references listed are for solutions.

[b] The relative receptivities D^p and D^C give measures of the intensities to be expected relative to those for the proton and ^{13}C, respectively. They are proportional to $\gamma^3 I(I+1)x$.

[c] Revised value—*see Pure Appl. Chem.* **80** (2008) 59, freely available at http://pac.iupac.org/publications/pac/pdf/2008/pdf/8001x0059.pdf.

[d] Highly toxic. Do not handle directly under any circumstances. Note that some other reference compounds are also toxic. Use of the unified scale for referencing, Ξ, is highly recommended for all such cases—*see Pure Appl. Chem.* **73** (2001) 1795 and *Pure Appl. Chem.* **80** (2008) 59.

[e] Corrected values—*see Pure Appl. Chem.* **80** (2008) 59.

THE SPIN PROPERTIES OF QUADRUPOLAR NUCLIDES[a]

Isotope	Spin	Natural abundance x/%	Magnetogyric ratio $\gamma/10^7$ rad s^{-1} T^{-1}	Quadrupole moment Q/fm^2	Frequency ratio Ξ/%	Reference compound	Relative receptivity[b] D^p	D^C
^{2}H	1	0.0115	4.106 627 91	0.2860	15.350 609	$(CD_3)_4Si$	1.11×10^{-6}	6.52×10^{-3}
^{6}Li	1	7.59	3.937 1709	-0.0808	14.716 086	LiCl	6.45×10^{-4}	3.79
^{7}Li	3/2	92.41	10.397 7013	-4.01	38.863 797	LiCl	0.271	1.59×10^3
^{9}Be	3/2	100	$-3.759\ 666$	5.288	14.051 813	$BeSO_4$	1.39×10^{-2}	81.5
^{10}B	3	19.9	2.874 6786	8.459	10.743 658	$BF_3 \cdot Et_2O$	3.95×10^{-3}	23.2
^{11}B	3/2	80.1	8.584 7044	4.059	32.083 974	$BF_3 \cdot Et_2O$	0.132	7.77×10^2
^{14}N	1	99.632	1.933 7792	2.044	7.226 317	CH_3NO_2	1.00×10^{-3}	5.90
^{17}O	5/2	0.038	$-3.628\ 08$	-2.558	13.556 457	D_2O	1.11×10^{-5}	6.50×10^{-2}
^{21}Ne	3/2	0.27	$-2.113\ 08$	10.155	7.894 296	Ne	6.65×10^{-6}	3.91×10^{-2}
^{23}Na	3/2	100	7.080 8493	10.4	26.451 900	NaCl	9.27×10^{-2}	5.45×10^2
^{25}Mg	5/2	10.00	$-1.638\ 87$	19.94	6.121 635	$MgCl_2$	2.68×10^{-4}	1.58
^{27}Al	5/2	100	6.976 2715	14.66	26.056 859	$Al(NO_3)_3$	0.207	1.22×10^3
^{33}S	3/2	0.76	2.055 685	-6.78	7.676 000	$(NH_4)_2SO_4$	1.72×10^{-5}	0.101
^{35}Cl	3/2	75.78	2.624 198	-8.165	9.797 909	NaCl	3.58×10^{-3}	21.0
^{37}Cl	3/2	24.22	2.184 368	-6.435	8.155 725	NaCl	6.59×10^{-4}	3.87
^{39}K	3/2	93.2581	1.250 0608	5.85	4.666 373	KCl	4.76×10^{-4}	2.79
$(^{40}$K)	4	0.0117	$-1.554\ 2854$	-7.3	5.802 018	KCl	6.12×10^{-7}	3.59×10^{-3}

(*continued on following page*)

Isotope	Spin	Natural abundance x/%	Magnetogyric ratio γ /10^7 rad s^{-1} T^{-1}	Quadrupole moment Q/fm^2	Frequency ratio Ξ/%	Reference compound	Relative receptivity[b] D^p	D^C
(^{41}K)	3/2	6.7302	0.686 068 08	7.11	2.561 305	KCl	5.68×10^{-6}	3.33×10^{-2}
^{43}Ca	7/2	0.135	−1.803 069	−4.08	6.730 029	CaCl$_2$	8.68×10^{-6}	5.10×10^{-2}
^{45}Sc	7/2	100	6.508 7973	−22.0	24.291 747	Sc(NO$_3$)$_3$	0.302	1.78×10^3
^{47}Ti	5/2	7.44	−1.5105	30.2	5.637 534	TiCl$_4$	1.56×10^{-4}	0.918
^{49}Ti	7/2	5.41	−1.510 95	24.7	5.639 037	TiCl$_4$	2.05×10^{-4}	1.20
(^{50}V)	6	0.250	2.670 6490	21.0	9.970 309	VOCl$_3$	1.39×10^{-4}	0.818
^{51}V	7/2	99.750	7.045 5117	−5.2	26.302 948	VOCl$_3$	0.383	2.25×10^3
^{53}Cr	3/2	9.501	−1.5152	−15.0	5.652 496	K$_2$CrO$_4$	8.63×10^{-5}	0.507
^{55}Mn	5/2	100	6.645 2546	33.0	24.789 218	KMnO$_4$	0.179	1.05×10^3
^{59}Co	7/2	100	6.332	42.0	23.727 074	K$_3$[Co(CN)$_6$]	0.278	1.64×10^3
^{61}Ni	3/2	1.1399	−2.3948	16.2	8.936 051	Ni(CO)$_4$	4.09×10^{-5}	0.240
^{63}Cu	3/2	69.17	7.111 7890	−22.0	26.515 473	[Cu(CH$_3$CN)$_4$][ClO$_4$]	6.50×10^{-2}	3.82×10^2
^{65}Cu	3/2	30.83	7.604 35	−20.4	28.403 693	[Cu(CH$_3$CN)$_4$][ClO$_4$]	3.54×10^{-2}	2.08×10^2
^{67}Zn	5/2	4.10	1.676 688	15.0	6.256 803	Zn(NO$_3$)$_2$	1.18×10^{-4}	0.692
(^{69}Ga)	3/2	60.108	6.438 855	17.1	24.001 354	Ga(NO$_3$)$_3$	4.19×10^{-2}	2.46×10^2
^{71}Ga	3/2	39.892	8.181 171	10.7	30.496 704	Ga(NO$_3$)$_3$	5.71×10^{-2}	3.35×10^2
^{73}Ge	9/2	7.73	−0.936 0303	−19.6	3.488 315	(CH$_3$)$_4$Ge	1.09×10^{-4}	0.642
^{75}As	3/2	100	4.596 163	31.4	17.122 614	NaAsF$_6$	2.54×10^{-2}	1.49×10^2
(^{79}Br)	3/2	50.69	6.725 616	31.3	25.053 980	NaBr	4.03×10^{-2}	2.37×10^2
^{81}Br	3/2	49.31	7.249 776	26.2	27.006 518	NaBr	4.91×10^{-2}	2.88×10^2
^{83}Kr	9/2	11.49	−1.033 10	25.9	3.847 600	Kr	2.18×10^{-4}	1.28
(^{85}Rb)	5/2	72.17	2.592 7050	27.6	9.654 943	RbCl	7.67×10^{-3}	45.0
^{87}Rb	3/2	27.83	8.786 400	13.35	32.720 454	RbCl	4.93×10^{-2}	2.90×10^2
^{87}Sr	9/2	7.00	−1.163 9376	33.5	4.333 822	SrCl$_2$	1.90×10^{-4}	1.12
^{91}Zr	5/2	11.22	−2.497 43	−17.6	9.296 298	Zr(C$_5$H$_5$)$_2$Cl$_2$	1.07×10^{-3}	6.26
^{93}Nb	9/2	100	6.5674	−32.0	24.476 170	K[NbCl$_6$]	0.488	2.87×10^3
^{95}Mo	5/2	15.92	−1.751	−2.2	6.516 926	Na$_2$MoO$_4$	5.21×10^{-4}	3.06
(^{97}Mo)	5/2	9.55	−1.788	25.5	6.653 695	Na$_2$MoO$_4$	3.33×10^{-4}	1.95
^{99}Tc	9/2	-	6.046	−12.9	22.508 326	NH$_4$TcO$_4$	–	–
^{99}Ru	5/2	12.76	−1.229	7.9	4.605 151	K$_4$[Ru(CN)$_6$]	1.44×10^{-4}	0.848

(*continued on following page*)

(*Continued*)

Isotope	Spin	Natural abundance $x/\%$	Magnetogyric ratio $\gamma/10^7$ rad s^{-1} T^{-1}	Quadrupole moment Q/fm^2	Frequency ratio $\Xi/\%$	Reference compound	Relative receptivity[b] D^p	D^C
^{101}Ru	5/2	17.06	−1.377	45.7	5.161 369	K$_4$[Ru(CN)$_6$]	2.71×10^{-4}	1.59
^{105}Pd	5/2	22.33	−1.23	66.0	4.576 100	K$_2$PdCl$_6$	2.53×10^{-4}	1.49
(^{113}In)	9/2	4.29	5.8845	79.9	21.865 755	In(NO$_3$)$_3$	1.51×10^{-2}	88.5
^{115}In	9/2	95.71	5.8972	81.0	21.912 629	In(NO$_3$)$_3$	0.338	1.98×10^3
^{121}Sb	5/2	57.21	6.4435	−36.0	23.930 577	KSbCl$_6$	9.33×10^{-2}	5.48×10^2
(^{123}Sb)	7/2	42.79	3.4892	−49.0	12.959 217	KSbCl$_6$	1.99×10^{-2}	1.17×10^2
^{127}I	5/2	100	5.389 573	−71.0	20.007 486	KI	9.54×10^{-2}	5.60×10^2
^{131}Xe	3/2	21.18	2.209 076	−11.4	8.243 921	XeOF$_4$	5.96×10^{-4}	3.50
^{133}Cs	7/2	100	3.533 2539	−0.343	13.116 142	CsNO$_3$	4.84×10^{-2}	2.84×10^2
(^{135}Ba)	3/2	6.592	2.675 50	16.0	9.934 457	BaCl$_2$	3.30×10^{-4}	1.93
^{137}Ba	3/2	11.232	2.992 95	24.5	11.112 928	BaCl$_2$	7.87×10^{-4}	4.62
^{138}La	5	0.090	3.557 239	45.0	13.194 300	LaCl$_3$	8.46×10^{-5}	0.497
^{139}La	7/2	99.910	3.808 3318	20.0	14.125 641	LaCl$_3$	6.05×10^{-2}	3.56×10^2
^{177}Hf	7/2	18.60	1.086	336.5	(4.007)	–	2.61×10^{-4}	1.54
^{179}Hf	9/2	13.62	−0.6821	379.3	(2.517	–	7.45×10^{-5}	0.438
^{181}Ta	7/2	99.988	3.2438	317.0	11.989 600	KTaCl$_6$	3.74×10^{-2}	2.20×10^2
(^{185}Re)	5/2	37.40	6.1057	218.0	22.524 600	KReO$_4$	5.19×10^{-2}	3.05×10^2
^{187}Re	5/2	62.60	6.1682	207.0	22.751 600	KReO$_4$	8.95×10^{-2}	5.26×10^2
^{189}Os	3/2	16.15	2.107 13	85.6	7.765 400	OsO$_4$	3.95×10^{-4}	2.32
(^{191}Ir)	3/2	37.3	0.4812	81.6	(1.718)	–	1.09×10^{-5}	6.38×10^{-2}
^{193}Ir	3/2	62.7	0.5227	75.1	(1.871)	–	2.34×10^{-5}	0.137
^{197}Au	3/2	100	0.473 060	54.7	(1.729)	–	2.77×10^{-5}	0.162
^{201}Hg	3/2	13.18	−1.788 769	38.6	6.611 583	(CH$_3$)$_2$Hg[c]	1.97×10^{-4}	1.16
^{209}Bi	9/2	100	4.3750	−51.6	16.069 288	Bi(NO$_3$)$_3$	0.144	8.48×10^2

[a] The table is taken from *Pure Appl. Chem.* **73** (2001) 1795, which should be consulted for the explanatory footnotes and for the precise conditions for the references. The full text of that article is freely available at http://pac.iupac.org/publications/pac/pdf/2001/pdf/7311x1795.pdf. The copyright, however, remains with IUPAC. Note that all the references listed are for solutions.

[b] The relative receptivities D^p and D^C give measures of the intensities to be expected relative to those for the proton and ^{13}C, respectively. They are proportional to $\gamma^3 I(I+1)x$.

[c] Highly toxic. Do not handle directly under any circumstances. Note that some other reference compounds are also toxic. Use of the unified scale, Ξ, for referencing is highly recommended for all such cases—see *Pure Appl. Chem.* **73** (2001) 1795 and *Pure Appl. Chem.* **80** (2008) 59, the latter also freely available at http://pac.iupac.org/publications/pac/pdf/2008/pdf/8001x0059.pdf.

APPENDIX C

LIOUVILLE SPACE, RELAXATION & EXCHANGE

C.1 INTRODUCTION TO LIOUVILLE SPACE

There are a number of important processes that cannot be described in terms of the evolution of a density operator under a Hamiltonian, as given by the Liouville–von Neumann equation (equation 4.50). The most important of these are relaxation of the nuclear spin coherences and chemical exchange. The reason why they cannot be accounted for in the nuclear spin Hamiltonian is their dependence on the external degrees of freedom that were discarded in reducing the total system Hamiltonian to one involving just the nuclear spins, as discussed in section 4.1. NMR relaxation, for instance, is driven by molecular motion, while the derivation of the nuclear spin Hamiltonian assumes molecular motion to be much faster than the NMR timescale and the time dependence could be ignored. Relaxation and other effects involving the non-spin coordinates could, of course, be included by working with the full Hamiltonian, but fortunately there is a much easier route. This section gives an overview of the theoretical tools required to treat these problems in solid-state NMR, with the emphasis on providing an intuitive picture of what is involved rather than a detailed treatment. See the Further reading for more thorough treatments.

The key to accounting for terms that cannot be included in the nuclear spin Hamiltonian is switching from the normal *Hilbert space*, where the number of degrees of freedom is the size of the Hamiltonian eigenbasis, to a much larger *Liouville space* in which the different coherences (or elements of the density matrix) effectively become the eigenbasis. Considering a single spin-$\frac{1}{2}$, for example, there are just two states, $|a\rangle$ and $|\beta\rangle$, in the Hilbert space, and the density matrix has four elements, $|a\rangle\langle a|, |a\rangle\langle\beta|, |\beta\rangle\langle a|$, and $|\beta\rangle\langle\beta|$. Hence the corresponding Liouville space has four degrees of freedom, with the basis $|a\rangle\langle a|, |a\rangle\langle\beta|, |\beta\rangle\langle a|$, and $|\beta\rangle\langle\beta|$, where $|j\rangle\langle k|$ represents the coherence between states j and k. A density matrix can then be represented in Liouville space as a simple vector of coefficients, for example, $(0, 1, 0, 0)$ would correspond to one unit of population in the $|a\rangle\langle\beta|$

coherence. Transformations in this Liouville space therefore involve 4×4 matrices, which may seem an excessive number of matrix elements to describe the dynamics of a single spin!

C.2   APPLICATION TO RELAXATION

The equivalent of Hamiltonian in Liouville space is the Liouvillian which is written $\hat{\hat{L}}$, the double hat being used to denote a *superoperator*. While a normal operator acts on a wavefunction, or (equivalently) a set of coefficients corresponding to the eigenbasis, a superoperator acts on the density operator itself and is therefore able to change the density operator in more radical ways. If the Hamiltonian operator is $\hat{H} = \Delta \hat{I}_z$, that is, rotation about z with frequency Δ, the corresponding Liouvillian has the matrix representation[1]

$$
\hat{\hat{L}} = \begin{array}{c} \\ |a\rangle\langle a| \\ |a\rangle\langle \beta| \\ |\beta\rangle\langle a| \\ |\beta\rangle\langle \beta| \end{array}
\begin{array}{cccc} |a\rangle\langle a| & |a\rangle\langle \beta| & |\beta\rangle\langle a| & |\beta\rangle\langle \beta| \\ \left(\begin{array}{cccc} 0 & 0 & 0 & 0 \\ 0 & 2\pi i\Delta & 0 & 0 \\ 0 & 0 & -2\pi i\Delta & 0 \\ 0 & 0 & 0 & 0 \end{array}\right) \end{array}
\tag{C.1}
$$

The equivalent of the Liouville–von Neumann equation in Liouville space is

$$
\dot{\hat{\rho}} = \hat{\hat{L}}\hat{\rho}
\tag{C.2}
$$

which has the solution

$$
\hat{\rho}(t) = \hat{\hat{U}}(t,0)\hat{\rho}(0) \quad \text{where} \quad \hat{\hat{U}}(t,0) = e^{\hat{\hat{L}}t}
\tag{C.3}
$$

$\hat{\hat{U}}$ is a propagator in Liouville space. Note how this acts directly on $\hat{\rho}$ rather than as a similarity transform (as is the case in Hilbert space). Since $\hat{\hat{L}}$ is already diagonal, the propagator is just:

$$
\hat{\hat{U}}(t,0) = \begin{pmatrix} 1 & 0 & 0 & 0 \\ 0 & e^{2\pi i\Delta t} & 0 & 0 \\ 0 & 0 & e^{-2\pi i\Delta t} & 0 \\ 0 & 0 & 0 & 1 \end{pmatrix}
\tag{C.4}
$$

---

[1] Formally $\hat{\hat{L}} = -2\pi i\, \hat{\hat{H}}$, where $\hat{\hat{H}} = [\hat{H},] = \hat{H} \otimes \mathbf{1} - \mathbf{1} \otimes \hat{H}$ is the superoperator form of the Hamiltonian. $\mathbf{1}$ is an identity matrix whose order matches the Hilbert space and $\otimes$ represents the Kronecker (direct) product.

The evolution of our initial density matrix, (0, 1, 0, 0), in Liouville space is simply

$$\hat{\rho}(t) = \hat{U}\,\hat{\rho}(0) = \left(0, e^{2\pi i \Delta t}, 0, 0\right)$$

C.5

that is, the $|a\rangle\langle\beta|$ coherence oscillates with a frequency Δ. Note how the evolution of the density matrix can be read directly from equation C.1: the populations, $|a\rangle\langle a|$ and $|\beta\rangle\langle\beta|$, do not evolve, while the $|a\rangle\langle\beta|$ and $|\beta\rangle\langle a|$ coherences evolve at frequencies of Δ and $-\Delta$, respectively.

The Liouville space treatment is a rather extravagant way of describing free precession. Consider, however, the Liouvillian and propagator:

$$\hat{\hat{L}} = \hat{\hat{R}} = \begin{pmatrix} -R_1 & 0 & 0 & 0 \\ 0 & -R_2 & 0 & 0 \\ 0 & 0 & -R_2 & 0 \\ 0 & 0 & 0 & -R_1 \end{pmatrix}$$

C.6

hence

$$\hat{\hat{U}}(t,0) = \begin{pmatrix} e^{-R_1 t} & 0 & 0 & 0 \\ 0 & e^{-R_2 t} & 0 & 0 \\ 0 & 0 & e^{-R_2 t} & 0 \\ 0 & 0 & 0 & e^{-R_1 t} \end{pmatrix}$$

C.7

Now the evolution of the density matrix will be

$$\hat{\rho}(t) = \left(0, e^{-R_2 t}, 0, 0\right)$$

C.8

that is, the $|a\rangle\langle\beta|$ coherence decays with rate R_2; this is simply T_2 relaxation.[2] Evolution under the Hamiltonian and relaxation can be naturally combined

$$\hat{\hat{L}} = -2\pi i \hat{\hat{H}} + \hat{\hat{R}}$$

C.9

where $\hat{\hat{R}}$ is a relaxation matrix as in equation C.6. Note how $\hat{\hat{H}}$ will give rise to oscillations of the coherence amplitudes (due to the factor of i), while $\hat{\hat{R}}$ will lead to exponential decays in the coherence amplitudes.

---

[2] Some subtle details about whether the "zero" density matrix refers to the equilibrium density matrix (equation 4.46) or its reduced version (equation 4.48) are being glossed over here.

C.3   APPLICATION TO CHEMICAL EXCHANGE

In problems involving slow exchange (i.e., dynamics that are much slower than the scale of the NMR interactions), it is often possible to avoid Liouville space treatments and simply consider the exchange of z magnetization during magnetization transfer or EXchange SpectroscopY-like experiments. If we consider, for example, a system in which z magnetization is exchanging between three sites, A, B, and C, with the following rates for individual transfers

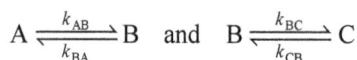

$$A \underset{k_{BA}}{\overset{k_{AB}}{\rightleftharpoons}} B \quad \text{and} \quad B \underset{k_{CB}}{\overset{k_{BC}}{\rightleftharpoons}} C$$

where k_{AB} is the rate of transfer *from* A to B. The overall rate of change of the magnetizations can be written down using simple chemical kinetics. For example,

$$\dot{M}_A = -k_{AB} M_A + k_{BA} M_B \qquad \text{C.10}$$

The set of overall rate equations for M_A, M_B, and M_C are elegantly summarized in matrix form

$$\dot{M} = KM = \begin{pmatrix} -k_{AB} & k_{BA} & 0 \\ k_{AB} & -k_{BA} - k_{BC} & k_{CB} \\ 0 & k_{BC} & -k_{CB} \end{pmatrix} \begin{pmatrix} M_A \\ M_B \\ M_C \end{pmatrix} \qquad \text{C.11}$$

where M is the vector containing the site magnetizations, M_k, and K is the *exchange matrix*. Note how the structure of K indicates which sites are exchanging, and how mass balance means that each column must sum to zero. Note also that K is not generally symmetrical, and care must be taken not to confuse rows and columns, forwards *vs.* backwards exchange rates, etc.

Equation C.11 has exactly the same form as equation C.2 and again has the solution

$$M(t) = e^{Kt} M(0) \qquad \text{C.12}$$

This is an elegant solution to an apparently complex problem involving coupled differential equations. These solutions will have the form

$$M_k(t) = \sum_j A_{jk} e^{\Lambda_j t} \qquad \text{C.13}$$

where Λ_j are the eigenvalues of K, and the coefficients A_{jk} are determined by its eigenvectors and the initial populations, $M(0)$. As a result of the mass balance, one of these eigenvalues must be zero, and the corresponding eigenvector gives relative populations at equilibrium, $t \to \infty$.

T_1 relaxation can be straightforwardly included by adding a relaxation matrix, $\boldsymbol{R}$, to the Liouvillian,[3] $\boldsymbol{L} = \boldsymbol{K} + \boldsymbol{R}$, that is,

$$\boldsymbol{R} = \begin{pmatrix} -R_A & 0 & 0 \\ 0 & -R_B & 0 \\ 0 & 0 & -R_C \end{pmatrix} \tag{C.14}$$

where R_k is the rate of T_1 relaxation for site k.

As a second example, we consider two-site exchange on the *intermediate timescale*, in which the motion occurs on the same timescale as differences in the NMR frequencies. The Hamiltonian and kinetic exchange matrix are

$$\hat{H} = \Delta_A \hat{I}_{zA} + \Delta_B \hat{I}_{zB} \qquad \boldsymbol{K} = \begin{pmatrix} -k_{AB} & k_{BA} \\ k_{AB} & -k_{BA} \end{pmatrix} \tag{C.15}$$

where Δ_A and Δ_B are the resonance frequencies of the two sites.

In this case, the xy magnetization is of interest, rather than the z magnetization. Since the coherence orders do not mix under free precession, we do not need the full Liouville space but only the observed $|a\rangle\langle\beta|$ coherences for the two sites. Hence the Liouvillian for this subspace is[4]

$$\hat{\hat{L}}_{|a\rangle\langle\beta|} = \begin{matrix} & |a\rangle\langle\beta|_A & |a\rangle\langle\beta|_B \\ |a\rangle\langle\beta|_A \\ |a\rangle\langle\beta|_B \end{matrix} \begin{pmatrix} 2\pi i\Delta_A - k_{AB} & k_{BA} \\ k_{AB} & 2\pi i\Delta_B - k_{BA} \end{pmatrix} \tag{C.16}$$

Simplifying the problem to symmetrical two-site exchange by writing $k = k_{AB} = k_{BA}$, the eigenvalues are:

$$\lambda = \pi i(\Delta_A + \Delta_B) - k \pm \sqrt{k^2 - \pi^2(\Delta_A - \Delta_B)^2} \tag{C.17}$$

In the slow exchange limit, $k \ll \pi\Delta$, the eigenvalues reduce to

$$\lambda_A = 2\pi i\Delta_A - k \quad \text{and} \quad \lambda_B = 2\pi i\Delta_B - k \tag{C.18}$$

---

[3] We are making the common assumption that the exchange process is instantaneous on the NMR timescale and does not directly influence the nuclear spin dynamics, that is, at one moment the nuclear spin is in site A and evolving under its Hamiltonian and relaxation parameters, and at the next moment it is evolving with the parameters of site B.
[4] This is identical to the matrix that would be obtained by simple modified Bloch equation treatments of the exchange of xy magnetization.

which corresponds to the A and B resonance frequencies damped by a decay rate k, that is, the lines broaden as the exchange rate increases.

FURTHER READING

Chapter 2 of *"Principles of nuclear magnetic resonance in one and two dimensions"*, R.R. Ernst, G. Bodenhausen & A. Wokaun, Oxford University Press, (1990), ISBN 0 19 855647 0.

APPENDIX D

INTRODUCTION TO SOLID-STATE NMR SIMULATION

The quantitative analysis of solid-state NMR experiments often requires the simulation of NMR data, for example fitting a set of spinning sidebands to anisotropy parameters, as discussed in section 8.2.3. Simulations are invaluable when developing more complex experiments to verify that the NMR response will be as expected.

This section sets out the general principles involved in the numerical simulation of solid-state NMR experiments and the various parameters involved. Simulation programs (a number of which are listed at the end of this appendix) come in a variety of forms: some are specialized for particular tasks, for example sideband analysis, while others are more general, but the same principles apply, and it should not be difficult to adapt the discussion to a particular program.

D.1   SPECIFYING THE SPIN SYSTEM

The first step in any simulation is specifying the set of nuclear spins and their NMR interactions needed to describe the problem at hand. Calculation times typically increase by almost an order of magnitude for each additional spin, and using the smallest spin system that reproduces the experimental results is vital to obtaining results in a reasonable length of time (as well as simplifying the resulting spectra). Fortunately it is rarely necessary to specify more than a couple of spins. For instance, the spectra of quadrupolar nuclei are dominated by the quadrupolar interaction and the coupling to other nuclear spins can often be ignored. Hence, a simple one-spin simulation involving just the quadrupolar interaction (and possibly the chemical shift anisotropy (CSA), depending on its size) should be sufficient to model the spectrum. Similarly, when fitting the ^{13}C CSA pattern, it is not necessary to consider coupling to surrounding protons. Although residual dipolar coupling will broaden the ^{13}C resonances, this does not need to be accounted for explicitly. Simulations involving abundant spins are considerably more challenging; calculating the lineshape associated with a set of coupled spins necessarily involves using a sufficiently large number of spins (typically about 10) in order to reproduce experimental results.

As an example, the spin system for a two-spin heteronuclear problem may be expressed (SIMPSON and pNMRsim) by

```
spinsys {
  nuclei 13C 1H
  dipole 1 2 -20e3 0 0 0
  shift 1 10p 50p 0.5 0 30 0
}
```

where `dipole` specifies the coupling between the spins to be 20 kHz, with its principal axis system aligned with the molecular axis system, that is with Euler angles $\Omega_{MP} = (0, 0, 0)$. The `shift` line specifies the isotropic chemical shift to be 10 ppm and the CSA to be 50 ppm with an asymmetry parameter of 0.5 and with a PAS tilted by 30° away from the molecular frame. Note how the anisotropy information, including relative orientations, must be fully specified for the calculation to be properly defined, which is another motivation for using the smallest possible spin system!

D.2 SPECIFYING THE POWDER SAMPLING

Most solid-state NMR experiments involve microcrystalline powder samples. Hence, it is necessary to average the results of calculations performed for a number of crystallite orientations. As the time taken is proportional to the number of orientations in the powder averaging, it is important to choose the number of sampling points appropriately, particularly if calculating a single response is time consuming. If insufficient points are used, however, individual subspectra from the different orientations will be observable and the sum spectrum will have an unphysical jagged appearance. The optimum number of powder steps is best determined via trial calculations—increasing the number of orientations until the calculated spectra are unchanged—but, as a rule of thumb, the number of orientations required is roughly proportional to the overall widths of the spectral features relative to the intrinsic linewidth. Hence many more orientations are required to obtain a satisfactory broad powder lineshape compared with obtaining, say, a satisfactory CSA spinning sideband pattern under moderately fast magic-angle spinning. In simulations, the intrinsic linewidth is generally added artificially by apodization of the simulated NMR signal. Stronger damping of the NMR signal will naturally decrease the spectral resolution, and will also reduce the number of crystallite orientations required to obtain a smooth powder lineshape.

There are several schemes for choosing which crystallite orientations to consider. Like any other numerical integration, there is no single pattern that converges most quickly to the correct integral for any function. A naive sampling based on equal angle steps of the α and β Euler angles is not particularly effective since equal time is spent on the small volume elements near the pole ($\beta \sim 0$) as points near the "equator" ($\beta \sim \pi/2$). Sampling schemes such as REPULSION distribute sampling points as evenly as possible across the unit sphere and so are invariably more efficient. The so-called ZCW sampling schemes in which the powder angles are incremented simultaneously are also efficient and widely used.

In contrast to static cases, problems involving powders under MAS generally require integration over all three Euler angles, Ω_{RM}, that connect the molecular and rotor axis frames, with the γ angle corresponding to the position of the crystallite around the rotor axis. In sufficiently simple cases (in particular if time-dependent radiofrequency is not involved), it is possible to perform the γ angle integration analytically (the "γ-COMPUTE" algorithm), reducing the requirement for explicit powder angle integration to just α and β. When this is not the case, the full three angle integration can be specified in terms of an integration scheme for α and β combined with a linear increment of γ over a specified number of steps, or, better, using a sampling scheme specifically designed to integrate over three angles. Hence in SIMPSON or pNMRsim.

```
crystal_file zcw132
gamma_angles 8
```

might be used to specify powder integration over 132 α, β pairs using a ZCW scheme, plus eight steps in the γ angle (over the range 0–2π). As a rule of thumb, the number of γ angles does not usually need to be much greater than the number of spinning sidebands in simple problems without RF. By contrast, the NMR response can be strongly dependent on the γ angle when time-dependent RF is present, and more angles may be required before the spectrum converges.

Algorithms such as γ-COMPUTE often require certain "synchronization" conditions to be met. In this case, the number of γ angle steps must be a multiple of the number of observation points (dwell time steps) per rotor cycle. In the example above, setting the spectral width to eight times the MAS frequency would ensure that the dwell time (one-eighth of the rotor period) will divide into the rotor period (i.e., that the spinning and observations periods are synchronized) and that the γ-COMPUTE synchronization condition is also met.

D.3   SPECIFYING THE PULSE SEQUENCE

Different simulation systems tend to use very different approaches to specify the pulse sequence. A simple example is given here to illustrate the principles; the details will need adapting to the system in use.

```
channels 1H
  ...
  start_operator 13C:x
  detect_operator 13C:+
  ...
  pulse 5 100e3 x
  pulse 5 100e3 -x
  store XiX
  ...
pulseq {
  acq 0 XiX
}
```

The different RF channels are specified in the channels directive, while the pulse directives are used to construct a pulse sequence, here the XiX sequence for heteronuclear decoupling, which consists of repeated pairs of π pulses with alternating x and $-x$ phase. The RF nutation rate on the ^{1}H channel has been set to 100 kHz and so duration of each π pulse is 5 μs. Once constructed, the sequence can then be used in `pulseq`, which simply consists of an acquisition (with a zero phase shift of the receiver) in the presence of the decoupling sequence. Note how the preparation of the initial ^{13}C magnetization has been skipped in favor of directly specifying the starting density operator (`start_operator`) to be x magnetization.

D.4   EFFICIENCY OF CALCULATION

As when specifying the spin system, keeping the pulse sequence as simple as possible simplifies the set-up of the simulation and increases the chances that it will run efficiently. For instance, ideal (delta function) pulses can be used whenever the effects of the finite amplitude of experimental RF pulses are not significant. Similarly there is little point in reproducing an experimental detail, such as two-pulse phase-modulated decoupling, when this is not of interest. (Because simple continuous-wave decoupling is not time dependent, it can be simulated with much greater efficiency than a modulated sequence.) In the example above, the initial state of the density operator was explicitly set to be x magnetization rather than starting with equilibrium (z) magnetization. This simplified the pulse sequence and potentially allowed the simulation to take advantage of extra symmetries that would not be present if the initial state of the system had been created explicitly using pulses. Interpretation of the simulation results is also simplified by factoring out aspects of little interest.

Avoiding unnecessary time dependencies is important, as a lack of "synchronization" between the different timescale scales present (e.g., spinning period, dwell time, cycle time of RF sequences) greatly complicates the simulation. For instance, if the dwell time, Δt, is 8 μs and the rotation period, τ_r, is 32 μs, then the signal is detected exactly four times per rotor period. This means that the evolution is completely defined by the four propagators $U(\Delta t, 0)$, $U(2\Delta t, \Delta t)$, $U(3\Delta t, 2\Delta t)$, and $U(4\Delta t = \tau_r, 3\Delta t)$ calculated over one complete rotor period. Efficient algorithms can be used to compute the free-induction decay (FID) or spectrum given this set of propagators. By contrast, if $\tau_r = 33$ μs, it is not possible to find a short common period, and it may be necessary to compute a new propagator for each point of the FID. Since computing the propagators is usually the time-limiting step, simulations where the synchronization conditions are not met will be much slower and may also be complicated to program.

A final parameter that determines the total time required for calculation is the time step, often called `maxdt`, which is used when determining propagators for time-dependent homogeneous Hamiltonians (see inset 4.5). In these cases, it is necessary to compute the propagators by breaking the evolution into small steps over which the Hamiltonian is assumed to be approximately constant. The propagators for these small steps are multiplied together to give the overall propagators. A coarse time step improves the speed of calculation, but too large a step will introduce numerical errors; the time step needs to be small enough to capture the time dependence of the density matrix. If, for instance, the evolution is driven by a time-dependent dipolar coupling of 20 kHz, then a time step of 10 µs should be adequate to calculate the density matrix evolution without significant numerical error. On the other hand, if the evolution is driven by time-dependent RF with amplitude of 200 kHz, a much shorter time step, say 1 µs, would be required.

FURTHER READING

"Numerical simulation of solid-state NMR experiments", P. Hodgkinson & L. Emsley, *Prog. Nucl. Magn. Reson. Spectrosc.*, **36** (2000), 201. DOI: 10.1016/S0079-6565(99)00019-9.

"Computer simulations in solid-state NMR", Mattias Edén, *Concepts Magn. Reson.*, **17A** (2003), 117; **18A** (2003), 1; **18A** (2003), 24. DOI: 10.1002/cmr.a.10061; 10.1002/cmr.a.10064; 10.1002/cmr.a.10065.

SIMULATION PROGRAMS

Simulation software is constantly evolving, so the list here represents a snapshot of general simulation programs at the time of writing. More specialist programs, for example for fitting spinning sideband data and 2D spectra are also available, such as **wSOLIDS** and **Dmfit**. All the packages can be found straightforwardly by a web search for the package name plus the keyword NMR.

GAMMA provides a C++ library (with Python interface) for building NMR simulations. Unlike other systems listed here, it provides a framework rather than a ready-to-use simulation system and requires significant programming skills, particularly as its functionality is mostly directed towards solution-state NMR and relaxation.

pNMRsim is the general simulation program used in the Durham NMR group and which is optimized for multi-spin problems.

SIMPSON is a popular general simulation program for solid-state NMR. SIMPSON is relatively easy to use and has a large, established user base.

SPINEVOLUTION is another general simulation program for solid-state NMR. Although specifying pulse sequences can be tricky, it handles multi-spin problems efficiently.

Index

265

Other Applied Science and General Titles Certain to Be of Interest

- Acoustic High-Frequency Diffraction Theory, Frederic Molinet
- Diffuse Scattering and the Fundamental Properties of Materials, R. I. Barabash, G.E. Ice, P.E.A. Turchi
- The Essentials of Finite Element Modeling and Adaptive Refinement: For Beginning Analysts to Advanced Researchers in Solid Mechanics, John O. Dow
- Polymer Testing: New Instrumental Methods, M. N. Subramanian
- Virtual Engineering, Joe Cecil
- Social Media for Engineers and Scientists, Jon DiPietro
- Professional Expression: To Organize, Write, and manage Technical communication, M.D. Morris
- The Engineering Language: A Consolidation of the Words and Their Definitions, Ronald Hanifan
- Reduce Your Engineering Drawing Errors, Ronald Hanifan

For more information, please visit www.momentumpress.net

The Momentum Press Digital Library
Engineering Ebooks For Research, Classrooms, and Reference

Our books can also be purchased in an e-book collection that features...

- *a one-time purchase that is owned forever; not subscription-based*
- *allows for simultaneous readers,*
- *has no restrictions on printing*
- *can be downloaded as a pdf*

The Momentum Press digital library is an affordable way to give many readers simultaneous access to expert content.

For more information, please visit www.momentumpress.net/library
and to set up a trial, please contact mptrials@globalepress.com

www.ingramcontent.com/pod-product-compliance
Lightning Source LLC
Chambersburg PA
CBHW082004190326
41458CB00010B/3066